Emergence in Complex, Cognitive, Social, and Biological Systems

Emergence in Complex, Cognitive, Social, and Biological Systems

Edited by

Gianfranco Minati
Italian Systems Society
Milan, Italy

and

Eliano Pessa
University of Pavia
Pavia, Italy

Kluwer Academic / Plenum Publishers
New York, Boston, Dordrecht, London, Moscow

Proceedings of the Second Conference of the Italian Systems Society, Associazione Italiana per la Ricerca sui Sistemi (AIRS), on Emergence in Complex, Cognitive, Social, and Biological Systems, held November 9–10, 2001, in Castel Ivano (Trento), Italy

ISBN 0-306-47358-5

233 Spring Street, New York, New York 10013

http://www.wkap.nl/

10 9 8 7 6 5 4 3 2 1

A C.I.P. record for this book is available from the Library of Congress

Printed in the United States of America

CONTRIBUTORS

1. Abram M. R.
2. Badalamenti F.
3. Bandinelli P. L.
4. Benedicenti E.
5. Bonfiglio N.S.
6. Bouquet P.
7. Brahms S.
8. Cambini R.
9. Camiz P.
10. Celani E.
11. Codetta A.
12. Codetta M.
13. Collen A.
14. D'Anna G.
15. Di Caprio U.
16. Di Gregorio S.
17. Falbo F. B.
18. Galgani L.
19. Iezzi A.
20. Magliocca L.
21. Mauri G.
22. Minati G.
23. Montesanto A.
24. Nardon M.
25. Not E.
26. Palma C.
27. Penna M.P.
28. Pessa E.
29. Petroni R.
30. Pianesi, F.
31. Pipitone C.
32. Puliti P.
33. Rabascini N.
34. Ricciuti A.
35. Ricciuti P.
36. Serra R.
37. Stara V.
38. Tascini G.
39. Terenzi G.
40. Tofoni M. C.
41. Trionfi M.
42. Trunfio G.A.
43. Vitiello G.
44. Zancanaro M.
45. Zandonà D.
46. Zandron C.

PREFACE

The systems movement is made up of many systems societies as well as of disciplinary researchers and researches, explicitly or implicitly focusing on the subject of systemics, officially introduced in the scientific community fifty years ago.

Many researches in different fields have been and continue to be sources of new ideas and challenges for the systems community. To this regard, a very important topic is the one of EMERGENCE. Between the goals for the actual and future systems scientists there is certainly the definition of a general theory of emergence and the building of a general model of it.

The Italian Systems Society, Associazione Italiana per la Ricerca sui Sistemi (AIRS), decided to devote its Second National Conference to this subject. Because AIRS is organized under the form of a network of researchers, institutions, scholars, professionals, and teachers, its research activity has an impact at different levels and in different ways. Thus the topic of emergence was not only the focus of this conference but it is actually the main subject of many AIRS activities.

Therefore the goal of this book, containing the proceedings of this conference, as well as of this short introduction, is to recall to the systems community an important challenge to be dealt with in the immediate future: the study and characterization of general features of what is commonly qualified as 'emergence', chiefly in complex systems such as biological and cognitive ones. Such a topic was a fundamental one at the very beginning of the systemic movement, and to it the founding fathers, such as Von Bertalanffy, Ashby and Von Foerster, devoted most efforts. In more recent times, however, the interests shifted towards an empirical study of systemic properties characterizing human organizations, and the subject of emergence was partly abandoned. Notwithstanding, the understanding of what is emergence, and of the circumstances which allow for its occurrence within a complex system, is of crucial importance for systemics. Namely all systemic properties – the ones which allow a system to behave as a whole and not as an aggregate of constituents – are just emergent properties.

In this regard, we remember that, starting from the sixties, many scientific disciplines, sometimes inspired by systems thinking, but most often completely unaware of it, produced a number of fundamental contributions to the topic of emergence. We can quote physics (the quantum theory of collective phenomena, the theory of symmetry-breaking phase transitions, the reformulation of quantum field theory, the theory of nonlinear phenomena, the study of classical and quantum chaos, the theory of dissipative structures, the birth of synergetics), biology (the birth of structuralism), neuroscience (the discovery of long range correlations in the brain, the discovery of role of chaotic processes in olfactory bulb, the birth of psychoneuroimmunology), cognitive science (the introduction of connectionist models of cognitive processing), artificial intelligence (the introduction of neural and neuro-fuzzy networks, of soft computing, of evolutionary algorithms, of artificial life), engineering (the birth of nanotechnology, of quantum computing, of self-designing machines), philosophy (the analysis of binding problems, of symbol grounding problems, and of concepts such as coherence and consciousness). All these developments evidenced how, in contrast to what was expected, the features of emergent phenomena, in a so wide range of different domains of inquiry, were not typical of the domain under study. In other words, emergence within the biological realm isn't intrinsically different from emergence, for instance, within the cognitive realm or within the physical realm. This allowed for a unified, and by its very nature transdisciplinary, treatment of emergence without a specific reference to a particular domain.

Here systemics enter into play. Namely, when dealing with systems, we must be aware of the existence of different levels of inquiry, which can be listed as follows:

a) the phenomenological level, in which we observe and describe the particular behaviors of a particular system, whose nature must be defined in advance in a detailed way, as a function of particular goals of the observer;
b) the modeling level, in which we build a specific model of a given system, or of a previously defined class of systems, and we use mathematics to derive from the model itself precise forecastings about system behavior;
c) the meta-modeling level, in which we reason about general properties of a wide class of models, aiming to discover general features characterizing each model of the class, independently from the domain to which it refers.

These different levels are connected both by bottom-up and top-down relationships. The bottom-up influences consist in the fact that, without experimental observations of particular systems, we cannot build models, and, in turn, without the previous existence of particular models, we cannot reason about model classes. The top-down influences are evidenced by the fact that, without having a model, not only experimental observations are devoid of any meaning, but we cannot even decide what observations should be made. Moreover, the decisions about what model to choose, between the virtually unlimited number of possibilities, and how to interpret its forecasting depend only on the results of general analyses done on large classes of models.

We claim that the main role of systemics is to give contributions at the meta-modeling level; and emergence is just a property pertaining to this level. Without this contribution every scientific progress will be blocked: the science and the engineering of the future will be almost exclusively the science and the engineering of emergent phenomena – in a word, of systems, and not of aggregates. Such a circumstance calls for a stronger engagement of systems community in a deep reflection about a number of questions relevant to emergence. Some of them are:

1. what are the possible definitions of emergence?
2. is emergence a feature only of systems described by quantum theory or even of systems described in classical terms?
3. is it possible to build many-level models, having more than the two usual levels of description (macroscopic and microscopic)?
4. is it possible to build a many-level model in which, given two immediately neighboring levels, their relationship be exactly the same as the one holding between any other different pair of immediately neighboring levels?
5. is it possible to observe two different levels at the same time?
6. how to deal with medium-size systems which are not infinitely large nor infinitely small?
7. up to what point is it possible to control the occurrence of emergence?
8. how would it be possible to build a general theory of the interaction between the observer and external world, able to contain, as a particular case, quantum theory, but without using the traditional formalism of quantum theory?

Systemics should contribute to the answering of these questions both in a conceptual and in a technical (that is mathematical) way. Namely every general proposal should be implemented in a particular model (which, of course, could be also a 'toy model'). It's time for systemics to continue a tradition lasting to researchers such as Mesarovic and Rosen, and to re-enter into the mainstream of actual scientific inquiry, by assuming a leading role.

We also want to take this opportunity to invite the entire systems and scientific community to find again the original pioneer attitude, having the ability to invent and propose new approaches, new projects, and new ideas. That's having the daring to try and propose paradigm shifts. This attitude is particularly required to systems scientists.

It is useful to add the following observations about the role of systemics.

1. Systemics is usually applied within specific disciplinary domains (maybe it is the only way to use it). A systems scientist should be very sensitive to detect and realize the systemic content of any disciplinary result. In such a way a systems scientist is able to give to theories and results a systemic interpretation and to develop an adequate systemic theorization able to generalize, amplify, and expand the disciplinary results and to design new disciplinary applications thanks to the systemic understanding.
2. Systemics is a very powerful approach, paradigm, methodology, and cultural resource, usable in any context. Nevertheless this must not be intended as an ideological or dogmatic assumption. Systems scientists must not only be ready to take into consideration, but welcome any new approach emerging from the current explicit or implicit systemic research.

The papers presented in these proceedings deal with different fields of systemics. The opening lecture by H. Haken refers to one of the more important and impressive systemic contributions, coming from a disciplinary application in physics, such as synergetics.

The conference was articulated in eight sessions dealing with the topic of emergence from different perspectives: modeling, cognitive science, theoretical issues in systemics, education, systemic approach and information technology, general systems, management, and systemic approaches to human care. Forty-six contributors authored thirty-two papers. The conference had a final panel on synergetics and systemics reported in this book under the form of short contributions by the invited participants.

We hope that systemic research will accept the challenge contained in the papers in this book trying to contribute to solve the previously quoted open problems thus preparing the basis for a new exciting phase of systemics.

Gianfranco Minati
AIRS President
Eliano Pessa
Co-Editor

Castel Ivano, November, 2001

ACKNOWLEDGMENTS

The Second Italian Conference on Systemics has been possible thanks to the contributions of many people that have accompanied and supported the growth and development of AIRS during all the years since its establishment, in 1985.

We have been honoured by the presence of Prof. Hermann Haken, by his opening lecture for this conference and for chairing the panel on synergetics and systemics.

We thank the Castel Ivano Association for hosting this conference and we particularly thank Prof. Staudacher, a reference point for the high level cultural activities in the area enlightened by its beautiful castle.

We thank also the Department of Philosophy of the University of Pavia, the Department of Information Sciences of the University of Milano (Prof. G. Mauri), the Department of Computer and Management Sciences of the University of Trento (Prof. A. Zorat), cultural sponsors of the conference.

Also, we thank all the authors who submitted papers for this conference and in particularly the members of the program committee and the referees who have guaranteed the quality of the event.

Finally we explicitly thank all the people that contributed during the conference, bringing ideas and stimuli to the cultural project of systemics and particularly M. Abram for helping so much in editing the book.

G. Minati, E. Pessa

CONTENTS

Emergence in Complex, Cognitive, Social, and Biological Systems

Opening Lecture

Synergetics - An Overview

HERMANN HAKEN
Institute for Theoretical Physics, Center of Synergetics
Pfaffenwaldring 57/4, D-70550 Stuttgart
E-mail: haken@theo1.physik.uni-stuttgart.de

1. COMPLEX SYSTEMS

As is witnessed by the program of the Italian Society for Systemics, as well as by the program of this congress, systemics deals with complex systems that are practically ubiquitous. Just to mention a few examples: schools, universities, and more generally education, hospitals, and more generally health care, factories, companies, cities, economy. In a more general frame, we may consider society. Globalization has been furthered considerably by the internet and the increasing traffic. A different field dealing with complex systems is legislation and still more pressing issues, such as ethics. Biological systems are outstanding examples of complex systems. Of particular interest is the brain. Its action is increasingly mimicked by computers as well as by computer nets. Robotics as well as multi-agent systems provide us with still more examples. There are, however, much simpler model systems in physics and chemistry, such as lasers, fluids and chemical reactions that again and again may play an important role in developing new ideas on how to cope with complex systems. A field of particular interest is quantum physics that will not be covered, however, in the subsequent talk.

Emergence in Complex, Cognitive, Social, and Biological Systems, Edited by
Minati and Pessa, Kluwer Academic/Plenum Publishers, New York 2002

2. SOME IMPORTANT APPROACHES

A number of important approaches have been developed that deal with complex systems from a unifying point of view. Let us mention some typical of them: General System Theory as developed by von Bertalanffy, Cybernetics I, developed by Norbert Wiener. His approach deals mainly with the control of biological and technical systems rather than with the self-organization taking place in complex systems. An important step into that direction was done by von Foerster in Cybernetics II. Another purely mathematical approach is Dynamic Systems Theory as initiated by Poincaré and others. Among more recent approaches, the one by Gianfranco Minati on his work on Collective Beings must be mentioned in particular.

In the following, I want to report on my own approach that I called synergetics.

3. SYNERGETICS

About thirty years ago, I coined the word synergetics in order to designate a field of research that - as I hoped - will develop. This field defined as "science of cooperation" starts from the following rather obvious observation: In nature, in technology, and also in sociology we are dealing with systems that are composed of many individual parts or elements. Because of their mutual interaction and their own properties, these systems are able to produce structures or functions at macroscopic scales. These structures or functions are not present at the microscopic level of the individual parts so that one speaks of the emergence of qualitatively new features. A few examples: In fluids, one may observe at the macroscopic level patterns, such as the formation of movement hexagons, spirals, etc., while at the microscopic level these fluids are composed of individual molecules. In the brain at the macro-level, we have phenomena, such as percepts, movement control, etc., while at the microscopic level the brain is composed by neurons and, perhaps, at a still more microscopic level by their constituents.

Finally, society while at the microscopic level composed of individual humans shows at the macroscopic level phenomena, such as language, the formation of social groups, etc.

The most striking feature of all these examples is the following fact: The macroscopic structures are not imposed on the system from the outside by specific means, rather these structures evolve via self-organization. Thus, synergetics may be considered as a theory of self-organization. It should be mentioned, however, that the close relationship between experiment and theory is a main aspect of synergetics. So far, about 90 volumes within the Springer Series in Synergetics have been devoted to this field. The basic question of synergetics is this: Are there general principles of self-organization irrespective of the nature of the parts? In spite of the great variety of the systems studied, this question could be answered in the positive provided we pay a prize, namely we confine our study on qualitative changes at macroscopic scales. Incidentally, in this way we also can unearth profound analogies between otherwise quite different systems.

4. BASIC CONCEPTS OF SYNERGETICS

The systems under consideration are exposed to external influences that are described by so-called control parameters. In a number of cases, also internally fixed parameters can be considered as such control parameters. The macroscopic structures or functions are described by so-called order parameters. These parameters play also a functional role, namely by means of the slaving principle they determine the behavior of the individual parts of the system. While the order parameters are small in number, the parts are still quite numerous. In this picture, the order parameters act like puppeteers. However, in synergetic systems, the relationship between order parameters and parts is more intricate, because the many parts determine the behavior of the order parameters and in a number of cases even define the order parameters. This leads to the concept of circular causality. The few order parameters determine the behavior of the individual parts, while the parts determine the behavior of the order parameters. When we follow up this circle in full, we have a picture in which the many variables or parts acquire a specifically ordered state, which might be described as kind of consensus between the individual parts. On the other hand, when we look at this circle from the point of view of order parameters, the behavior of the complex system is described by these rather few variables so that we obtain a pronounced information compression.

5. THE BASIC STRATEGY OF SYNERGETICS

A basic strategy, though not the only one of synergetics, is as follows: We study systems in situations, where the macroscopic structure or function changes qualitatively. In mathematical terms, we proceed as follows: We start from a stable state of the system, then change one or several control parameters and study how the system becomes destabilized. In such a situation, fluctuations act as triggers for the system so that it can acquire a new state. During such a transition period, in particular critical fluctuations may occur. When the control parameters are further changed, different collective behaviors may emerge that are described by order parameters. These order parameters may compete, cooperate, or coexist. Finally, the system gets into a new stabilized state, whereby, in general, several realizations are possible - a phenomenon known in physics as symmetry breaking.

6. PATTERN RECOGNITION BY THE SYNERGETIC COMPUTER

The concept of the synergetic computer is based on three ingredients: First of all, according to Steinbuch and later to Kohonen, pattern recognition is based on the action of an associative memory. An example is provided by a telephone book that complements the name of a person with his or her telephone number. Thus, an associative memory serves for the completion of a set of data. Furthermore, in the spirit of synergetics, we try to realize an associative memory by the dynamics governed by a set of coupled differential equations.

Last but not least, we conceive pattern recognition as pattern formation. This is illustrated by the following example: In a fluid heated from below a stripe pattern of up- and downwelling fluid can emerge. When in a circular vessel, we prescribe one upwelling stripe, eventually a full stripe pattern in that direction develops. When we prescribe two upwelling stripes in different directions, the computer simulation shows that, eventually, a specific stripe pattern evolves that points in the direction of the originally stronger upwelling stripe. In terms of synergetics, the following has happened: The initial state calls upon order parameters that compete with each other, one order parameter wins the competition and via the slaving principle brings the whole system into the corresponding ordered state. The same principle is invoked for recognition of patterns, where few features are initially given. These features call upon the corresponding order parameters that compete. One order parameter wins the competition and finally brings the system into the completely recognized state. An important ingredient of this computer concept is the appearance of what we call attention parameters that allow the computer to analyse scenes. After some preprocessing that makes the recognition process invariant against translations, rotations, and scalings, the computer can recognize a scene, for instance composed of five faces in the following way: It first recognizes one face, then the corresponding attention parameter is put equal to zero, then it recognizes the second face, and so on. The synergetic computer can also recognize depth in stereovision. In all these cases, a specific type of equations occur, namely equations that describe the competition between order parameters. I personally believe that the human brain operates along similar lines. This may be substantiated by experiments that are described in the next section. We suggest that in pattern recognition, or with respect to the brain, in perception, the dynamics is governed by order parameters. Invoking the phenomenon of symmetry breaking, an order parameter can adopt two different states (bistability). Such bistability of perception can be experienced when we look at some of the pictures by the medieval Italian painter Arcimboldo (Fig. 1), where we recognize for instance a face, but when looking more closely, we realize that the face is nothing but a bunch of fruits and vegetables.

A further example of the behavior of order parameters is hysteresis. Depending on the history, the system may acquire two entirely different states in spite of the fact that the external conditions (control parameters) are precisely the same. This can be exemplified by looking at Fig. 2, either top-down or bottom-up. In these cases, the switch from one interpretation, CHAOS, to the other one, ORDER, occurs at a different line. A third example of the behavior of order parameters is that of oscillations that can be easily realized when looking at Fig. 3, where our percept oscillates between face and vase. In Italy, these processes have been quantitatively experimentally studied by Borsellino and his co-workers and the mathematical model developed by my former co-worker Ditzinger and myself are an excellent agreement with these findings.

The synergetic computer is a beautiful example of the fruitfullness of the concepts of synergetics, for instance to reduce the complexity of a problem. In fact, the recognition task requires the solution of coupled nonlinear differential equations, whose number equals that of the number of pixels. This is in analogy to other approaches, e.g. Hopfield nets. In the present case, it is possible, however, to rigorously reduce the dimension of the problem to the - in general much smaller - number of stored prototype patterns that can be represented by their order parameters. They obey a set of rather simple nonlinear differential equations that describe the competition between order parameters and also allow for hardware realizations by semiconductors and lasers.

7. DECISION MAKING

We may go one step further and interprete decision making as pattern recognition. In both cases, an incomplete pattern in the sense of an image or in the sense of a set of abstract data must be complemented in a specific way. Following up the analogy with pattern recognition, we may expect the following kinds of decision making:

1. Unique recognition: unique decision.
2. Oscillations between two percepts: oscillations between two decisions.
3. Hysteresis. Perception depends on history: decisions depend on previous experience. This is a phenomenon that I called the manager syndrom. A manager makes a decision similar to one he has done before where he has been successful, in spite of the fact that the situation may have changed. The hypothesis that decision making acts similarly as pattern recognition, which in turn is based on similarities, leads us to propose specific learing strategies, namely to relate material that has to be learnt anew to material that has already been learnt by means of similarity relations.

8. ASSIGNMENT PROBLEMS AND MULTI-AGENT SYSTEMS

The concept to describe systems by means of order parameters that obey competition equations has been applied by us to a number of further problems that I want to sketch in the following. In assignment tasks, the following problem has to be solved: In a factory, a set of machines is given and another set of tasks. The machines, however, are differently suited for the fulfillment of such tasks, i.e. the relationship between tasks and machines may be described by cost functions. The problem is to find such an assignment that the minimum of costs is reached. A still more complicated problem that is considered to be NP-hard is the following: There are three sets given, namely machines, tasks and workers. Their mutual assignment must be arranged in such a way that the costs are again minimized. As was shown by Starke, Schanz and myself, such problems can be, at least approximately, solved by means of competition equations similar to the ones used in section 6. An explicit problem of this sort has been treated jointly with Portugali and Daffertshofer, where we studied the distribution of citizens over houses or flats in a city. In this way, one may also study the self-organization of a city, which leads to specific patterns of habitation. A whole book by Portugali is devoted to the problem of self-organization and the city. The behavior of the individual citizens can be simulated by a computer approach that realizes multi-agent systems. Here an important task is the choice of appropriate hierarchies of order parameters, for instance the order parameter of the city, the order parameters of the individuals, and finally their neurons as the basic individual parts.

Here we are dealing with interacting agents, whose goal is the solution of specific tasks. Such tasks may be externally prescribed or internally defined, for instance "survival" of the system (e.g. on another planet). An agent interacts with its surrounding (which includes other agents) by perception and action. An agent possesses memory and, e.g., decision making algorithms. It can be realized by a computer program or - invoking principles of embodiment and situatedness - by specific robots. Synergetics may offer hints on how to construct such multi-agent systems by invoking hierarchies of order parameters. A detailed account is beyond the scope of my presentation. An example may suffice here. In the above mentioned example of occupation of houses, we have three levels: The order parameter of the city that enslaves the individual humans, and the order parameters of their brains that enslave the individual neurons. Important aspects here are time-scale separation and circular causality.

9. SOME GENERAL CONCLUSIONS

From our experience on the mechanisms of self-organization in a variety of systems ranging from physics to sociology, we may draw several quite general conclusions. Self-organizing systems are characterized by the fact that the behavior of the individuals or individual parts is not directly steered from the outside, rather the system finds its organization or function by its own. Changes between different kinds of functions or structures can be induced from the outside by adequate changes of control parameters. In general, however, the destabilization of the old state by such a control parameter state can lead to basically several new states. Some of them may be wanted, some of them may be unwanted. Therefore, some guidance during the initial phase at such a transition point may be necessary. On the other hand, the use of control parameters means a considerable facilitation, because only few measures are necessary to steer a complex system. The transformation of a system from one state to another state is, of course, of vital importance in psychotherapy and a number of basic principles of synergetics have found applications there, for instance in books and articles by Schiepek.

Similar problems arise also in management when the concept of self-organization is realized, for instance by introducing a flat hierarchy. Studies of that sort can be found in works by Beisel as well as in articles by the present author.

10. ETHICS

Let us conceive ethics as a system that obeys the laws of synergetics. As all other systems, that system is also subject to the influence or changes of control parameters. Basically, there are two situations:

1. The system changes smoothly, if control parameters are changed.
2. It undergoes qualitative changes.

Let us consider the second perhaps more tantalizing situation with respect to ethics. In fact, presently we observe a dramatic change of control parameters, i.e. external conditions to which are societies exposed, for instance by dramatic changes in technology and especially biology, such as the question, on the one hand, of stem cell research and, on the other, on - to put it harshly - death control. When we consider humankind as a self-organizing system, we surely have to face the possibility of the evolution of different kinds of ethics that may act like order parameters in the sense of synergetics, namely, on the one hand, they may compete or coexist, and simultaneously they determine the behavior of the corresponding individuals. We surely will be confronted with the question whether there are principles that at a higher level determine ethics uniquely. I don't want to dare at an answer here, but rather would like to leave this question for later discussions.

11. CONCLUDING REMARKS

In my talk I have tried to give a brief outline of basic principles of synergetics and to indicate how they could be made operational in particular in mathematical terms that allow us to treat a number of problems ranging from pattern recognition to city organization. Synergetics also allows for a number of qualitative considerations, where a brief outline was given in sections 10 and 11. There is an example of how we can apply synergetics to sociology. We just mention that we can easily identify a number of order parameters, such as language, the system of laws, corporate identity, paradigms in the sense of Thomas S. Kuhn, and so on. Thus, I hope that the reader will find my article useful for his own future research.

REFERENCES

Beisel, R., 1996, *Synergetik und Organisationsentwicklung. Eine Synthese auf der Basis einer Fallstudie aus der Automobilindustrie.* (2. Auflage), Hampp, München.

v. Bertalanffy, L., 1968, *General system theory.* Braziller, New York.

Daffertshofer, A., Haken, H., and Portugali, J., 2001, Self-organized settlements. *Environment and Planning B: Planning and Design* **28**: 89-102.

Haken, H., 1981, *Erfolgsgeheimnisse der Natur. Synergetik: Die Lehre vom Zusammenwirken.* DVA, Stuttgart.

Haken, H., 1990, *Synergetic computers and cognition.* Springer, Berlin.

Haken, H., 1996, *Principles of brain functioning. A synergetic approach to brain activity, behavior and cognition.* Springer, Berlin.

Haken, H., Schanz, M., and Starke, J., 1999, Treatment of combinatorial optimization problems using selection equations with cost terms. Part I. Two-dimensional assignment problems. *Physica D* **134**: 227-241.

Starke, J., Schanz, M., and Haken, H., 1999, Treatment of combinatorial optimization problems using selection equations with cost terms. Part II. NP-hard three-dimensional assignment problems. *Physica D* **134**: 242-252.

Minati, G., 2001, *Esseri colletivi. Sistemica, fenomeni collettivi ed emergenza.* Apogeo, Milano.

Portugali, J., 1999, *Self-organization and the city.* Springer, Berlin.

Figure 1. Face or fruits and vegetables? Bistability in perception.

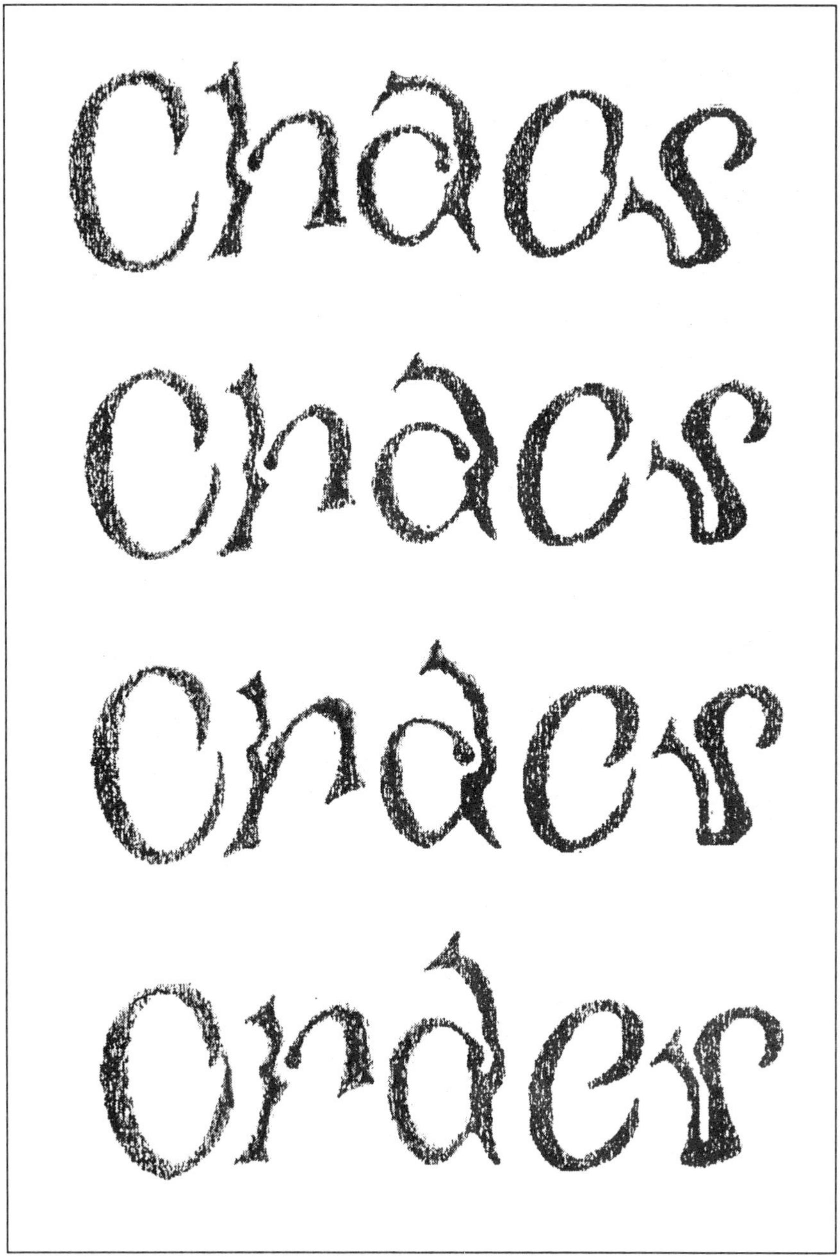

Figure 2. From CHAOS to ORDER and back again. Hysterisis in perception.

Figure 3. Oscillations between the percepts "faces" and "vase".

Modelling

A First Cellular Automata Model of Red Mullet Behaviour

FABIO BADALAMENTI*, GIOVANNI D'ANNA*, SALVATORE DI GREGORIO#, CARLO PIPITONE*, and GIUSEPPE A. TRUNFIO#
*Marine Biology Lab., IRMA-CNR, via G. da Verrazzano, 17
91014 Castellammare del Golfo, Italy - E-mail: {fbadala, danna, pipitone}@irma.pa.cnr.it;
#Department of Mathematics, University of Calabria, Arcavacata, 87036 Rende, Italy
E-mail: toti.dig@unical.it

1. INTRODUCTION

Cellular Automata (CA) are a paradigm of parallel computing; they are good candidates for modelling and simulating acentric systems, i.e. dynamical systems, whose evolution depends exclusively on the local interactions of the constituent parts (Di Gregorio and Serra, 1999).

A CA involves a regular division of the space in cells, each one characterised by a state that represents the actual conditions of the cell. The state changes according to a transition function that depends on the states of neighbouring cells and of the cell itself; the transition function is identical for all the cells. At the time $t = 0$, cells are in states, describing initial conditions and the CA evolves changing the state of all the cells simultaneously at discrete times, according to the transition function.

CA were successfully utilised to simulate fish school reactions for herring banks, predicting anti-predator behaviour as observed in nature (Vabø and Nøttestad, 1997).

We have already adopted this paradigm to model the behaviour of the greater amberjack Seriola dumerili, a carnivorous fish, during the first year of life. Further information on this model, named CANNIZZU, is not pointless, because it permits to understand the genesis of a more ambitious

Emergence in Complex, Cognitive, Social, and Biological Systems, Edited by
Minati and Pessa, Kluwer Academic/Plenum Publishers, New York 2002

project. Our first model of the red mullet behaviour is a small part of this project.

Juveniles of S. dumerili aggregate under the shade of floating objects (antropic waste or vegetal materials) that drift in the currents. If they are displaced to coastal areas, they may meet the FADs (i.e., Fish Attracting Devices used by fishermen) and may be "attracted" by them. Once "attracted" by the shade of FADs, they can be easily fished (shade fishing) (Badalamenti *et al.*, 1995; Pipitone and Andaloro, 1995; Badalamenti et al., 1998; D'Anna *et al.*, 1999). The results of the simulations with CANNIZZU show a phenomenon development that fits significantly with the data obtained in the Gulf of Castellammare and the first model validations can be considered positive. That means that CANNIZZU may be used to study optimisation processes in the disposition of FADs (Piscitelli, 2000; Piscitelli *et al.*, 2001).

Basing on those positive results we start to develop new, more general, CA models of behaviour and interaction of several fish species in the area of the Gulf of Castellammare. CA permit easily to melt models for different species, just defining their interactions in local terms and correcting the statistical part in the case that it accounts for those interactions.

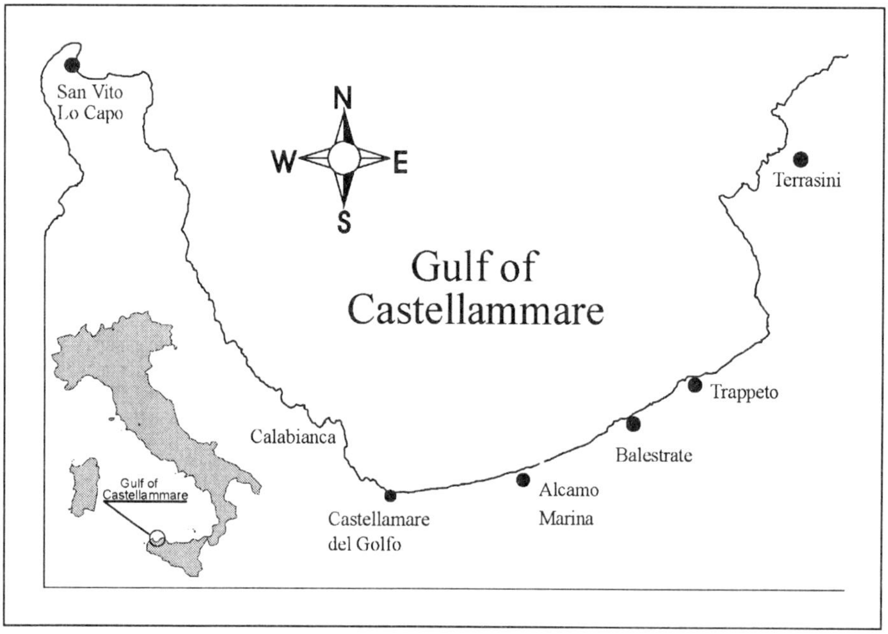

Figure 1. Gulf of Castellammare and localization.

The present work aims at modelling the distribution and abundance patterns of red mullet in the Gulf of Castellammare. The aim on the long run is to obtain an effective simulation tool for investigating the consequences of trawling on fish distribution and abundance.

In fact the Gulf of Castellammare has been interdicted to trawlers since 1990, resulting in an increase of fish biomass (Pipitone *et al.*, 1996, 2000a). A modelling tool able to forecast changes in the fish population size and structure could provide a useful support for different management options that could possibly arise in the future.

We start by the consideration that an aquatic ecosystem is a very complex system; using computer simulation and effective models, it could be possible to predict, with enough reliability, the effects of a significant variation of externals actions (e.g., fishing increase), obtaining a deeper understanding of a "global" phenomenon. The Gulf area was partitioned in Elementary Sampling Units (ESUs), each one characterized, for our purposes, by average sea depth and sea bottom quality. Every ESU is numbered according to the bathymetric band in which it falls (see fig. 2). The sea bottom quality is related to the biocoenosis concept, which considers the species assemblage (animals and plants) and the environment where they live (i.e., the substrate). Such data are expressed by conventional acronyms (Peres and Picard, 1964). Examples are: VTC that refers to the biocoenosis of coastal terrigenous mud; SFBC, that refers to the biocoenosis of fine well sorted sand. As shown in figure 3, every ESU is characterized by a specific biocoenosis. Such data are relevant because the biocoenosis affects considerably the behaviour of bottom-dwelling fish species, mainly for feeding reasons.

The first species included in our CA model is the red mullet (Mullus barbatus), that is the most abundant fish species in the Gulf (Pipitone *et al.*, 2000b) and plays a 'template' role, so that the implementation of other significant species, with similar behaviour, will require only tuning of a few parameters. This species, which is characterized by a gregarious behaviour, is a diurnal feeder that feeds continuously during the daytime period on small invertebrate prey (Lipari *et al.*, 1998). The sexual maturity is reached approximately after one year from birth; the spawning period is May-June. Red mullet larvae and pre-recruit juveniles (below 5-6 cm of length) are pelagic; juveniles live at depths smaller than 10 meter. Afterwards, the main mechanisms that determine the grown-up red mullet behaviour are related to feeding and reproductive activity, the latter only for sexually mature fish in the spawning period.

We have used knowledge of fish behaviour derived from the scientific literature (Badalamenti *et al.*, 1993; Lipari *et al.*, 1998), but our model is mainly based on the analysis of a vast amount of data available (i.e., trawl survey data collected by the IRMA staff), over the demersal (=bottom-dwelling) species living in the Gulf (Pipitone *et al.*, 2000b).

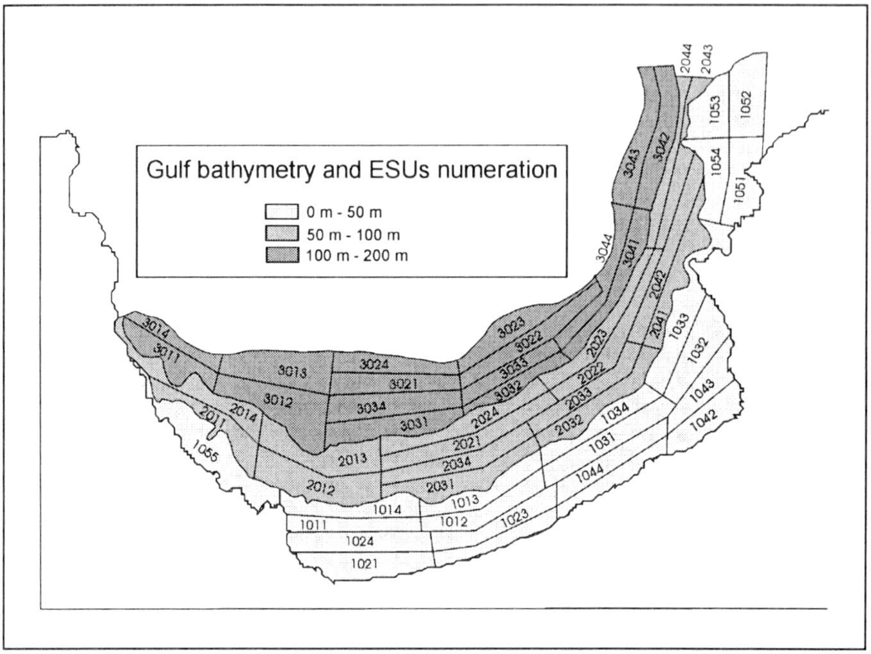

Figure 2. Gulf of Castellammare: bathymetry, ESUs subdivision and numeration.

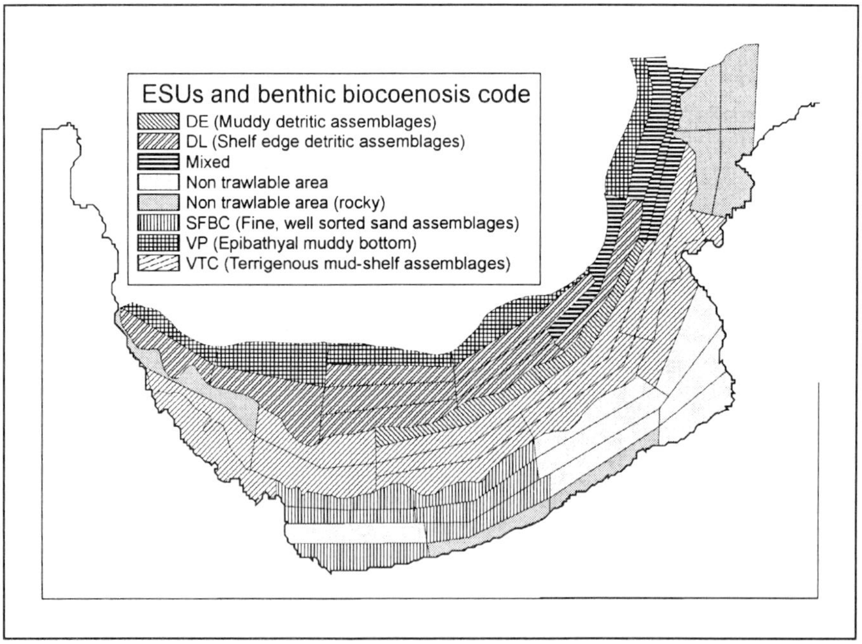

Figure 3. Gulf of Castellammare: ESUs biocoenosis code.

The biomass of the target species in the total study area, by season and year-round, was estimated by the "swept area" method (Sparre and Venema, 1992) and was expressed in terms of density (kg/km^2).

Table 1. Fully grown red mullet density (Kg/km^2) from survey data relative to year 1998 (n.s. = not sampled)

ESU	SPRING	SUMMER	AUTUMN	WINTER	ESU	SPRING	SUMMER	AUTUMN	WINTER
1011	1530	1877	1508	1692	2011	378	544	249	12
1012	n.s.	885	n.s.	1170	2012	711	470	793	559
1013	n.s.	n.s.	720	n.s.	2013	n.s.	262	510	122
1014	1817	2000	1752	n.s.	2021	202	224	n.s.	n.s.
1021	47	n.s.	211	328	2022	155	222	n.s.	16
1023	177	269	n.s.	77	2023	114	n.s.	202	67
1033	n.s.	n.s.	354	139	2024	n.s.	n.s.	114	70
1055	642	1887	n.s.	300	2031	n.s.	n.s.	967	366
					2032	n.s.	693	n.s.	163
					2033	376	547	441	n.s.
					2034	n.s.	388	305	393
					2041	46	294	224	92
					2042	201	n.s.	137	n.s.
					2043	n.s.	17	n.s.	n.s.
					2044	19	n.s.	n.s.	11

To estimate the swept area (i.e., the bottom area swept by the trawl net during one tow) the length of the tow and the wing spread of the trawl net were used. The mean density (i.e. biomass per Km^2) was estimated dividing the catch per haul (=CPUE) by the corresponding swept area. The density of fully grown red mullet as obtained from the 1998 surveys is reported in table 1. The same data are shown in fig. 4, in which grey tone intensity is proportional to the density of fish.

The analysis of trawl survey data allows to relate the concentration and distribution of red mullet to their sex and age and to the time of year, using the biocoenosis code of the bottom and the sea depth value. This makes the basis of our CA's evolution rules.

The model, according to the approach adopted by Vabø and Nøttestad (1997) and Piscitelli *et al.* (2001), supposes that red mullet reacts only to the local conditions defining simple interaction rules, deduced from studies and observations made in the Gulf of Castellammare. The application of these simple rules generates the global complex behaviour of the red mullet, i.e. birth, growing and death processes, furthermore the moving is based on attractive ESU features for feeding and/or reproduction.

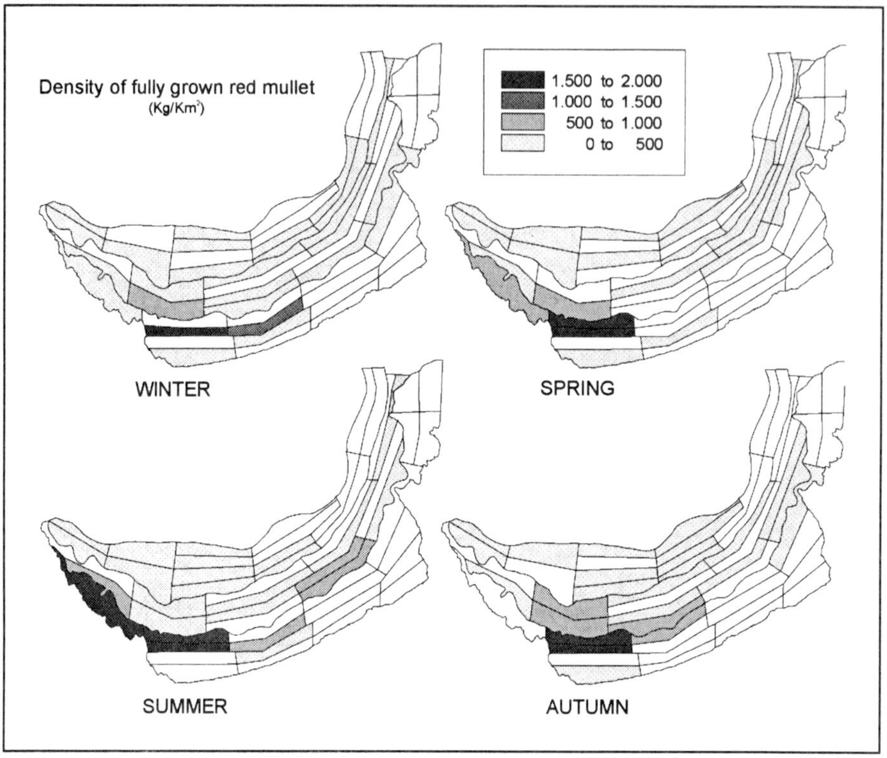

Figure 4. Gulf of Castellammare: map of fully grown red mullet survey data relative to year 1998 (blank ESUs were not sampled).

Modelling is based on a CA empirical method (Di Gregorio and Serra, 1999): each characteristic relevant to the evolution of the system and relative to the space portion corresponding to the cell is individuated as a substate. The values associated to such components can vary depending on interactions among substates inside the cell (internal transformation) and local interactions among cells.

Such a model is sketched in the next section; the transition function is specified in the third section; features of the model implementations and preliminary results are in the fourth section; at the end there are comments and the conclusion.

2. THE PROTOTYPE OF CA MODEL

Our two-dimensional CA with hexagonal cells is represented by the sextuple:

$$CA = \langle E, X, S, P, \sigma, I, \Gamma \rangle$$

where:

- E is the set of hexagons in the finite region, where the phenomenon evolves; each cell corresponds to a territory portion, mostly a sea portion.
- The set $X = \{0..6\}$ identifies, by pointers at the neighbour cells, the geometrical pattern of the cells which influences the change in the state of each one (see fig. 5).

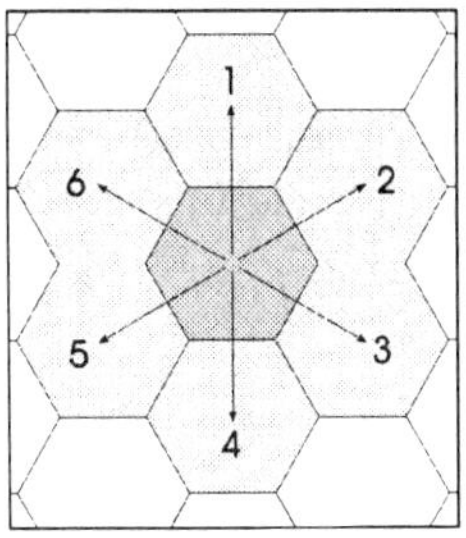

Figure 5. Hexagonal cell with her neighbours.

- The finite set S of the states is: $S = S_{FISH} \times S_{ENV}$, where S_{FISH} is the set of substates associated to the presence of the fishes in cell and S_{ENV} is the set of substates that accounts for bio-physical environment properties, relevant to the fish behaviour. Two further subdivisions in substates are introduced:

$$S_{FISH} = S_{FNB} \times S_{FJU} \times S_{FGU} \times S_{FFG} \times S_{LNB} \times S_{LJU} \times S_{LGU} \times S_{LFG}$$

S_{FNB}, substate "newborn fishes", takes the value of newborn fishes number in the cell;
S_{FJU}, substate "juvenile fishes", takes the value of juvenile fishes number in the cell;
S_{FGU}, substate "grown-up fishes", takes the value of grown-up fishes number in the cell;
S_{FFG}, substate "fully-grown fishes", takes the value of fully-grown fishes number in the cell;
S_{LNB}, S_{LJU}, S_{LGU}, S_{LFG} substates "leaving fishes" takes, respectively for each previously defined age's classes of fishes, the value of fish number that are leaving the cell (six components for the six neighbour cells);

$$S_{ENV} = S_{TSF} \times S_D$$

S_{TSF} substate "kind of sea bottom" refer to biocoenosis concept. As previously explained, every sea cell fall on an ESU that is characterized by a benthic biocoenosis code. So the substate S_{TSF} takes the value in the finite set {VTC, SFBC, DE, DL, VP, AR, MIXED, ROCKY}. The value of S_{TSF} substate contribute to determine juveniles, grown-up and full-grown fish behaviour.

S_D substate "sea depth" contributes to determine newborns and juveniles fish behaviour. It takes the zero value on the cell that fall in the land (this able to identify the coast line; the coast cells have $S_D = 0$ and at the least a neighbour cell with S_D different from zero).

- P is the finite set of global parameters of the CA, which effect the transition function, but are invariant in space: $P = \{A, \Delta t, t, L_{LNB}, L_{LJU}, L_{LGU}, L_{LFG}, R_{LNB}, R_{LJU}, R_{LGU}, R_{LFG}, V_{max}, V_{min}, W_{max}, k, k_1, k_2\}$ respectively the area of the hexagonal cell, the time corresponding to a CA step, the current time (month, day, hour and minute), the four fixed average length for each age class, the four vectors of rating scale for benthic biocenosis quality of sea bottom, the maximum and minimum average speed of fish, the maximum total weight of fish admissible on the cell, concerning the flows of fish between cells and parameters concerning the growth rate. Such parameters will be explained in the next section.
- $\sigma : S^7 \rightarrow S$ is a part of the transition function, it accounts for growth, death and outflows (inflows are deduced trivially by the outflows) of fish in the cell.
- $I \subset E$ individuates the cells, where external temporary actions effecting the states of selected cells (e.g., fishing).
- $\Gamma = \{\gamma_1, \gamma_2\}$ is the set of functions representing an external influence on CA. They are computed at each step before the application of σ.

  $\gamma_1 : \Delta t \times t \rightarrow t$ determines the current time of the CA step (the behaviour changes according the season);

  $\gamma_2 : N \times I \rightarrow S_{FISH}$ accounts for external actions (e.g., fishing) to cells of I at prefixed steps, N is the set of natural numbers identifying the CA steps.

3. THE TRANSITION FUNCTION

3.1 The Preparatory Phase

The preparatory phase fixes the value of substates "kind of sea bottom" and "sea depth" by simple import from available GIS data. The sea cells are initialised in the following way:

- for each cell the fish number substates S_{FNB}, S_{FJU}, S_{FGU} and S_{FFG} takes the initial value according to a statistical analysis of the real observed behaviour;

– substates "leaving fish" takes the value zero.

3.2 The Evolution Phase

The evolution phase, specified by the transition function, is decomposed in two internal transformations: (growth process in the cell, death process in the cell) and in the local interaction fish diffusion. The computation is applied to each cell according the following stages.

In our model the growth process is discrete and consists in transition from one age's class to her upper. Therefore the fish number substates S_{FNB}, S_{FJU}, S_{FGU} and S_{FFG} are first updated accounting on the growth processes. The current length of the fish is obtained by the growth law of Von Bertalanffy (1968): the length $L(t)$ at time t is depending on the maximum length L_{max} that can be reached by the fish, on the length L_1 at time t_1 and on a parameter k.

$$L(t) = L_{max} - (L_{max} - L_1)\, exp\, [-k\, (t - t_0)]$$

Comparing $L(t)$ with average length that characterizes any age class we compute current values of fish number substates. Since length of the fish is related to the weight $W(t)$, is then possible to compute new value of total weight of fish on the cell (this is necessary in order to keep count of maximum weight admissible for the cell);

Therefore the fish number substates S_{FNB}, S_{FJU}, S_{FGU} and S_{FFG} are updated accounting on the death processes. The rate of natural mortality at any time t, which we shall denote by $\Delta N / \Delta t$, depends on the number N of fish present at that time, and we assume:

$$\Delta N / \Delta t = - M\, N$$

where M is the natural mortality coefficient. This formula, used by several authors, is taken to imply that natural death is due to a large number of causes acting randomly, and that the probability of a particular fish dying between any time t and time $t + \Delta t$, is constant. We may expect the natural mortality coefficient to vary with age of fish: it decreases when the fish grows, because its surviving capacity increases. The following formula for M is derived from experimental data, using multiple regression analysis, and accounts for the previous considerations:

$$log_e\, M = k_1 + log_e\, L_{max} + k_2\, log_e\, k$$

where L_{max} is the maximum length that can be reached by the fish, k_1, k_2 are opportune constants.

The computation of new values of substates "leaving fish" considers one case for each of first three age's class, and two cases for fully-grown fishes (one for reproduction time of the year and one for the normal feeding activity). The main idea is that the red mullet, that pastures the sea bed, spends a time proportional to the abundance of food on the sea bed; the presence of some food items could be dramatically altered by trawling. All rules are mainly based on the computation of average speed of fish in cell V_a from P and S_{ENV}. The former include rating by fish of both sea bottom (depending from biocenosis code) and sea depth. Then, by geometric and probabilistic considerations, we derive substates "leaving fish" according to the following formula:

$$F = N\,a\,V_a\,\Delta t\,/\,A$$

where F is the value of fish number that are leaving the cell, A is the area of the cell, a is the length of cell's edge, Δt if CA time step and N is the fish number in the considered age class. Correction is necessary in order to consider the maximum weight of fish admissible on the cell. Note that for each cell the net value of flow at every step is the outcome of algebraic sum of outgoing flows and incoming flows.

The fish number substates S_{FNB}, S_{FJU}, S_{FGU} and S_{FFG} are up-dated accounting on the new values of substates "leaving fish". These stages are cyclically executed; actions of the interactive phase may be inserted at the beginning of each cycle.

3.3 The Interactive Phase

At the moment simulation of fishing with bottom trawl nets is in a preliminary phase. We will improve this phase introducing better modelling of trawling effects on the sea bottom.

4. IMPLEMENTATION AND PRELIMINARY APPLICATIONS

The preliminary implementation of CA was developed in the programming language C++ using a RAD environment (see fig. 6). Result is an effective simulation tool that, at the present, is mainly used for improving and validating our algorithm.

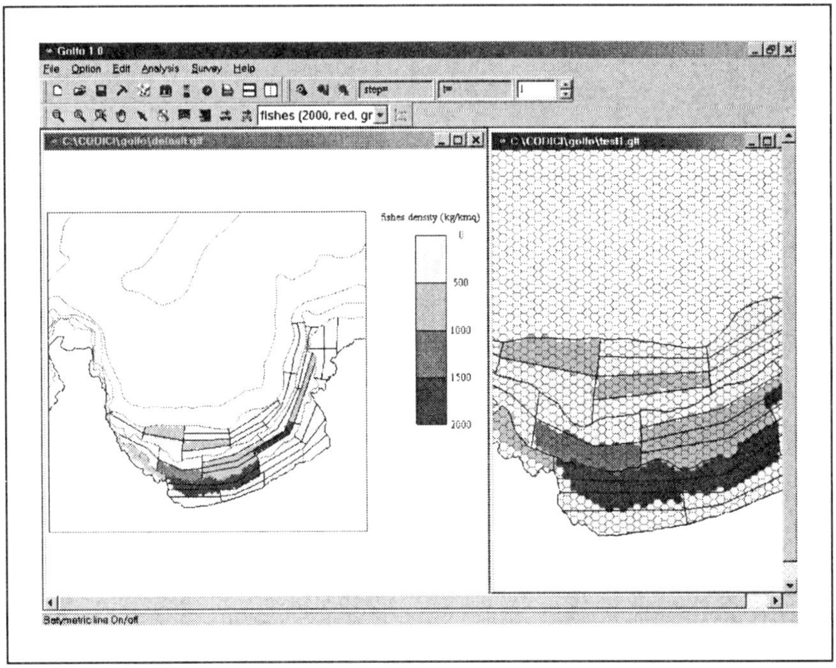

Figure 6. Screen shot of simulation program with initial configuration loaded (winter red mullet distribution from survey data).

The simulated area (the Gulf of Castellammare in our case) is visualised at each step by a matrix, whose cells represent the total fish biomass in the cell with colours or grey tones. The simulation can be interrupted at any given moment for introducing the external actions (e.g., trawling). Data may be recorded during the simulation and comparison may be executed with the statistical data available.

Many simulations were performed considering different initial conditions or different conditions related to the interactive phase. Some parameters, whose values cannot be obtained accurately by observations, have been tuned on the basis of the results of the simulations. So the model was improved appreciably.

Figure 7 refer to an example of simulation in which initial configuration is represented in figure 6 (that fit with winter survey data). We have used a 80x54 matrix of hexagonal cell each one of area equal about to 0.12 km^2. Starting day is 15/01 and each step corresponds to one hour. Simulation is carried out until summer in 5328 steps.

In table 2 we compare the simulation results concerning the red mullets average density with the corresponding available survey data.

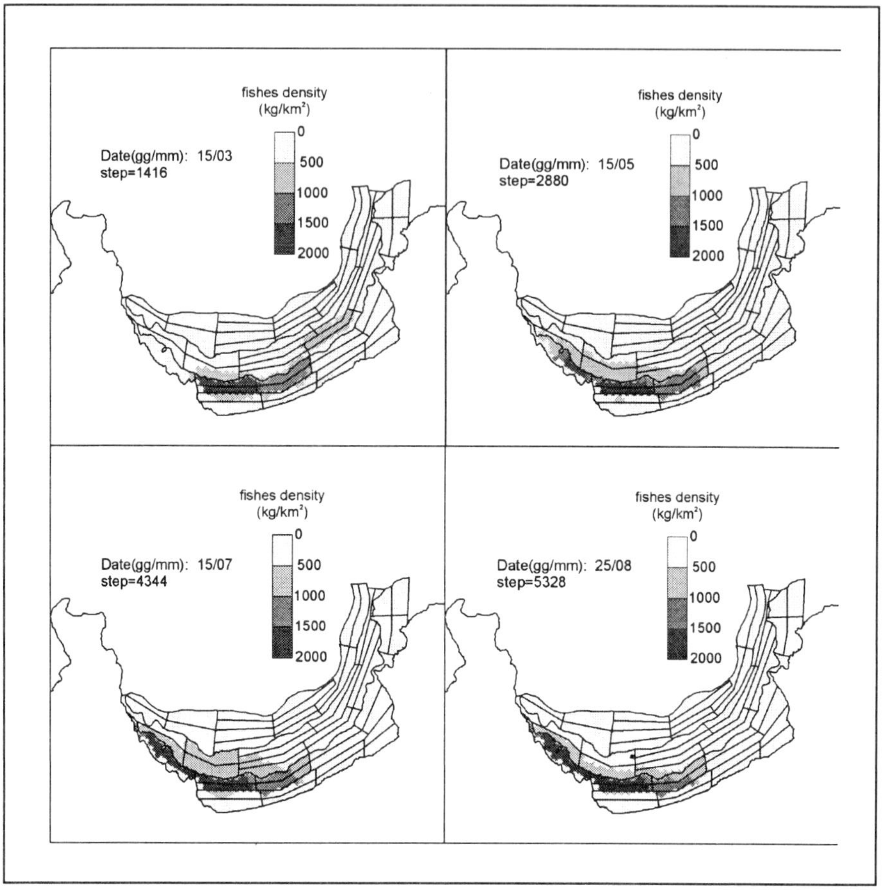

Figure 7. Example of fish density evolution during simulation (grown-up red mullets move from the shallow zone to VTC and SFBC biocoenosis zone).

As shown in figure 7 and in table 2, first simulations account for one of the main characteristics of the phenomenon: the diffusion of fishes. To compare real data with simulation data needs a statistical elaboration of available survey data. That's because observations were performed in different times and each observation did not cover all the ESUs area. Therefore it is necessary to consider groups of near observations, that cover all the ESUs, but a statistical processing is always necessary because of the data superpositions (the same ESU may be involved in more observations of a group).

A comparison is performed in table 2 between average density of fish in simulation final configuration (corresponding about to middle summer) and survey data relative to summer 1998. We judge that approximately 10%

difference between real data and simulation data represents an acceptable result on the base of the previous considerations.

The results are compatible with the experimental data except two cases: ESU 2043 and ESU 2032. The discrepancy for ESU 2043 is justified by the particular shape of the ESU, de facto divided in two parts and by the low values of red mullet density. The lower discrepancy for ESU 2032 is more difficult to understand; we suspect that some component of the red mullet behaviour is not appropriately introduced in the model.

Table 2. Outline comparison between average density of fish in final configuration of simulation and survey data relative to summer 1998 (cells with depth less than 20 meters were excluded in the computation of the mean)

ESU	CA Result (Kg/m^2)	Survey Data (Kg/m^2)	Err(%)	ESU	CA Result (Kg/m^2)	Survey Data (Kg/m^2)	Err(%)
1011	1838	1877	2	2011	569	544	5
1012	967	885	9	2012	509	470	8
1014	1989	2000	1	2013	279	262	6
1023	282	269	5	2021	237	224	6
1055	1876	1887	1	2022	246	222	11
				2032	494	693	29
				2033	488	547	11
				2034	420	388	8
				2041	282	294	4
				2043	24	17	41

5. CONCLUSION

The results of the simulations show a phenomenon development that fits not only qualitatively but also quantitatively with fish behaviour deduced by data obtained in the Gulf of Castellammare. Although abundant, such data are partial because of the great extension of the territory under control. The first simulations can be considered positive in the limits of a first approach to the problem. They account for the many characteristics of the phenomenon; some parameters have been tuned in the simulation phase; so the model has been improved on the basis of the results of the simulations. Smaller hexagonal cells could improve the simulation precision.

This model, improved and validated with further observations, could be the first significant step towards a more general model extended to other species for investigating the consequences of trawling on fish distribution and biomass.

REFERENCES

Badalamenti, F., D'Anna, G., Fazio, G., Gristina, M., and Lipari, R., 1993, Relazioni trofiche tra quattro specie ittiche catturate su differenti substrati nel Golfo di Castellammare (Sicilia N/O). *Biol. Mar., Suppl. al Notiziario S.I.B.M.* **1**: 145-150.

Badalamenti, F., D'Anna, G., Lopiano, L., Scilipoti, D., and Mazzola, A., 1995, Feeding habits of young-of-the-year greater amberjack *Seriola dumerili* (Risso, 1810) along the N/W Sicilian coast. *Sci. Mar.* **59** (2-3): 317-323.

Badalamenti, F., Fazio, G., Zagami, G., and D'Anna, G., 1998, Ciclo alimentare giornaliero e selezione delle prede di *Seriola dumerili* associata a FADs di superficie nel Golfo di Castellammare. *Biol. Mar. Medit.*, **5** (1): 71-79.

D'Anna, G., Badalamenti, F., and Riggio, S., 1999, Traditional and experimental floating fish aggregating devices in the gulf of Castellammare (N/W Sicily): results from catches and visual observations. *Sci. Mar.* **63** (3–4): 1-10.

Di Gregorio, S., and Serra, R., 1999, An empirical method for modelling and simulating some complex macroscopic phenomena by cellular automata. *Future Generation Computer Systems* **16**: 259-271.

Lipari, R., Badalamenti, F., and D'Anna, G., 1998, Relazioni trofiche e selezione alimentare di Mullus barbatus L. 1758) nella comunità a sabbie fini del Golfo di Castellammare (Sicilia N/O). *Biol. Mar. Medit.* **5** (1): 513-516.

Peres, J. M., and Picard, J., 1964, Nouveau manuel de bionomie benthique de la Méditerranée. *Rec. Trav. Stn. Mar. Endoume* **31**: 1-137.

Pipitone, C., and Andaloro, F., 1995, Food and feeding habits of juvenile greater amberjack, *Seriola dumerili* (Osteichthyes, Carangidae) in inshore waters of the central Mediterranean sea. *Cybium* **19**(3): 305-310.

Pipitone, C., Badalamenti, F., D'Anna, G., and Patti, B., 1996, Divieto di pesca a strascico nel Golfo di Castellammare (Sicilia nord-occidentale): alcune considerazioni. *Biol. Mar. Medit.* **3** (1): 200-204.

Pipitone, C., Badalamenti, F., D'Anna, G., and Patti, B., 2000a, Fish biomass increase after a four year trawl ban in the gulf of Castellammare (N/W Sicily, Mediterranean sea). *Fish. Res.* **48**: 23-30.

Pipitone, C., Badalamenti, F., D'Anna, G., Whitmarsh, D., Pickering, H., and James, C., 2000b, Trawling ban in the gulf of Castellammare: effects on the small scale fishery economics and on the abundance of fish. *Final report of Study 97/063, EC DG XIV.*

Piscitelli, M., 2000, Laurea graduation thesis (in Italian) *Sviluppo di un modello ad Automi Cellulari per la simulazione del comportamento e dello sviluppo della Seriola dumerili.* Dept. of Mathematics, Univ. of Calabria.

Piscitelli, M., Badalamenti, F., D'Anna, G., and Di Gregorio, S., 2001, A Cellular Automata Model of Fish-Aggregating-Devices Performance. In *Proceedings of EUROSIM 2001 (The 4th International EUROSIM congress "Shaping the future with simulation"),* June 26-29 2001, Delft, The Netherlands, CdRom ISBN: 90-806441-1-0.

Vabø, R., and Nøttestad, L., 1997, An individual based model of fish school reactions: predicting antipredator behaviour as observed in nature. *Fish. Oceanogr.* **6**(3): 155-171.

Von Bertalanffy, L., 1968, *General systems theory.* Bartiller, NewYork.

Simulations of Forest Fires by Cellular Automata Modelling

ERMINIA BENDICENTI[*], SALVATORE DI GREGORIO[*], FRANCESCO M. FALBO[#], and ANGELA IEZZI[*]
[*]*Department of Mathematics, University of Calabria, Arcavacata, 87036 Rende, Italy*
[#]*Comando Provinciale Vigili del Fuoco CS, Viale Repubblica 56, 87100 Cosenza, Italy*

1. INTRODUCTION

Forest fires represent a serious environmental problem, whose negative impact is becoming day by day more worrisome.

Forest fires are very complex phenomena; that need an interdisciplinary approach. The adopted method to modelling involves the definition of local rules, from which the global behaviour of the system can emerge. The paradigm of Cellular Automata was applied and the model ABBAMPAU was projected to simulate the evolution of forest fires.

Cellular Automata (CA) are a paradigm of parallel computing; they are good candidates for modelling and simulating complex dynamical systems, whose evolution depends exclusively on the local interactions of their constituent parts (Di Gregorio and Serra, 1999).

A CA involves a regular division of the space in cells, each one characterised by a state that represents the actual conditions of the cell. The state changes according to a transition function that depends on the states of neighbouring cells and of the cell itself; the transition function is identical for all the cells. At the time $t = 0$, cells are in states, describing initial conditions, and the CA evolves changing the state of all the cells simultaneously at discrete times, according to the transition function.

The CA features seem to match the system "forest fire"; the parameters, describing globally a forest fire, i.e. propagation rate, flame length and

Emergence in Complex, Cognitive, Social, and Biological Systems, Edited by Minati and Pessa, Kluwer Academic/Plenum Publishers, New York 2002

direction, fireline intensity, fire duration time, etc. are mainly depending on some local characteristics i.e. vegetation type (live and dead fuel), relative humidity, fuel moisture, heat, territory morphology (altitude, slope), etc. The only global characteristic is given by wind velocity and direction, but wind velocity and direction is locally altered according to the morphology; therefore wind has also to be considered at local level.

The fundamental studies of Rothermel (1972, 1983) point out these characteristics; its forest fire spread model is based on dynamic equations, that allow the forecast of the rate of spread and the reaction intensity knowing certain properties of the fuel matrix (fuel particle, fuel arrangement) and environmental conditions (slope, wind, fuel moisture) in which the fire occurs.

Some models were previously developed in terms of CA: Green (1983a,b) considered a two-dimensional square CA, where each cell is identified by fuel content and its combustion duration. The fire-spread mechanism is shortly the following: each burning cell generates an ellipse with a focus at the cell centre; the ellipse dimension and orientation is depending on the wind direction and strength. If a cell with fuel content is inside an ellipse, it will burn at the next steps.

Gonçalves and Diogo (1994a,b) developed a more complex probabilistic CA model based on the Rothermel's (1972, 1983) equations of fire rate spread. A burning cell determines, according the fire rate spread toward the direction of neighbouring cells, an ignition probability.

Malamud and Turcotte (2000) developed a statistical model of comprehensive effects in a forest for the two opposite phenomena of trees taking root in free areas and trees burnt by the fire.

The proposed CA model ABBAMPAU (Iezzi, 2000) is deterministic, it is related to Rothermel's studies (1983) and is based on an empirical method (Di Gregorio and Serra, 1999) that may be applied to some macroscopic phenomena in order to produce a proper CA model. The main points of the method are here illustrated shortly:

- The state of the cell must account for all the characteristics, relevant to the evolution of the system and relative to the space portion corresponding to the cell; e.g. the altitude. Each characteristic must be individuated as a substate. The substate value is considered constant in the cell.
- As the state of the cell can be decomposed in substates, the transition function may be also split in many components. We distinguish two types of components: internal transformation, depending on the substates of the cell and local interactions, depending on the substates of the cells in the neighbouring. Local interactions may be treated in terms of flows of

some quantity or propagation of properties towards the neighbouring cells, in order to reduce "unbalance" conditions, e.g. fire propagation.
- Special functions must supply the history of "external influences" overall on the CA cells and direction.
- Some cells represent a kind of input from the "external world" to the CA; it accounts for describing an external initial influence, which cannot be described in terms of rules; e.g. the fire starting points.

This paper illustrates in the next section the model ABBAMPAU, the CA transition function is described in the successive section; the fourth section treats the model implementations and first results of applications; at the end comments conclude the paper.

2. THE CA MODEL ABBAMPAU

A CA was developed for simulating forest fire; its was named **ABBAMPAU** (**AB**ating **B**urning danger through **A** **M**odel **P**rovided by cellular **AU**tomata - to be read "ahb-bahm-'pah-oo", as the acronym was devised to mean "it blazed up" in Calabrian and Sicilian language).

A first release of ABBAMPAU was developed (Iezzi, 2000); improvements have been added in order to better simulate real events, in particular, fire jumps were not considered. The following model represents our more complete version that was partially implemented on a real case.

ABBAMPAU is a two-dimensional CA with square cells:

$$\text{ABBAMPAU} = \langle R, X, S, P, \sigma, I, \Gamma \rangle$$

where
- $R = \{(x, y) | \ x, y \in N, 0 \leq x \leq l_x, 0 \leq y \leq l_y\}$ is the set of points with integer co-ordinates in the finite region, where the phenomenon evolves; N is the set of natural numbers; each cell corresponds to a territory portion.
- The set $X = \{(i, j) \mid -5 \leq i, j \leq 5\}$ identifies the geometrical pattern of the cells, which influences the change in the state of each one. In the fact, the Moore neighbouring, respectively the central cell and the surrounding ones ($\{(i, j) \mid -1 \leq i, j \leq 1\}$) is sufficient for all the local interactions, except that regarding the fire jumps.
- The finite set S of the states of the cell is:

$$S = S_A \times S_V \times S_T \times S_H \times S_C \times S_D \times S_{WD} \times S_{WR} \times S_J \times (S_{FS})^8$$

S_A, substate "altitude", takes the altitude value of the cell;

S_V, substate "vegetation", specifies the type of vegetation, relatively to the properties of catching fire and burning;

S_T, substate "temperature", takes the temperature value of the cell;

S_H, substate "humidity", takes the relative humidity value of the cell;

S_C, substate "combustion" for the two possible fire types in the cell: "surface fire" and "crown fire"; it takes one of the values "non-inflammable", "inflammable", "burning" and "burnt" for each type.

S_D, substate "duration", takes the value of the duration rest of the fire in the cell;

S_{WD}, substate "wind direction", takes the values of the wind directions (the eight directions of the wind rose) at ground level (that could be different from the free wind direction);

S_{WR}, substate "wind rate", takes the values of the wind rate (Km 0-60) at ground level (that could be different from the free wind rate);

S_J, substate "fire jump", takes the values 0, 1, 2, 3 depending on the possible maximum extension of a fire jump from the cell;

S_{FS}, substate "fire spread", accounts for the fire spread from the central cell to the other neighbouring cells; S_{FA}, substate "fire acquire", is just a renaming of S_{FS} and individuates the fire propagation to the central cell from the other neighbouring cells.

- P is the finite set of global parameters, which affect the transition function; they are constant overall in the cellular space:

  p_e, the size of the cell edge; in this case the size of the cell is 20 meters.

  p_s, the time corresponding to an ABBAMPAU step; the step of ABBAMPAU is 3 minutes.

  P_v, the set of parameters concerning catching fire and burning, depending on the type of vegetation; they are also constant in time.

  Other parameters are:

  p_t, the current time (month, day, hour and minute).

  p_w, the weather conditions depending on the sun and concerning a fixed characteristics: exposure and characteristics, varying during the day according the season: reference values of humidity and temperature (such values are correct considering the type of vegetation, the wind and in the case of a previous raining episode as far as a month.

  p_{wd}, the free wind direction.

  p_{wr}, the free wind rate.

  Note that the function γ_1 keeps p_t up-to-date, the function γ_2 supplies the history of p_w, p_{wd} and p_{wr} at each CA step.

- $\sigma : S^{121} \rightarrow S$ is the transition function, that will be sketched in the next section; it accounts for the following aspects of the phenomenon: effects of combustion in surface and crown fire inside the cell, crown fire triggering off; surface and crown fire spread, determination of the local wind rate and direction, fire jumps.

- $I \subset R$ individuates the cells, where the fire starts.
- $\Gamma = \{\gamma_1, \gamma_2, \gamma_3, \gamma_4\}$ is the set of functions representing an external influence on ABBAMPAU. They are computed at each step before the application of σ.

  $\gamma_1 : p_s \times p_t \rightarrow p_t$ determines the current time of the CA step;

  $\gamma_2 : N \rightarrow p_w \times p_{wd} \times p_{wr}$ supplies the weather conditions related to sun, the external wind direction and rate, N is the set of natural numbers identifying the ABBAMPAU steps;

  $\gamma_3 : N \times I \rightarrow S_C$ accounts for external setting fire to cells of I at prefixed steps;

  $\gamma_4 : N \times R \rightarrow S_C$ accounts for firemen intervention at prefixed steps.

3. THE TRANSITION FUNCTION

The transition function involves the following internal transformation, concerning the effects of combustion in surface and crown fire inside the cell:

$$\sigma_{T1} : S_V \times S_T \times S_H \times S_C \times S_D \times S_{WD} \times S_{WR} \times P_v \times p_w \times p_t \rightarrow S_T \times S_H \times S_C \times S_D.$$

The following local interactions that compute, in order, the wind direction and rate at the cell altitude, the change of combustion conditions cell, the fire spread toward the neighbouring cells:

$$\sigma_{I1} : (S_A)^9 \times p_w \times p_t \times p_{wd} \times p_{wr} \rightarrow S_{WD} \times S_{WR}$$

$$\sigma_{I2} : (S_{FA})^8 \times S_C \rightarrow S_C$$

$$\sigma_{I3} : (S_A)^9 \times S_V \times S_C \times S_H \times S_T \times S_D \times S_{WD} \times S_{WR} \times p_s \times p_e \times P_v \rightarrow (S_{FS})^8$$

$$\sigma_{I4} : (S_A \times S_{WD} \times S_{WR})^{121} \times S_J \times S_C \times S_D \times p_s \times p_e \times P_v \rightarrow S_C \times S_J$$

The internal transformation and the local interactions are computed in the same order of the presentation. the transition functions is computed

T1) When the substate S_C is not "burning", then it doesn't change together the substate S_D, while the substates S_H and S_T vary their previous values on the basis of the weather change (p_w), the day hour (p_t), the wind (S_{WD} and S_{WR}) and the vegetation type (S_V and P_v). When the substate S_C is "burning", then the substates S_T, S_H, S_C and S_D depend on the previous values of S_H, S_D and S_T. The value 0 for S_D determinates the change of S_C to "burnt". The conditions for the ignition from fire surface to crown fire are applied in T1.

I1) The computation of the substates S_{WD} and S_{WR}, "wind direction and rate" depends on p_{wd} and p_{wr} that represent the values of the free wind. Such values are reduced in relation to the altitude of the cell according to an empirical table of Rothermel (1983), obtaining the wind vector, related to the cell altitude. A corrective vector wind is computed, considering the angle between the slope relative to the cell and the free wind direction. It accounts on that, in absence of free wind, convective currents form along the maximum slope directions; their rate depends also on the weather and the hour of the day. Empirical tables of Rothermel (1983) indicate the value of such vector. The new values of S_{WD} and S_{WR} are obtained adding the altitude vector to the corrective vector.

I2) This functions tests if the fire is spreading toward the central cell from the other cells of the neighbouring. If the combustion substate S_C is "inflammable", then it changes to "burning".

I3) Three computation steps are considered, if the state S_C is "burning": a) it is calculated a circle, whose radius is proportional to the fire intensity, depending on substates S_V, S_H, S_T, S_D and the set of parameters P_v; of course the time corresponding to the step of ABBAMPAU p_s and the size of p_e effect also the size of the circle; b) such circle is transformed in ellipse with a focus in the centre of the cell, considering the effects of the wind (substates S_{WD} and S_{WR}); c) the fire can propagate towards the neighbouring cells inside the ellipse, i.e. S_{FS} takes the value "true".

I4) Each cell is tested if it can be reached by incandescent material from other cells of neighbouring. The "fire jump" conditions depend on the wind rate and direction, morphology and combustion of the cell: the cell fire intensity is computed; if it is larger than a threshold, then the slope along the wind direction is considered; if the wind rate is enough strong, then the incandescent material can deposit or fly over the area of the cell. In the former case the cell will became "burning", in the latter case, the cell can be the origin of a jump fire.

4. IMPLEMENTATION AND PRELIMINARY APPLICATIONS

The implementation of ABBAMPAU was developed in the programming language Delphi. Minor features of the model were not implemented: cells are considered burning indefinitely (lack of substate S_D and value "burnt" of S_C), but this absence does not affect the simulation; the parameter p_w, is not considered to be influenced by the wind (S_{WD} and S_{WR}), because a satisfying more precise correlation is not significant at this level of the model.

A first validation of this new release of ABBAMPAU was tested on a real case of fire forest (fig. 1) in the territory of Villaputzu, Sardinia island, August 22nd, 1998, (SALTUS, 1998-2000). This forest fire was a case treated and monitored in the milieu of European Research Program SALTUS, Environment and Climate, (1998-2000). The beginning time was very nearly 12.30 and lasted less than 9 hours. The wind was in direction SE at a rate 20-30 Km/h. The burnt area is approximately Km^2 3.

The forest fire area in Fig. 1 is evidenced by grey tones, whose labels specify the different type of burnt vegetation:

- Label "A" indicates a vegetation of cistus, (on the left top, at the starting point of forest fire);
- Labels "Y_1" indicate the prevalent vegetation, maquis with cistus, phillyrea and lentiscus;
- Labels "Z" indicate a forest of conifers with some oaks;
- Label "C" indicates a vegetation of maquis with junipers.

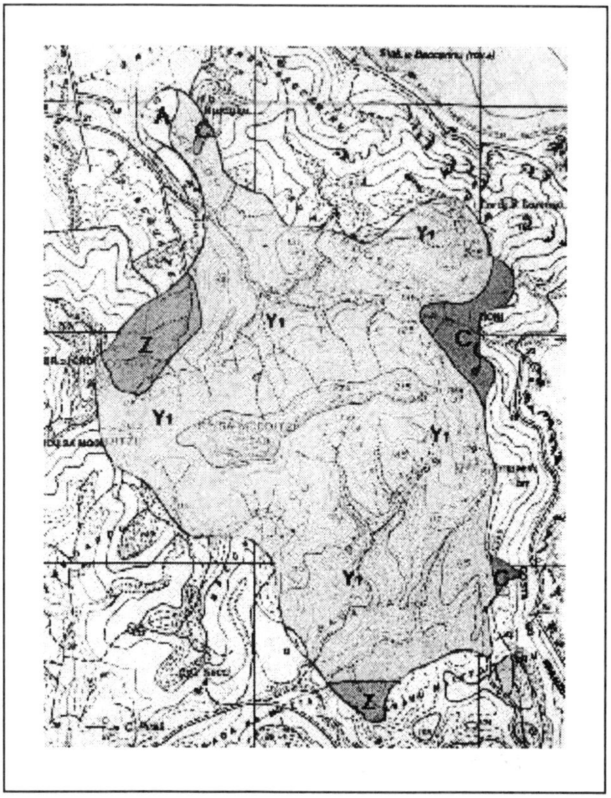

Figure 1. The forest fire of Villaputzu, August 22nd, 1998.

The dotted lines individuate the roads. The roads on the left and on the right of the initial area of the fire represents a barrier, because the human intervention to limit the fire spread is easily performed in the zones adjacent to roads.

The simulated area (fig. 2) is visualised at each simulation step by a square matrix with 168 rows and 131 columns; the cell edge is m 20; the sea is light-grey, the burning cells are grey, (the darker grey colour is the surface fire, the less dark grey colour is the crown fire) and natural or artificial fire barriers (also determined by the firemen intervention) are black.

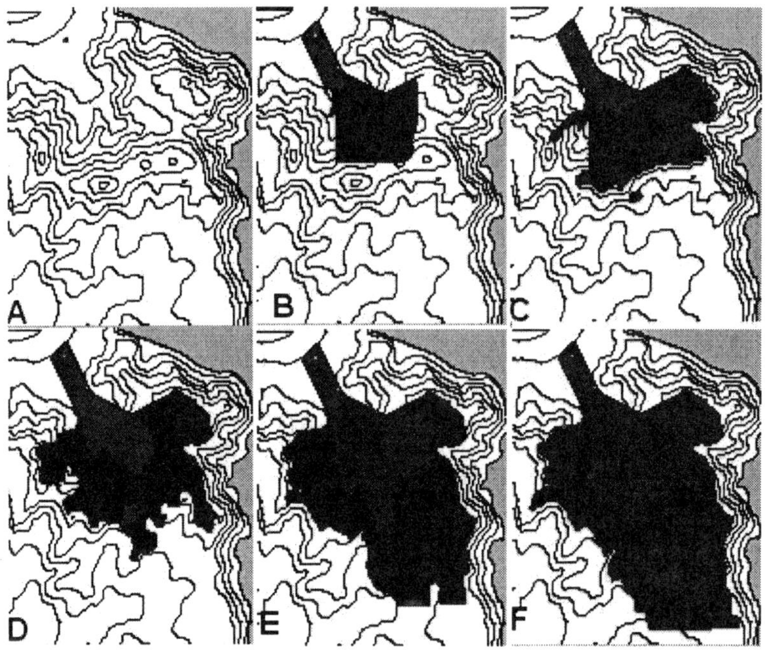

Figure 2. Sequence of six simulation steps: figures A, B, C, D, E, F correspond respectively to steps 3, 70, 130,147, 162, 175.

The forest fire started on the left top of fig. 1; the third step of the simulation (fig. 2-A, less than 10 minutes) individuates the first cells burning. In the step 70 (fig. 2-B, 3.5 hours), the fire is reaching rapidly the peaks of mountain; a barrier is introduced in the fire beginning area, because an easy human intervention was performed there. The fire propagates along the peaks and a fire jump has occurred after the left peak (fig. 2-C, 6.5 hours).

The fire is propagating according the wind direction. The positive slope enlarges also the fire toward E, NE; (fig. 2-D, less than 7.5 hours). Then the propagation toward the wind direction is almost exhausted, its rate slows with negative slope (fig. 2-E, almost 8 hours). The only propagation directions are now toward SW and the fire died out (fig. 2-F, less than 9 hours).

Fig. 3 represents the superimposition of the figures of real fire on simulated fire (left) and vice versa (right).

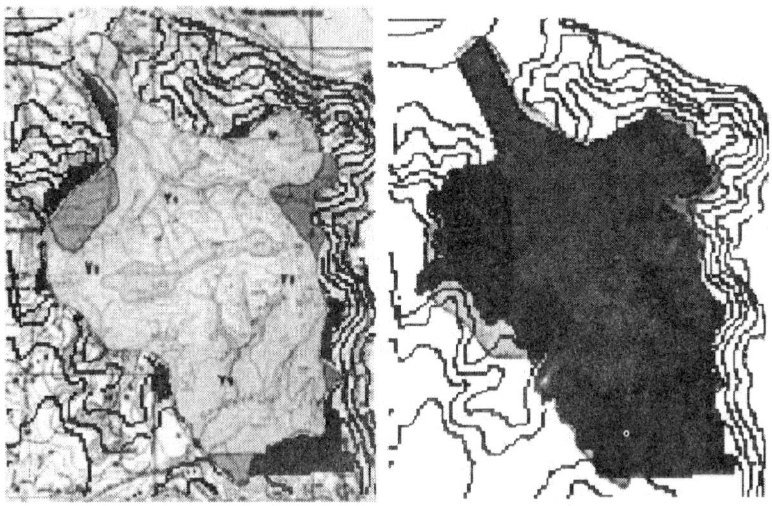

Figure 3. Simulated forest fire vs real forest fire of Villaputzu.

The result of simulation looks good, few areas of the real fire don't appear to be burnt in the simulation and few areas, which were not burnt, are computed to be "burnt" in the simulation.

5. CONCLUSION

The results of the simulations show a phenomenon development that fits significantly with real data but it is necessary to stress some problems:

- standard fire records don't account for all the data necessary to the simulation; e.g. firemen intervention was inserted in simulation according to good sense considerations;
- vegetation data are usually incomplete;

- Rothermel tables (1972, 1983) must be modified and completed according the features of European vegetation; they must be improved in consideration of local humidity;
- the spurious symmetries of the square tessellation can effect negatively the simulation in flat areas; our final results are good because the presence of flat areas is not relevant.

A first improvement of the model involves the introduction of hexagonal cells; the wind directions must be increased; fires with more complete data

ACKNOWLEDGEMENTS

The authors thank the components of European Research Program SALTUS, Environment and Climate, (1998-2000) for the precious data utilised in this research.

REFERENCES

Di Gregorio, S., and Serra, R., 1999, An empirical method for modelling and simulating some complex macroscopic phenomena by cellular automata. *Future Generation Computer Systems* **16**: 259-271.

Gonçalves, P., and Diogo, P., 1994a, Geographic Information Systems and Cellular Automata: A New Approach to Forest Fire Simulation. In *Proceedings of The European Conference on Geographical Information Systems (EGIS 94)*, Paris France, April 1994.

Gonçalves, P., and Diogo, P., 1994b, Forest Fire Modelling: A New Methodology Using Cellular Automata and Geographic Information Systems. In *Proceedings of 2nd International Conference on Forest Fire Research*, Coimbra 21/24 Nov. 1994.

Green, D. G., 1983a, Shapes of simulated fires in discrete fuels. *Ecological Modelling* **20**: 21-32.

Green, D. G., 1983b, Fires shapes and the adequacy of fire spread models. *Ecological Modelling* **20**: 33-45.

Iezzi, A., 2000, Laurea thesis (in Italian) *Sviluppo di un modello ad Automi Cellulari per la simulazione di incendi agro-boschivi*. (*A Cellular Automata model for forest fire simulation*). Department of Mathematics, University of Calabria.

Malamud, B. D., and Turcotte, D. L., 2000, Cellular Automata models applied to natural hazard. *Computing in Science and Engineering* **2**(3): 43-51.

Rothermel, R, 1972, *A Mathematical Model for predicting fire Spread in Wildland Fuels*. Res. Pag. INT-115, Department of Agriculture, Forest Service, Intermountain Forest and Range Experiment Station, Ogden, UT, U.S.

Rothermel, R., 1983, *How to Predict the Spread and Intensity of Forest Fire and Range Fires*. Gen. Tech. Rep. INT-143, Department of Agriculture, Forest Service, Intermountain Forest and Range Experiment Station, Ogden, UT, U.S.

SALTUS Reports, (1998-2000) from European Research Program SALTUS, Environment and Climate.

The Dynamic Usage of Models (DYSAM)

GIANFRANCO MINATI[*], and SABRINA BRAHMS[#]
[*]*AIRS - Associazione Italiana per le Ricerca sui Sistemi - E-mail : gianfranco.minati@iol.it*
[#]*M.A.; 368 Museum Drive, Los Angeles, CA, 90065, USA – E-mail: sabejams@pacbell.net ;*
Consulting Faculty, Phillips Graduate Institute, Los Angeles, CA, USA;
Doctoral Student, Saybrook Graduate School, San Francisco, CA, USA

1. INTRODUCING THE DYSAM APPROACH

Usually a (dynamic or non dynamic) model[1] is formulated to understand, simulate, manage, forecast the behavior of a system or phenomena of different kinds (i.e. physical, biological, chemical, social). The models are formulated on the basis of theoretical assumptions, for instance suitable to be used to produce software models and mathematical descriptions.

The idea that the more effective model must exist and must be unique is based on an a classic objectivistic approach. This approach is not anymore the only one accepted in science, for instance because of the fundamental role of the observer, with its cognitive processing, as in the processes of emergence. Depending on the interest of the observer and on the nature of the context one model may be more suitable than another one.

For instance, a model based on the symbolic approach may be more suitable than one based on the sub-symbolic one when coping with problems characterized by events and explicit knowledge.

Because both (1) the system to be modelled may be changing, not only in time as usual for dynamic systems, but in *structure* (Cruchtfield, 1994) and (2) the observer may be interested in different aspects of the phenomena, in order to be effective it is important the availability of different kinds of

[1] http://pespmc1.vub.ac.be/ASC/Model.html

Emergence in Complex, Cognitive, Social, and Biological Systems, Edited by
Minati and Pessa, Kluwer Academic/Plenum Publishers, New York 2002

models. This is particularly interesting for emergent processes where *intrinsic emergence* takes place, in which not only the occurrence of a particular behavior cannot be predicted in advance (even if it is compatible with model assumptions), but its appearance gives rise to a deep modification of system's structure, in such a way to require the formulation of a new model of the system itself (Minati, 2001a).

The concept of *Collective Being* (Minati, 2001b) has been introduced to model *multiple systems* emerging from the behavior of agents having the same cognitive model. A Collective Being is intended to be a system oscillating among different systemic aspects, having components simultaneously belonging to many systems and to be suitably modeled by using a DYnamic uSAge of Models (DYSAM) (Minati, 2001c). A Collective Being is intended to

- emerge from the behavior of agents provided with cognitive models and having no internal functionally different roles (i.e. flocks, swarms, herds);
- have components belonging to different systems, even *simultaneously*;
- present dynamic, intelligent collective behavior (i.e. hunting, escaping a predator);
- be oscillating among different systemic aspects and simultaneously requiring focus on the behavior of single agents to be managed (i.e. classrooms), *showing in time different predominant aspects and not only a standard one.*

The agents elements of a Collective Being are not having different functions as subsystems. Example is a behave versus a swarm. The assumption to deal with agents provided with cognitive systems is crucial because they may do not interact following standardized rules as properties of physic or biologic components. They are able to modulate, to change, to design their way to interact having awareness (Anderson, 1985; Benjafield, 1992). A social system may be very different in time because of the change in the way the components interacts.

The example of the crowd may be considered. It may be constituted of

Agents, like
people walking, standing, silent, singing, clapping, etc.: single agents may be interested in specific actions, like shopping, watching, listening to.

Depending on the interest of the observer the

Collective Being *Crowd*
may be modelled with reference to security problems, traffic circulation, money distribution in shopping, pollution generation, noise generation, etc.

Another example takes place when the observer is interested in different aspects of the phenomena, considered physically constant and then suitable to be dealt with by using the same model. The need to use different models comes from the different, changing interests of the observer. Like the lava erupted by a volcano may be interesting for its effects on people, on territory, on buildings. Or tide interesting for fishing or for its impact on damaging buildings.

The system *system-observer* to be modeled may be so changing that the observer may need:

- To have different models available depending on the approach required;
- To have different models available until a more complete and effective one is identified. Two frameworks may be distinguished: (1) when the more suitable model is not easily identifiable, but the dynamic use of models is expected to *converge* to the more effective and more suitable one; (2) when the problem is not to identify the more suitable one because the continuous changing of the phenomenon to be modelled. The problem in this case is to adapt as soon as possible to a changing context. It may be an effective approach to model emergence processes having unexpected evolutions.
- To have different models to be used *alternately* in consequence to continuous changing;
- To have different models to be processed *in parallel* (through parallel processing), by using strategies to decide the one to be used depending on the time available to take action;
- To have different interacting models to be processed *in parallel* (through parallel processing) and simultaneously used to take decisions.

The DYSAM approach makes reference to the last point when the observer-user system is expected to keep *memory* of usage, allowing *learning* processes.

Can DYSAM be itself considered a model? It is a way to design the use of different models simultaneously, interactively, by allowing learning processes to manage phenomena having emerging properties. As we will discuss it may be considered a model of emergence and its *dynamics* (model selection, interactions, parallel processing, learning and decision processes) may be considered as an index of emergence with reference also to the ergodicity of the system (Minati, 2002).

2. A GENERAL SCHEMA OF THE DYSAM APPROACH

The DYSAM approach may be considered articulated in five components:

a) A data base of symbolic, and non-symbolic, deterministic and non-deterministic, dynamic and non-dynamic, context- and non context-sensitive, etc., models;
b) Contextual information about the system/phenomenon to be modeled;
c) *Interaction among models* using cross-results, giving meaning to errors and to different results depending on *memory* about the usage of models (context, interactions, consequences) and *learning* processes: a Dynamic Processing of Models;
d) Representation of the case to be modeled suitable to be processed by using models;
e) Use of models to simulate, to decide, to manage.

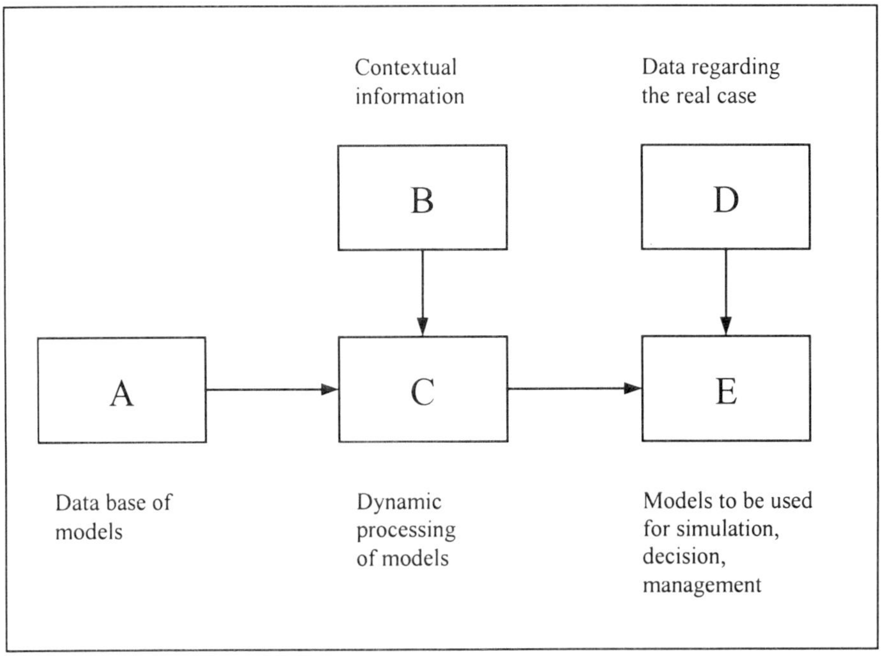

Figure 1.

The core of the DYSAM approach is well represented by the component c). In this phase very complex tasks are processed: an example is introduced in the next paragraph where the abstract problem of recognizing handwritten digits is discussed.

3. EXEMPLIFYING THE DYSAM APPROACH: HANDWRITTEN DIGITS AUTOMATIC RECOGNITION

An introductory example regarding handwritten digit recognition is introduced in (Minati, 2001b). This example is discussed in order to better illustrate the DYSAM principles. In fact, real application is at this time, not so interesting being the Optical Character Recognition (OCR) technology very effective without the need to have a dynamic usage of models.

In this case dynamics in using models is required by the unexpected and unknown user writing the digit to be recognized.

a) Available models to be applied may regard:
   - how many holes there are in the sign to be interpreted;
   - if crossing lines are present;
   - number of the pixels composing the image to be interpreted;
   - overlapping degree with other images in the database related to shapes to be recognized;
   - position of "holes" with reference to axis (up, down, left, right);
   - and so on, by using well known sophisticated interpreting techniques.

   They may be linked by a neural network able to propose to step c) the more suitable models to be applied in reference to a previous training process. The ability to get suitable information from errors takes place both in this step and in step c).
b) Contextual information may regard geographical place where the recognition is attempted: numbers are written differently in US, Europe, etc., additional user information (sex, age, instruction, right or left writing) from company's personal data records, kinds of documents where the digit has been written (getting information about physical context of writing: office, home, standing), filling into a form digit by digit or by writing a complete number, supposed written at the same time or in different times, like amount on a check, etc., knowing the meaning of the complete number: amount, date, code, age, zip code, weight, phone number, etc. Is it possible to establish some expectances about the next field?
c) This layer is expected to process the selections proposed by layer a), depending on contextual information b) and previous learning processes. Different strategies, tactics and styles may be applied. A strategy regarding automatic reading of a form may regard to keep data consistent, checking their consistency by accessing external database and by applying control rules. A tactic may be to try recognition step by step: digit per digit, then field by field, then form per form, then slot per slot by assuming some possible homogeneity. A possible style is to

continuously show the process and the decisions assumed by the system, for instance in a window, making the user able to interact, to make selection or not. Another strategy may be to apply some graphoanalysis models. In this case data must regards a full form or a sufficient number of data to be recognized authored by the same user. Applications may regard signature authentication and handwritten forging.

d) Data may be images of the handwritten digit to be recognized proposed pixel by pixel, red by an optical reader, a video camera, etc.
e) The system may at this time propose to use of
   - model #1, if and how many holes there are in the sign to be interpreted;
   - model #2, if and how many crossing lines are present;
   - model #3, how many pixels are composing the image to be interpreted;
   - model #4, position of "holes" with reference to axis (up, down, left, right);

   simultaneously, by considering proposed results as a whole (if #1 says X, and #2 says yes and $k_1 > Y > k_2$, and ...) proposing different compositions. The model is also context-sensitive as introduced in step (c).

The model applied at time T may be used to recognize digits and complete numbers, but different models may be required even for the same form having different fields when it is compiled by different people, or by the same person in different times, or by using different pencils, or by writing in different positions. When a sufficient number of items refer to the same user a tentative user model may be proposed.

In this context it's also evident the difference between adopting DYSAM for (1) converging, heuristic processes oriented to identify the more suitable model for a specific phenomenon having some *stationary states* (for instance the same interacting user for some time) and (2) to adapt to an highly changing, unpredictable phenomenon.

4. APPLICATIONS

4.1 Applications of DYSAM in the Fields of Marriage and Family Therapy and Conflict Resolution

4.1.1 Marriage and Family Therapy

Parents operate with a cognitive model that permits them to instruct their children throughout the day and through their childhood years. It would not be unusual to hear a parent say, "make you bed", "brush your teeth", "pick up your clothes from the floor". Within the same cognitive model it would not be appropriate, however, to give the same kinds of directives to one's spouse. If a wife finds herself speaking to her husband in similar ways that she speaks to her children, the couple may experience marital dissonance.

The following is another example of marital discord that may be experienced as a result of inappropriately using only one cognitive model. A woman works as a high-powered executive in a corporation. In this position, she is accustomed to making important decisions and delegating responsibilities. If when she comes home, she continues this role, while her husband operates with a cognitive model that places her in a subservient role to his decisions and authority, the couple may face tension and strife.

Transactional Analysis (Stewart, 1989; Goulding and Goulding, 1997) is a form of psychotherapy that addresses the idea of cognitive models although calls them "scripts". In this method of therapy a client becomes aware of the unconscious scripts that motivate him through life. "Be successful at any cost" may be one such life script. Another may be, "don't disappoint others". Living with these unconscious stories may become debilitating but through the therapeutic process, the client has the opportunity to change these limiting cognitive models.

In Transactional Analysis, the emphasis is on replacing the limiting cognitive models with a more expansive model. According to DYSAM, it would not be as important to replace our limiting cognitive models but instead to be aware of when they are useful and when they are not, using the appropriate model when necessary. DYSAM does not encourage the replacement of models, but the addition of new models. This way of thinking seems more realistic because it is difficult to imagine that someone is able to completely rid himself of a model that has been in use for most of his life. It is also a more gentle approach, allowing the limiting model to remain but only as one aspect of many models that may be used in any given situation.

4.1.2 Conflict Resolution

In order to describe the possible uses of DYSAM in the arena of conflict resolution, the example of the Israeli-Palestinian conflict will be used, but the issues addressed here are relevant in many situations of conflict resolution. In the case of Palestinians and Israelis, each group operates from a different cognitive model about the reasons for the conflict between the two groups. Each has a different version of history, specifically when it comes to a sense of entitlement to the land. Each has a different set of religious doctrines as well as a different view of the value of conflict. Although both peoples now live in the Middle East, Palestinians carry a cultural view that is indigenous to the region, while many Israelis are from European descent and therefore have more of a Western cultural view. All these factors influence which cognitive models are employed in consideration of the other. On a psychological level, each group answers the following questions differently. Who is "us"? Who is the "other"? How do we respond to members of our community; how do we respond to members of the enemy community? Thus a member of one community may be very giving and loving toward his brethren while having no compunctions about killing members of the outside community. This illustrates the core feature of DYSAM, that many models may be present in one system.

The complexity of the dynamic interactions between these multiple cognitive models can certainly be seen in a series of peace negotiations. Each member of the negotiating team is operating with a number of models that may change, depending on the context. The complexity is increased by the interplay between many different people, each with his own complex set of operating models. Acknowledging some of these cognitive models through an active process of disclosure may begin to create a productive foundation for actual negotiations. In other words, if team members from each group are asked to identify the cognitive models of which they are aware, to team members from the other group, a common understanding of the presiding assumptions may begin. And when negotiations come to an impasse, it may be possible to return to the operating cognitive models to examine whether they are best serving the group or its purposes.

In general, it seems to be essential for human systems to increase awareness of the multiple cognitive models that impact thoughts and decisions. This awareness is the starting point for choosing the most appropriate model for each occasion.

4.2 An Educational Lab based on DYSAM: teaching to a Collective Being is different than teaching to a system of students, as a classroom is

An educational lab based on interactive computerized systems and using the DYSAM approach may allow the following functions.

The teacher presents the lecture as it happens during normal lessons. But the teacher is supposed to have previously developed a model for each attending student. With the help of a computer supported editor he designs *multiple* lessons *taking place at the same time* (Mingers, 1998) His/her verbal lesson is supposed to be the same for everybody, but on the individual screen different messages are shown depending on the student and on the model the teacher has of him/her. Different examples, different information, different stimuli (analogies, problems, correspondences, questions, considerations, relationships, in different language also) are shown during the lesson. This part must be synchronized with the verbal presentation. Anyway it must be possible for the student to process this individual lesson personally: replaying any part, interacting, comparing data and having different simultaneous presentations (pie charts, histograms, etc.), highlighting, zooming, having audio input using ear-phones also. The second part of the lesson may be dedicated to individual interactions through the computer systems. During the first and second part of the lesson personal interactions between students and teacher are expected and welcome.

An automated, intelligent, version of the Educational-DYSAM consists in having a system able to automatically translate, transform the original lesson in the different versions depending on the student models.

The attendants *oscillate* from establishing a system (classroom) and a Collective Being (an emerging system dealing for multiple strategies corresponding to individual needs and characteristics).

4.3 Discussion with multiple simultaneous speakers

This case has been already introduced by presenting the concept of logical openness as an extension of the classic thermodynamic openness (Minati, 1998). The speaker assumed to defend proposals in a debate following his/her presentation must select the model that better fits with the single questions. Then the questioner and the proposer carry recursively out model (Allen, 1990) of the model of each other and their interactions are modulated by using a hierarchy of models. The model used at each time is assumed to be an improvement of the one used in the first interaction, being the result of a convergent process activated by learning on the basis of increasing information.

4.4 Adaptive Books based on DYSAM

An adaptive book is an E-book able to organize itself, to assume a structure *depending on the reader profile*.

The text of the book is not assumed to be static, equal for any kind of reader. Depending on the reader profile, on the reader model, the system is able to organize the text, previously produced by the author, in order to match the reader skills and expectances. The book is presented to the general reader readership by using title, keywords and some standard text. Then the reader may have access to the table of contents as proposed by the author, as it usually happens consulting a book in a bookstore. If the potential reader is interested s/he may ask the system for interacting with him/her and to carry out his/her model. Depending on this model a new version of the book is produced. If the new personalized version of the table of content is satisfying the potential reader then a new complete new version is generated.

The book is sold with a software to be used with other E-books, able not only to browse the text finding words, but even taking track of the reader notes, crossing different texts and notes. The book is not only red by the reader, but processed and in time the book is produced more by the reader than from the original author. Different solution, hardware and software, are already available in the market to read E-books. A new generation of intelligent and adaptive tools, as mentioned in this paragraph, are expected to be available in the future.

4.5 A Virtual Supermarket based on DYSAM: from offering products to a market intended as a system of anonymous interacting agents to sell to single components of a market

This Virtual Supermarket is able not only to offer on line products like the client was effectively shopping, just offering as additional service home delivery. By following the DYSAM approach, the client who is entering the Virtual Supermarket is asked to present the shopping list. This includes preferences and goods to be excluded for any reasons, the budget (max expense), general requirements (how many in the family, how long what bought is expected to be sufficient for), how big the refrigerator is, availability of micro wave oven, pantry, dish and laundry washing machines, etc. The system is asked to look for special offers.

The client profile is detected during a previous interaction.

The client is not asked to just select among offered products: the client presents the shopping list and the system is expected to match offer and demand. *At the end, the system presents its selection and asks the client to confirm item per item allowing the system to learn about this client.* Clients may interact with the systems manager giving suggestions and judgements. Clients may interact among them too sharing considerations. Clients may participate in games, looking for prizes and suggesting new clients to shop in this virtual supermarket getting discount and special offers.

Same approach may be assumed in banking selecting the best way to invest for customers, depending on their model. Or by a telephone company offering the best mix of rates to customers depending on their model. Models are assumed to be adaptive depending on the customer's behavior.

5. CONCLUSION

The DYSAM approach introduces a conceptual shift from averaging, generalizing, standardizing to multi-modeling by keeping specific models on line simultaneously. To adapt, to learn, to be context sensitive, to be interactive, to have memory of usage by using Cognitive Processing in interfacing are fundamental abilities to deal with continuously changing environment avoiding the classic expensive and averaging approach based on standardization, requiring to be the user to adapt. In this case there is loss of information, as it happens when bio diversity is substituted by single, divided sources like in monocultures vs. polycultures (The Land Institute, 2001). The DYSAM approach relates to emergence processes.

Thanks to the DYSAM approach it is possible to design devices able to avoid standardization as the "best" solution to deal with the problem to interact and manage different kinds of simultaneous users.

REFERENCES

Allen, R. B., 1990, User model: theory, methods, and practice. *International Journal of Man-Machine Studies* **32**: 511-543.

Anderson, J., 1985, *Cognitive Psychology and its Implication*. 2nd Edition, W. H. Freeman and Co, New York, N Y.

Benjafield, J. G., 1992, *Cognition*. Prentice-Hall, Englewood Cliffs, NJ.

Cruchtfield, J. P., 1994, The Calculi of Emergence: Computation, Dynamics and Induction. *Physica D* **75**: 11-54.

Goulding, M. M., and Goulding, R. L., 1997, *Changing lives through redecision therapy*. Grove Press, New York.

Krippendorff, K., Definition of Model. In *Web Dictionary of Cybernetics and Systems* (F. Heylighen, ed.). This dictionary is a combination (with permission) of the ASC Glossary, Krippendorff's Dictionary and Hornung's Glossary. The original texts were compiled and converted to hypertext by Francis Heylighen, helped by An Vranckx. http://pespmc1.vub.ac.be/ASC/IndexASC.html.

Minati, G., 2001a, The Concept of Emergence in Systemics. In, *General Systems Bulletin, Vol. 30*, International Society for the Systems Sciences, Cookeville, USA, pp 15-19.

Minati, G., 2001b, *Esseri Collettivi*. Apogeo, Milano, Italy (Collective Beings, 2001, in progress).

Minati, G., 2001c, Experimenting with the DYnamic uSAge of Models (DYSAM) approach: the cases of corporate communication and education. In *Proceedings or the 45th Conference of the International Society for the Systems Sciences (ISSS)*, USA.

Minati, G., 2002, Emergence and ergodicity: a line of research. In *Proceedings of Second Italian Systems Conference*, Kluwer Academic.

Minati, G., Penna, M. P., and Pessa, E., 1997, A conceptual framework for self-organization and merging processes in social systems. In *Systems For Sustainability*, Plenum Press, New York.

Minati, G., Pessa, E., and Penna M. P., 1998, Thermodynamical and logical Openness. *Systems Research and Behavioral Science* **15**: 131-145.

Mingers, J., and Gill, A., (eds), 1998, *Multi methodology: Towards Theory and Practice and Mixing and Matching Methodologies*. Wiley & Sons, U.K.

Stewart, I., 1989, *Transactional analysis counselling in action*. Sage Publishing, Newbury Park, CA.

The Land Institute, 2001, http://www.landinstitute.org.

Cognitive Science

The Emergence of Strategies in Cognitive Development

MARIA PIETRONILLA PENNA
Facoltà di Scienze della Formazione, Università degli Studi di Cagliari
E-mail: npenna@hotmail.com

1. INTRODUCTION

The topic of emergence is of crucial importance in studying human cognitive development. Namely, the latter is characterized by the sudden appearance of new abilities, which involve a number of new correlations (and synchronizations) between actions and behaviors already present before within subject behavioral repertory. It is tempting to think of these phenomena as of occurrences of new coherent states of cognitive system, emergent as a consequence of some cognitive "phase transition" (see, to this regard, Pessa and Penna, 1996). However, it is to be underlined that such a view of cognitive development is rather vague and qualitative. We need, therefore, a more precise definition of "cognitive emergence", which, however, can be given only within well-defined (and whence limited) contexts. In this paper we will take into consideration the particular context of cognitive strategies used by children in solving arithmetic problems, more precisely the ones relative to the operation of multiplication.

To this regard we must start from a fundamental fact, evidenced by a number of experimental researches (cfr. Schneider and Pressley, 1989): children have at disposal a priori a number of different strategies for solving a given kind of arithmetic problems (in our case related to the multiplication). Besides, the strategy effectively chosen by children depends on their age and level of learning. A possible picture of such a situation can be easily obtained if we adopt the well known framework of Newell and

Emergence in Complex, Cognitive, Social, and Biological Systems, Edited by
Minati and Pessa, Kluwer Academic/Plenum Publishers, New York 2002

Simon about human problem solving: each strategy is nothing but a class of *equivalent paths* (from the starting state to the goal) within *problem space*, all paths belonging to the same equivalence class being characterized by the same kind of *coherence* between the choices made in order to go from a problem state to the next problem state. Thus, the emergence of cognitive strategies can be considered as equivalent to the choice of a path equivalence class as the fundamental one. In other words, emergence consists in a transition from a situation in which a particular class is the stable equilibrium state of the cognitive system to another situation in which the old equilibrium becomes unstable and a different class becomes the new stable equilibrium state. The next problem is: what mechanism produces such a change of stability? In this paper we will investigate such a question both experimentally and theoretically, in the case of multiplication problems.

2. MECHANISMS OF STRATEGY CHOICE

According to a widely diffused theory (cfr. Flavell, 1981), there should exists a *metacognitive* system, deputed to control the choices made by the cognitive system. The metacognitive system should decide, in the case of arithmetic operations, what strategy to choose, on the basis of its cognitive load, of its effectiveness, of its pleasantness, and, chiefly, of the opinion the subject itself has about his skills. In the Figure 1 it is possible to see a schematic picture of how a model of this kind works.

Such models were implying a set of cause-effect relationships, in which the metacognitive knowledge was determining the choice of the cognitive strategy, whereas the latter, in turn, was determining the cognitive performance of the subject as observed in the experiments and in the normal school activity.

Such models, however, were not able to account for all experimental data which evidenced how, in many cases, the metacognitive knowledge was not so important for the choice of a cognitive strategy. For instance, in many cases the children, even if they know that a higher-level strategy is very effective and, in principle, are able to implement it, continue to use a lower-level strategy and resolve to abandon the latter only after a number of successes obtained in using the high-level strategy. Starting from these considerations, in the Eighties and Nineties Siegler and coworkers (Siegler, 1987; Siegler and Shipley, 1995; Siegler, 1996) proposed a new model in which the choice of a strategy was directly depending both on the success obtained by the children when implementing it, and on the cognitive load required by its implementation.

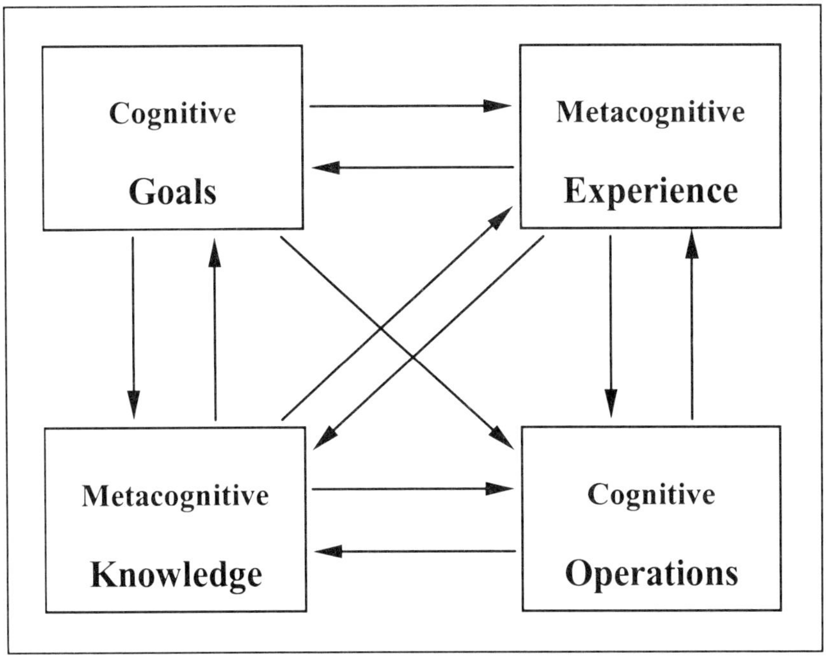

Figure 1. A model of cognitive performance based on metacognition.

In this model there was no need for an explicit intervention of a metacognitive system and, analogously to what occurs in phase transitions in physical systems, the emergence of a cognitive strategy was determined by a change of the constraints experienced by the cognitive system. The research described in this paper had just the goal of testing the ideas of Siegler, both through an experiment on children solving multiplication problems and a computer simulation of a model inspired by this framework.

3. THE EXPERIMENT

The experimental phase was implemented to test the validity of two main hypotheses:

1. children solve multiplication problems by using a number of different strategies;
2. as the learning level increases, the frequency of use of each strategy undergoes a change, that is the percentage of use of memory recall strategy increases, whereas the percentages of use of the other strategies decrease.

3.1 Subjects

The subject sample included 41 children (mean age 7.5 years) attending the elementary school (second grade).

3.2 Method

The subjects participated, one at time, from the end of February up to the end of April, to three successive test sessions, with an interval between the sessions of four weeks. Each test was consisting of 81 multiplication problems, whose results were always lesser than 100, chosen in a random way, but allowing the possibility that a problem contained in a given test session could be present even in the next session. In each test we monitored the percentage of correct answers, the kinds of strategy used, and their frequency of use. As regards the kind of strategy used, it was deduced directly from the observation of subject's behavior, without resorting to specific questions.

3.3 Results

The children used six different strategies:

- recall from memory (R)
- repetition aloud of multiplication tables (T)
- count on their fingers (AD)
- repeated addition (A)
- problem decomposition (SC)
- answering "I don't know" (N)

The Table I contains the percentages of use of these strategies as a function of the serial number of the test session. The latter, of course, is related to children learning level: a higher serial number denotes a higher learning level.

Table 1. Percentage of use of the different strategies as a function of the serial number of test session

	% USE		
	1	2	3
R	48.20	65.11	72.90
T	11.54	11.38	13.56
A	8.94	1.92	0.54
SC	7.69	6.58	3.18
AD	4.17	2.37	2.07
N	16.14	10.84	7.20

These results are in agreement with the previous hypotheses, even if there is a little discrepancy as regards the strategy of repetition aloud of multiplication tables. Such a discrepancy, however, could be explained in a number of different ways, by resorting to occasional factors.

4. THE MODEL

In order to account for the previous experimental results, we formulated a model, which was partly inspired by the *Distribution of Associations Model*, already proposed by Siegler and Shrager (cfr. Siegler and Shrager, 1984). The operation of our model was characterized by two different phases: the *learning* phase, and the *test* phase. During the learning phase we had a comparison between the effective answers, given by the model to particular multiplication problems, and the correct answers to the same problems. The results of such a comparison gave rise to a change of the values of model's dependent variables and parameters. In the test phase, on the contrary, every change of this type was impossible, learning procedure didn't work, and the model was able only to output, when dealing with a particular multiplication problem, the answer to it, the strategy and the number of trials used to obtain such an answer.

The model was based on the fundamental assumption that the child memory system be structured in such a way as to contain, for each multiplication problem (that is, for each possible factor pair), a set of *associations* between the factors and the possible numerical answers (within a given interval) to this problem. Each association is characterized by a numerical *weight*. The latter gives a measure of the probability of giving, as output, the associated answer. Whereas these weights can vary during the learning phase, they are kept constant during the test phase.

Within our model the initial distribution of the weights was given *a priori* for each strategy and given, for each problem, by a Gaussian distribution, centered on the correct answer to the problem itself. Besides, we hypothesized that the variances of these distributions were decreasing in a progressive way, from the more difficult to the less difficult strategy. The model operation was stochastic: given a particular multiplication problem, the answer was generated in a random way, but according to a probability given by the weight distribution. However, if the answer effectively generated was associated to a probability whose value was lesser than a given threshold value (the *criterion*), then the answer was not accepted and the model tried to generate a new answer. The maximum number allowable of trials for each problem and for each strategy was specified by a user-determined parameter. The model, for each problem, started always by choosing the most difficult strategy and, if the maximum number of allowable trials was exceeded, it turned to the immediately next strategy in decreasing order of difficulty. When, using such a procedure, the model gave a correct answer to a particular multiplication problem, the weight associated to this answer was increased of a fixed quantity. In the contrary case of an incorrect answer it was decreased of fixed quantity. After the list of the multiplication problems to be solved was exhausted, the model updated the variances of the weight distributions for each strategy, by resorting to a best fitting procedure based on a Gaussian function.

5. COMPUTER SIMULATIONS

In order to start the computer simulations of our model, we assumed that the order of difficulty of the strategies, from the more difficult to the less difficult, were the following: recall from memory (R), repetition aloud of multiplication tables (T), count on one's fingers (AD), repeated addition (A), problem decomposition (SC), answering "I don't know" (N). The adopted values for the model parameters were:

- initial variance for the most difficult strategy: 0.9;
- difference of variance between a strategy and the immediately following one in order of difficulty: 0.04;
- criterion value : 0.1;
- maximum number of allowable trials for each strategy : 1;
- weight increase in the case of a correct answer : 0.5;
- weight decrease in the case of a incorrect answer : 0.25;
- number of learning steps : 1000.

The simulation, done with the same multiplication problems used in the experiment, required three different phases:

- *first learning*, corresponding to test session 1 of the experiment with children;
- *learning continuation*, corresponding to test session 2;
- *test*, corresponding to test session 3.

The obtained results of computer simulations are shown in the Table 2.

Table 2. Percentages of use of each strategy in computer simulations

	% First learning	% Learning Continuation	% Test.
R	12.5	11.0	13.5
T	9.47	9.34	10.26
A	11.18	11.97	9.47
SC	26.7	25.7	27.8
AD	20.3	17.89	18.6
N	21.5	20.5	18.2

These results evidence a trend similar to the one observed in experiments (cfr. Table 1), even if there are differences with respect to this latter both in the actual values of percentages, and, chiefly, in the fact that our model appears to have favoured the strategy of problem decomposition, contrarily to what occurred in children.

6. CONCLUSIONS

The results so far obtained confirmed our hypotheses and showed that the mechanism of choice of strategies by children could, in principle, be explained by a model lacking of any metacognitive module. Even if this model should be improved, for the reasons quoted in the previous section, nonetheless its existence suggests the feasibility of a theory of emergence of cognitive strategies based on the effect induced by external (and inner) constraints on a set of already available solution behaviors. Such a theory can, at least, be used to explain some facts connected to the acquisition of arithmetical skills.

An important challenge for the future research will be to understand how a new strategy can originate from a previous state in which the subject was totally unaware of its existence. We feel that, to this regard, the theory of emergence will give new fruitful insights.

REFERENCES

Flavell, J. H., 1981, *Development of Comprehension Monitoring and Knowledge about Communication*. University of Chicago Press, Chicago.

Pessa, E., and Penna, M. P., 1996, Can Learning Process in Neural Networks be Considered as a Phase Transition? In *Neural Nets. Wirn Vietri-95* (M. Marinaro and R. Tagliaferri, eds.), World Scientific, Singapore, pp. 123-129.

Schneider, W., and Pressley, M., 1989, *Memory development between 2 and 20*. Springer, New York.

Siegler, R. S., 1987, Strategy choices in subtraction. In *Cognitive process in mathematics* (J. Sloboda and D. Rogers, eds.), Oxford University Press, Oxford, pp. 81-106.

Siegler, R. S., 1996, *Emerging minds: The process of change in children's thinking*. Oxford University Press, New York.

Siegler, R. S., and Shipley, C., 1995, Variation, selection, and cognitive change. In *Developing cognitive competence: New approaches to process modeling* (T. Simon and G. Halford, eds.), Erlbaum, Hillsdale, NJ:, pp. 31-76.

Siegler, R. S., and Shrager, J., 1984, A model of strategy choice. In *Origins of cognitive skills* (C. Sophian ed.), Erlbaum, Hillsdale, NJ, pp. 229-293.

Emergence of the Symbolic Function

GRAZIANO TERENZI
Dipartimento di Filosofia – Sezione Psicologia, Università degli Studi di Pavia
Piazza Botta, 6 (Palazzo S. Felice) 27100 Pavia, Italy - E-mail: g.terenzi@inwind

1. INTRODUCTION

On the surface, *natural language* appears to be embodied in an extremely sophisticated *communication system* that, through the mutual recall of meanings accomplished by using suitable triggering stimuli, allows distinct cognitive agents to perform adaptive behaviors; besides, such stimuli are typically thought of in terms of a set of *symbols*, i.e. meaning recalling entities, associated to a set of *processes* for their manipulation. However, the nature of both communication and symbols as well as the nature of their manipulation processes isn't still clearly understood.

In this paper we argue that the need for an explanation of the appearance of languages, endowed with the features of *stability*, coherence and *structural complexity* typical of natural languages, forces us to face preliminarily the *problem of emergence*, that is the general problem of the explanation of the appearance of new and ever more complex patterns of organization in nature. The goal of the present work is then to show a tighter link between the problem of emergence and the nature of language. To this aim we argue that emergence can be better understood as the transition, taking place within *logically open systems*, such as artificial neural networks (ANN), from unordered collective behaviors to ordered collective behaviors, i.e. *phase transitions*. By appealing to the latter analysis we illustrate a neural network model which catches some features of the problem of the emergence of symbols and of the resulting problem concerning the grounding of symbols in external environment (cfr. Harnad, 1990). The model consists of a neural architecture which is based essentially on two interconnected modules: one deputed to symbol categorization and another deputed to associate thematic roles to sentence constituents. The operation of the model is based on the idea according to which the emergence of entities endowed with semantic content (i.e. symbols) depends both on the formation of a suitable mapping from linguistic stimuli to perceptive meanings (and the other way down), and on the further semantic processing to which such entities undergo within a speaker's cognitive system. Clearly the centrality of the notion of emergence suggests the adoption of a systemic approach to the scientific study of natural language.

2. TOWARDS A SYSTEMIC APPROACH TO NATURAL LANGUAGE

2.1 Language ...

The highly ordered patterns exhibited by linguistic phenomena, from coordination of communicative processes taking place between distinct individuals who try to attain some goal, to the syntactic and semantic complexity of signals used thereby, mandatorily calls for a systemic explanation. In fact, were the linguistic competence of speakers completely and univocally determined by innate constraints (Chomsky, 1980) or were it not (Elman, 1999), in both these circumstances the problem remains of explaining the appearance of a phenomenon which exhibits such a degree of order. In the first case, the problem amounts to explain the appearance of such constraints under adaptive pressure and to demonstrate how that would suffice to account for the complexity of linguistic phenomena. In the second case, instead, the problem is that of accounting for the appearance of extremely general, adaptive structural constraints and learning mechanisms, underlying the functioning of cognitive systems, in such a way as to explain how they concur, through the essential interaction with an environment, to the production of the same class of phenomena. Clearly, in both the circumstances the problem of emergence is in the background.

Since linguistic phenomena take place within the context of a communication system emerging in a population of communicative agents which are endowed with suitable cognitive capacities and which interact with each other in a particular environment, then the notion of communication system seems to play an important role in the explanation of language. As it stands, the latter notion is strictly tied to Information Theory. However, as evidenced by very influential authors (cfr. Haken, 1988), whereas the characterization thereby contained catches properly the syntactic features of the communication of information, it is lacking as regards its semantic and pragmatic features. This limit, due to the lack of a suitable treatment of the structure of receiver and emitter, calls for an appropriate extension of the aforementioned notion. To this aim, one is forced to take into account both the structure of communicative agents and the semantic content associated to communicative signals. On that account, a communication system could be conceptually qualified as the set of the following things:

- *signal*, i.e. the external coding of sent information
- *receiver*, i.e. the cognitive structure of the agent receiving the signal
- *emitter*, i.e. the cognitive structure of the agent emitting the signal
- *channels*, i.e. physical medium for the transmission of signal
- *content*, i.e. the meaning of the signal.

Clearly, the receiver, the emitter and the channels are given at start and, once the cognitive structure of agents has been determined, their definition is quite unproblematic. On their side, signal and content are quite more problematic entities. Indeed, external embodiments of symbols (i.e. the set of admissible syntactic constructs) are a subset of the set of all possible signals. Content is an even more problematic notion, since it calls for external relations between representations and physical inputs as well as for internal relations holding between representation themselves (cfr. Chalmers, 1994). However, the internal component of content, bound to a set of semantically admissible constructs, could be seen, on a first approximation, as a part of the structure of the cognitive agents. If, as we do, we cannot a priori know what part of the signal space or of the content space corresponds to the set of admissible syntactic constructs or, respectively, to the set of admissible semantic constructs, then both signal and content are better conceivable as entities emerging through interactions between agents and their environment. This means that linguistic signals emerge in the channel in which the signal space is embedded whereas content emerges largely within the representational space embedded in the cognitive structure of each agent. Thus, Syntax and Semantics are better conceived as emergent features of any evolutionary communication system.

2.2 ...and Emergence

But what more exactly we refer to when we talk of emergence? Why and how such spatio-temporal patterns of organization do emerge? At a first glance emergence seems to be an ubiquitous phenomenon in nature. It is essentially characterized by the appearance of *novelties* within the causal organization of a system which can give rise to an *ordered pattern* of behavior quite different from the one that was initially displayed by the system. On a first approximation, we can refer to emergent phenomena as the appearance, within a system, of collective behaviors that are not only unpredictable, but cannot in any way be derived by using the information incorporated within the model that describes it; and this is due to the fact that the latter does not catch essential features of the modeled system (cfr. Heylighen, 1989; Crutchfield, 1994). On this view, the model itself is incomplete and has to be revised by including the relevant information. Since a fixed model is not sufficient to account for the complexity of a system that is able to display emergent behaviors, then we are forced to appeal to a meta-model, that is a model that would allow the construction of an entire ensemble of models which together approximate the behavior of the system under study.

An instance of such models is given by Artificial Neural Networks. As it is well known, ANN are networks of primitive functions (Rojas, 1996). In the simplest case, each primitive function is located at some node of the network and computes its output in a way dependent on its input signals and on the parameters associated with its afferent connection lines. An ANN is then identified by a network function corresponding to a particular composition of primitive function. The same network can give rise to quite different behaviors depending on the fact that it is defined in a particular region of its space of parameters rather than another one.

Behaviors that result from the application of the function to one region can be much different from behaviors that result from the application of the same function to another region of the space. Whereas, on the one region, the network might not exhibit global ordered patterns of behavior that are meaningful for an observing system, on the other one it might well do. The latter region comprises just the values for parameters that would lead the ANN towards an appropriate solution to the processing task at hand. Obviously, such a solution is expressed through the activation value of a suitable subset of the set of the computation units, which in turn depends on network parameters and on activation values of linked units. On their side, the values of connection weights (i.e. some of the network parameters) are determined by means of suitable learning algorithms, which constrain behaviors of the system within a certain region, and, more precisely, the one corresponding to the region of the solutions to the problem at hand.

In this context, emergence is tied to the fact that the same ANN that starts from an unordered pattern of behavior, ends up, throughout the adjustments due to the operation of the learning algorithm, in a ordered pattern of behavior that is meaningful from the point of view of the observing system. Here, we are not dealing with a single fixed model (i.e. a single network function), on the contrary we are dealing with an entire set of different models (i.e. different network functions) that succeed one another in such a (complementary) way as to give rise to an ordered behavior within the system's state space. On this view, we could not be allowed to account for emergence simply by appealing to one fixed model, but only by resorting to a sequence of complementary models, each of which is developed depending on 'external' information. In the case of ANNs it is the learning algorithm that, as a function of external information, modifies connection weights in such a way as to maximize the performance.

The considerations just unfolded clearly suggest that systems supporting emergent behaviors cannot be described by means of *logically closed models*. As a matter of fact, a logically closed model is thought of in terms of *Formal System*. Briefly, a Formal System is a, possibly infinite but decidable, set of formulas, expressed by means of a Formal Language, which correspond to

1. *logical axioms*, *special axioms* of the model (e.g. evolutive laws, constitutive laws, constraints, initial conditions and boundary conditions) and of other theoretical domains,
2. *theorems* derived from axioms through the application of the inference rules of the system.

Clearly, a logically closed model is relevant only to the description of logically closed systems. As regards these latter systems, information needed to their description could be algoritmically specified ab initio. Indeed, if a set of behaviors is recursively denumerable (i.e. produced as output by a computable function) but it is not also recursive (i.e. unpredictable because of undecidability), then this same set of behaviors could be, in one sense, new for an observer, but not for the system itself; for this reason, such circumstances should be ruled out from the definition of emergent phenomena. But, as we have seen, a fundamental property of emergence is its innovating (and possibly supervenient) character. Due to its finitistic nature, a formal system cannot, in isolation, give rise to genuine novel patterns of organization, even if not all of their properties are decidable (i.e. predictable). Thus, any formal system and, in turn, any logically closed model cannot provide us with an appropriate description of systems supporting emergent behaviors. In order to come up with such an account, we should rather resort to *logically open models* (cfr. Minati *et al.*, 1998). A logically open model is not a formal system in that the set of formulas that make it is not recursive (i.e. undecidable), that is it does not exist any algorithm for establishing what information is, or is not relevant to the description of the system's behavior; and such a circumstance can be a consequence of the fact that information needed is provided by the environment in an non algorithmic manner, or, equivalently, as pointed out by Minati *et al.* (1998), the relationships between the system and its environment are only implicitly described and it does not exist any method for making them explicit. In other words, logically open systems are systems for which no complete description exists.

According to our intuitions, the ability to support novel behaviors is not the only relevant property of systems that could give rise to emergent processes; indeed, such novel behaviors should display *macroscopic order* too. Briefly, such systems must posses the ability to support transitions from unordered macroscopic states to ordered macroscopic states, namely *phase transitions*. However, this is one of the most difficult issues as we are forced to face the difficulty of finding a *principled way* to account both for what the macroscopic states (order parameters) are that drive the functioning of the lower level as a whole, and for how they effectively emerge. On this account, the arguments put forward by Pessa (1998), Pessa and Vitiello (1998), according to which Quantum Field Theory (where phase transitions are a consequence of the theory itself) would constitute an "ideal" framework for the explanation of emergence seem to be highly cogent.

Summing up, the conclusion we draw from the preceding discussion is that emergence can take place only within *logically open systems* which display *phase transitions*. Therefore, if language is an emergent entity, as it appears to be, it also must be rooted in logically open systems which display phase transitions.

3. A NEURAL NETWORK MODEL OF THE EMERGENCE OF THE SYMBOLIC FUNCTION

3.1 Symbols and symbolic mediation

The model being illustrated in this section was originally put forward as a neural approach to the Symbol Grounding problem, namely the problem of understanding how communicative agents autonomously connect symbols to meanings (see Pessa and Terenzi, 2001, for the details). However, this latter can be considered as a direct consequence of the more general problem of Symbol Emergence, namely the problem of understanding how, in the first place, some events internal to cognitive systems end up to correlate systematically, and in specific ways, with environmental situations (cfr. Plunkett *et al.*, 1992).

The problem is further complicated by the need, originally pointed out within Semiotics and Philosophy of Language of distinguishing between different, non equivalent notions of "symbol" (more properly, *sign*). Leaving aside the other ones, the subject of interest here is the more problematic notion of *linguistic symbol*. By reformulating a proposal originally put forward by Pearce (1978), the latter notion can be characterized, in the simplest way, as follows: a physical stimulus S, corresponding to the linguistic stimulus, should be considered as a symbol *if and only if* its occurrence is correlated with the occurrence of entities that belong to the same equivalence class M, corresponding to the meaning of the symbol S, by means of a mediation performed by the cognitive system of the speaker.

The problem here amounts to establish how can such a symbolic mediation actually take place within the speaker's cognitive system. We refer to this problem as the problem of the *emergence of the symbolic function*. According to a view dating back to De Saussure (1922) the symbolic mediation is better understood as a complex relation which is incorporated within the speaker's mind and which holds between (a) a class of auditory representations of linguistic stimuli and (b) a class of semantic representations (perhaps, concepts) relating to (or triggered by) suitable physical events. However, the situation is complicated by the following circumstances

1. there are, at least, two different kinds of semantic representations: the low-level *perceptual categories* to which physical events belong, and the higher-level categories, which catch correlations between different low-level perceptual categories;
2. the relationship between auditory representations and semantic representations isn't topology-preserving: this means that two linguistic stimuli very close one to another from the auditory point of view can be related to quite different semantic representations
3. whereas, on the one hand, semantic representations are *causally related* to the environmental events they refer to, on the other one they depend, under many respects, on the internal structure of the cognitive system that makes the symbolic mediation.

3.2 Architecture of the model

The architecture we are going to illustrate was meant to solve the two following issues:

1. given a set of categorical representations of acoustic signals and given a set of physical stimuli that correspond to visual images belonging to distinct categories, how can it happen that, from one certain instant of time on, the former ones get systematically associated to the latter ones (and conversely), within the speaker's cognitive system?
2. Moreover, how do semantic representations, determined this way, both bear on and depend on the execution of specific semantic tasks of the processing system?

Besides, the idea underlying the model can be stated as follows: the "natural" association between symbols and physical inputs results from a suitable adaptation process in which a neural network architecture learns to perform a semantic task by progressively adjusting the form of the relationship between symbols and inputs as a function of the success obtained in performing the task itself. As a consequence of this, semantic representations should be *not fixed* in advance by the modeler, but *autonomously developed* by the cognitive agent itself throughout the learning process. The performance of the model was assessed by means of a semantic task consisting of associating thematic roles (such as agent, patient, instrument and the likes) to sentence constituents. The overall architecture (see Fig. 1) of the model is constituted by three main components:

1. *Global Symbol Lexicon* (GSL): which contains semantic representations of words and is operated on by the remaining components
2. *Symbol Categorization Module*: which (a) associates each physical stimulus (e.g. here identified with a visual image) to the code of a suitable linguistic stimulus (i.e. a word), as well as the low-level and high-level category to which the physical stimulus belongs; (b) encodes each linguistic stimulus (word) as a suitable low-level and as an high-level categorical representation associated with the corresponding physical stimulus; in turn it decodes low-level representations as the associated words; (c) sets the initial structure of GSL, whose representations undergo successive modifications depending on the output error made by the Semantic Processing Module
3. *Semantic Processing Module*: which performs the semantic task consisting of associating thematic roles to sentence constituents.

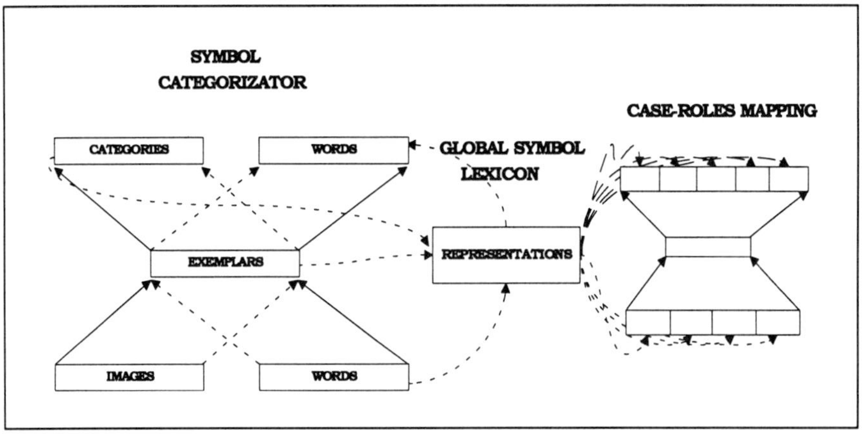

Figure 1. Overall architecture of the model including both symbol categorizator and FGREP module.

3.2.1 Symbol Categorization Module

As regards the symbol categorization module, it is composed by three distinct sub-modules:

1. *main categorizator*, which associates visual images to low-level categories, and, in turn, associates these latter to high-level categories
2. *encoder*, which associates input words to low-level categories
3. *decoder*, which associates low-level categories to output words.

The main categorizator was implemented through a RBF network, inspired by ALCOVE model (Kruschke, 1992, 1993). Its input layer is activated by a grey-level coding of the pixels of the visual image that corresponds to a suitable physical stimulus. On its side, each hidden-layer unit corresponds to a particular (low-level) perceptual category; for this reason, each hidden unit is associated to a particular *prototypical* visual image. As regards the encoder and the decoder, both were implemented through simple two-layer Perceptrons.

The connection weights of the Symbol Categorization Module were determined through a supervised learning procedure consisting of two successive phases. In the first one the main categorizator was trained on a set of 32 different visual images, to associate each image to its correct high-level category. Widrow-Hoff ruel was used to find the values of connection weights from hidden layer to output layer. In the second learning phase the encoder and the decoder were trained separately. The encoder was trained to associate each input vector (coding a particular word, like "*dog*") to the correct pattern of activation, on hidden units of the main categorizator, that corresponded to the presentation as input to this latter of a particular image (e.g. the image of a *dog*). In this case, we adopted an hypothesis of *timely coincidence* between the presentation of a visual image and the presentation of an acoustic stimulus; this means that each target vector of the encoder was determined through the simultaneous presentation, during the learning phase, of the stimulus to which the word is to correspond. On its side, the decoder was trained to associate each particular pattern of activation, on hidden units of the main categorizator (such as the one resulting from the presentation of the image of a *dog*), to an output vector coding a particular word (e.g. "*dog*"). Again, Widrow-Hoff rule was used.

3.2.2 Semantic Processing Module

As we have anticipated above, the Symbol Categorization Module is also deputed to set the *initial structure* of GSL, which contains representations of words used throughout the successive semantic processing.

The method we adopted to connect such a structure with the performance in the semantic task at hand, was the so called FGREP method (Forming Global Representations through Extended back-Propagation), already introduced by Miikkulainen and Dyer (1991). The simplest FGREP module is based on two main components: a three-layer Perceptron, and a Global Lexicon containing explicit vector representations of the symbolic entities being used (words, in our case). The system is trained to perform a specific semantic task typically by taking as external input the surface vector coding of the symbolic entities being used throughout the task. Within the Lexicon, such a coding gets associated with vectors of initially randomly set values. Later on, representations of the same items are taken out of Lexicon and possibly given as input and as target to the Perceptron, in a way dependent on the task; error signal is then backpropagated to the input layer, changing, along the path, both weights and input representations as if they were an extra layer of weights. The resulting representations replace the old ones in the Lexicon. As regards the case-roles mapping task at hand, the operation of the system can be summarized as follows:

1. *forming input representations*: for each component of the input sentence, *get* the associated representation from Lexicon and *concatenate* them within a single input vector
2. *forming target representations*: for each component of the target sentence (corresponding to the correct assignment of thematic roles), *do* the same as in 1. except for concatenating representations within a single target vector
3. *learning the processing task*: for each couple of vectors in the training set, *propagate* input, *back-propagate* target and *update* weights
4. *updating representations in Lexicon* (Pessa and Terenzi, 2001).

3.3 Numeric Simulations

By using the previously described architecture, we performed a number of different experiments in order to assess both the level of performance of the model and the possible emergence of the symbolic function.

In particular, we tried to evaluate the hypothesis according to which the performance of the system crucially depends on the specific form of the relationship holding between physical inputs and their inner representations, as it is learned throughout the processing task and as a function of the particular processing conditions. The processing conditions we have examined comprised the following situations:

- GSL starting with random representation vectors only
- GSL starting with high-level categorical vectors only
- GSL starting with low-level categorical vectors only

– GSL starting with mixed random + high-level categorical vectors.

The results obtained from numerical experiments demonstrated that the system was effectively able to perform the aforementioned task and that a systematic relationship between visual stimuli and acoustic stimuli effectively emerges. A meaningful result is that the highest performance in the association task (i.e. 88.5 % of correct assignments) is achieved when the semantic representations used to determine the initial structure of GSL are chosen from low-level categories, instead of high-level ones. As a consequence of this circumstance, the evaluation of the performance of the model showed that the dependence of semantic representations on perceptive processing bears on the execution of the particular semantic task at hand.

4. CONCLUSION

The discussion just carried out suggests, essentially, that models based on artificial neural networks could give rise to collective behaviors that are relevant as to the emergence of the symbolic function and to semantic processing as well. As a matter of fact, instances of this kind show how logically open systems, supporting transitions from unordered collective states to ordered collective states, can give rise to emergent phenomena, in a sense made clear above. As an examination of the previously illustrated model makes clear, the identification of the appearance of symbolic structures, and of the successive operations they undergo, is strictly linked to the existence of an observer who recognizes some collective behaviors of the system as meaningful from his own point of view, besides the point of view of the system itself.

However, many are the questions out of reach of our solution. It is in fact lacking under many fundamental respects. Firstly, it is not based on a general theory of emergence, providing us with general principles and tools both for identification and synthesis of emergent phenomena; in our case, identification is let to the observer in an unprincipled way. Secondly, by holding an hypothesis of timely coincidence between given visual inputs and given linguistic stimuli, it leaves unexplained how such simultaneity effectively takes place and how the stimuli, that are given in the model, emerge as regularities in their respective channels. Finally, as regards the specific problem at issue, it leaves out a number of other fundamental issues within the theory of meaning such as the hierarchical structure of categorical representations used by human subjects and the grounding of symbols to internal parameters (besides the external ones) that would serve as suitable directions for the system in selecting different kinds of information within the same pattern of stimulation. Whereas the two latter problems are domain-specific and pertain to a particular class of system (namely the one supporting the phenomena labelled as "natural language"), the first problem is one of the most general nature and therefore bears on the whole field of Systemics. And if we are correct by considering language as an emergent entity, then it will be of paramount importance to have a general theory of emergence that would clarify its general features and its underlying principles.

REFERENCES

Chalmers, D. J., 1994, *The Components of Content*. PNP Technical Report 94-04, Whashington University.

Chomsky, N. A., 1980, On Cognitive Structures and their Development. In *Language and Learning: the Debate between Jean Piaget and Noam Chomsky* (M. Piattelli- Palmarini, ed.), Harvard University Press, Cambridge, MA.

Crutchfield, J. P., 1994, The Calculi of Emergence: Computation, Dynamics and Induction. *Physica D* **75**: 11-54.

De Saussure, F., 1922, *Cours de Linguistique Generale*. Editions Payot, Paris.

Elman, J. L., 1999, The Emergence of Language: A Conspiracy Theory. In *Emergence of Language* (B. MacWhinney, ed.), LEA, Hillsdale, NJ.

Haken, H., 1988, *Information and Self-Organization: A Macroscopic Approach to Complex Systems*. Springer Verlag, Berlin-Heidelberg-New York.

Harnad, S., 1990, The Symbol Grounding Problem. *Physica D* **42**: 335-346.

Heylighen, F., 1989, Self-Organization, Emergence and the Architecture of Complexity. *Proceedings of the First European Conference on Systems Science*, AFCET, Paris.

Kruschke, J. K., 1992, ALCOVE: An Exemplar-based Connectionist Model of Category Learning. *Psychological Review* **99**: 22-44.

Kruschke, J. K., 1993, Human Category Learning: Implications for Back Propagation Models. *Connection Science* **5**: 3-36.

Miikkulainen, R., and Dyer, M. G., 1991, Natural Language Processing with Modular Neural Networks and Distributed Lexicon. *Cognitive Science* **15**: 343-399.

Minati, G., Penna, M. P., and Pessa, E., 1998, Thermodinamical and Logical Openness in General Systems. *Systems Research and Behavioral Science* **15**: 131-145.

Peirce, C. S., 1978, *Collected Papers. Vol. II: Elements of Logic*. (C. Hartshorne and P. Weiss, eds.), Belknap, Cambridge, MA.

Pessa, E., 1998, Emergence, Self-Organization and Quantum Theory. In *Proceedings of the First Italian Conference on Systemics*, Apogeo, Milan.

Pessa, E., and Terenzi, G., 2001, A Neural Solution to the Symbol Grounding Problem. In *WIRN – Vietri* (M. Marinaro and R. Tagliaferri, eds.), in press.

Pessa, E., and Vitiello, G., 1998, Dynamical Generation of Symmetry Patterns and Coherence in Quantum Fields Theory. *Proceedings of the First Italian Conference on Systemics*, Apogeo, Milano.

Plunkett, K., Sinha, C., Moller, M., and Strandsby, O., 1992, Symbol Grounding or the Emergence of Symbols? Vocabulary Growth in Children and a Connectionist Net. *Connection Science* **4** (3/4): 293-312.

Rojas, R., 1996, *Neural Networks: A Systematic Introduction*. Springer-Verlag, Berlin-Heidelberg-New York.

Synaptic Development as a Variational Problem

PAOLO CAMIZ
Department of Physics, Università di Roma "La Sapienza", Piazzale A. Moro,5 – Roma, Italy
INFN Sezione di Roma; ECONA

1. INTRODUCTION

A model for synaptic development in natural perceptive systems based on perturbation theory has been presented during the First Italian Conference on Systemic (Camiz, 1998a; Camiz, 1998b), and some further refinements and applications have been discussed in Fourth Science European Congress (Camiz, 1999) and in International Conference on Spatial Cognition 2000 (Camiz, 2000). The underlying idea for the perturbative neural network (PNN) has been suggested by the observation that some synaptic weights of visual system can be interpreted as eigenvectors of a suitable hermitian matrix, which turns out to be the correlation matrix of input vectors.

A reasonable choice of the perceptive scenario allows a very satisfactory reproduction of synaptic weights of some sections of visual system and also a prediction (to be verified by neurological tests) about the synaptic structure of labyrinthic system. Despite the success of PNN, the mathematical algorithm governing the updating of the network has scarce probability for an effective implementation on a natural perceptive system: indeed for some terms of the relevant formulae a neurological interpretation is very difficult, up to my knowledge.

On the other hand it is well known that the eigenvalue equation for a selfadjoint operator (hermitian matrix) is equivalent to a variational problem, namely the search for a set of vectors that make stationary the expectation value of the operator; the eigenvalues play the role of Lagrange multipliers ensuring the eigenvectors normalization.

Emergence in Complex, Cognitive, Social, and Biological Systems, Edited by
Minati and Pessa, Kluwer Academic/Plenum Publishers, New York 2002

In this paper I intend to make a critical analysis of PNN structure, to determine which functions are reasonable candidates for a variational problem, and to explore alternative processes, with respect to the rigorous perturbative method, capable to produce the same result, maybe in a longer time but with more realistic ingredients.

2. THE STRUCTURE OF THE PNN

I recall briefly the relevant features of the network: it consists of two layers of n neurons, the first one directly exposed to input vectors $x_i(t)$; $i = 1,2 \ldots n$, the second one connected to the first one by synaptic vectors ψ^i_α ; $\alpha = 1,2 \ldots n$ so that the response of second layer units is given by

$$\xi_\alpha = \psi^i_\alpha x_i(t) .$$

In the following, synaptic vectors ψ_α will be also represented in Dirac notation $|\, a >$ or $< a \,|$.

If the network must describe synaptic development during a perceptive experience, synaptic vectors are also functions of time, and their asymptotic value is such that

$$G_{ij}\psi^j_\alpha = \lambda_\alpha \psi^i_\alpha , \tag{1}$$

where

$$G_{ij} = \frac{1}{T}\sum_{t=1}^{T} x_i(t)\, x_j(t) = \frac{1}{T}\sum g_{ij}(t)$$

is the correlation matrix of the stimulating vectors, and λ_α are its eigenvalues, whose sum is 1.

According to perturbation theory for eigenvalues and eigenvectors of hermitian matrices the updating is given by the following rules:

$$\delta\lambda_\alpha(t) = f\psi^i_\alpha g_{ij}(t)\psi^j_\alpha = f\xi^2_\alpha \tag{2}$$

$$\delta\psi^i_\alpha(t) = f\sum_{\beta\neq\alpha}\psi^i_\beta \frac{\psi^k_\beta\, g_{kj}(t)\,\psi^j_\alpha}{\lambda_\alpha - \lambda_\beta} = f\sum_{\beta\neq\alpha}\psi^i_\beta \frac{\xi_\beta\xi_\alpha}{\lambda_\alpha - \lambda_\beta} \tag{3}$$

where f is a small learning parameter ensuring that each new stimulus can be considered as a small perturbation by the system. At each step of the updating procedure, normalization conditions for λ's and ψ's are imposed.

Furthermore a reorthogonalizing step, in the form of

$$|\alpha> \;\rightarrow\; |\alpha> - \frac{1}{2}\sum_{\beta\neq\alpha} |\beta><\beta|\alpha> \tag{4}$$

is necessary in order to prevent a dimensional collapse, namely the tendency of synaptic vectors to parallelize to the principal one.

Each eigenvalue represents the probability that the corresponding eigenvector is contained in the average stimulating vector.

The behaviour of the PNN is largely independent from the initial values of λ's and ψ's, provided that the formers are all different ant the latters are orthogonal. Asymptotic values are reached in shorter time for large eigenvalues, in longer time for small ones; the numerical value of the learning parameter f can be chosen in order to reach a reasonable compromise between a rapid learning (large f) and a stable one (small f). It has been verified that the network converges to the right values even if the function $1/(\lambda_\alpha - \lambda_\beta)$ in formula (3) is changed into a wide variety of odd functions of the same argument.

Is it possible that a similar procedure may occur in a natural perceptive system? A careful inspection to formulae (2-4) indicates that x's, ξ's, ψ's are actual quantities, whereas λ's are only virtual ones, which are not accessible to the system, unless there is some memory available in which current values of λ's can be stored. In absence of such a storage the PNN has to be modified in order to use only actual quantities in the updating procedure.

3. THE VARIATIONAL PROBLEM

The eigenvectors ψ_α of the correlation matrix G have the property that its average values $G_{\alpha\alpha} = <\alpha|G|\alpha>$ are stationary with respect to first order infinitesimal variations of ψ_α; indeed only the two eigenvectors belonging to the largest and the smallest eigenvalue of G correspond respectively to a maximum and a minimum of the average value of G, whereas the remaining ones are saddle points in a suitable vector space. During the learning process G is not available to the system: only the perturbing contribution $g(t)$ is an actual quantity, whose average value on ψ_α corresponds to the squared excitation of the second layer unit $\xi^2{}_\alpha$; its time average is the average value of G.

If the stimulating scenario is such that each of these quantities undergoes small fluctuations around the average or, in other words, if the scenario is relatively stable, the stationarities of g and G are equivalent.

A new problem now arises, because a stationarity condition is difficult to impose to a living system: better to impose to maximize (or minimize) a function, otherwise the requirement might be ambiguous. There are two possible solutions to this problem: the first one takes into account the fact that eigenvectors of a matrix are zeros for its root mean square deviation ΔG since $<\alpha|G^2|\alpha> = <\alpha|G|\alpha>^2$; therefore the condition to be imposed is to minimize ΔG.

The second solution is suggested by the previously described time behaviour of eigenvalues, and by the observation of visual performances of living systems during early stages of life: it is known that visual acuity increases during growth, provided that a suitable visual stimulation occurs. Increase of visual acuity means the ability to detect finer details of a given pattern or, in mathematical terms, to register the presence of higher spatial frequencies in its Fourier transform. It turns out, as can be shown by numerical simulation, that low (high) frequency waves correspond to larger (smaller) eigenvalues of the correlation matrix obtained with input vectors presented to a model of retina, and, as we have said, that large eigenvalues reach their asymptotic value before smaller ones.

This circumstances suggest that the (neurobiological) rule describing the solution of the variational problem could be the following:

- in correspondence of each stimulus adjust weights of the winning unit of the second layer in order to increase its response;
- once these weights have reached a reasonable stability proceed in the same way with the second winning unit, keeping its synaptic vector orthogonal to the previous one;
- continue with the following units until the last one has reached stability.

In this way the stationarity requirement (ambiguous) is converted into a requirement of absolute maximum in a hierarchy of orthogonal spaces. Of course the procedure could be applied in reversed order: first minimize the response of the losing unit, then the one of the second losing unit in an orthogonal space, etc. It is only the circumstance that the principal eigenvector requires a shorter time to reach its asymptotic value that suggest to maximize the response.

4. THE PNN DRIVEN ONLY BY ACTUAL QUANTITIES

As I have suggested in the preceding section the substitution of virtual quantities (λ's) with actual ones (ξ's) makes the PNN more realistic. Moreover, since λ's are not only the eigenvalues of the correlation matrix G, but also asymptotic values of the quadratic response of the second layer units, the new version of updating procedure concerns only synaptic vectors and:

$$\delta | \alpha > = f \sum_{\beta \neq \alpha} | \beta > \xi_\beta \xi_\alpha \, odd(\xi_\alpha^2 - \xi_\beta^2), \tag{5}$$

where $odd(x)$ is an odd function to be chosen arbitrarily.

Looking to the formula (5) we see at glance that a correlated response of two units α and β ($\xi_\alpha \xi_\beta > 0$) produces a "transfer" of synaptic weight from the losing unit to the winning one of the pair, as if synaptic weights could be considered as a fluid; an anticorrelated response produces an opposite behaviour.

The only feature difficult to be accepted from a neurological point of view is the reorthogonalizing process: nevertheless its presence is completely coherent with the need, for any perceptive system, to extract the largest amount of information from input vectors.

The orthogonalization step (4) is not necessary in the simplest case of 2 dimensions, where an analytic solution can be given, as I shall discuss in the next section.

5. ANALYTIC SOLUTION IN 2 DIMENSIONS

Let $| a >, | b >$ be two orthonormal two components vectors

$$| a > = \cos \alpha, \ \sin \alpha, \qquad | b > = \sin \alpha, \ - \cos \alpha,$$

λ, $1 - \lambda$ the corresponding eigenvalues and

$$G = \lambda | a > < a | + (1 - \lambda) | b > < b |$$

the correlation matrix, whose explicit form is

$$G_{aa} = \sin^2\alpha + \lambda\cos 2\alpha \ ,$$
$$G_{ab} = G_{ba} = \left(\lambda - \frac{1}{2}\right)\sin 2\alpha \ ,$$
$$G_{bb} = \cos^2\alpha - \lambda\cos 2\alpha \ ; \qquad (6)$$

the angle α can be considered as a constant or a random variable.

Assume that initial values of synaptic vectors are

$$\psi_1 = (\cos\beta, \sin\beta); \qquad \psi_2 = (\sin\beta, -\cos\beta) \ ;$$

in this basis matrix elements of G are given by:

$$G_{11} = \xi_1^2 = \lambda\cos^2(\alpha-\beta) + (1-\lambda)\sin^2(\alpha-\beta)$$
$$G_{22} = \xi_2^2 = (1-\lambda)\cos^2(\alpha-\beta) + \lambda\, sin^2(\alpha-\beta)$$
$$G_{12} = G_{21} = \xi_1\xi_2 = -\left(\lambda - \frac{1}{2}\right)\sin 2(\alpha-\beta)$$
$$G_{11} - G_{22} = (2\lambda-1)\cos 2(\alpha-\beta). \qquad (7)$$

The updating formula (5) can be transformed into a first order differential equation, in which we have only to define the arbitrary odd function. Let put first $odd(x) = x$: the equation reads

$$\frac{d\psi_1}{dt} = f\psi_2\left(\lambda - \frac{1}{2}\right)^2 \sin 4(\beta(t) - \alpha) \ , \qquad (8)$$

where α, for the moment, is a constant which can be put equal to zero without loss of generality. Equation (8) and the analogous one for ψ_2 are equivalent to the following eq.

$$\frac{d\beta}{dt} = -f\left(\lambda - \frac{1}{2}\right)^2 \sin 4\beta \ , \qquad (9)$$

which can be rewritten as $d\theta / dt = -\gamma\sin\theta$ where $\gamma = 4f(\lambda - 1/2)^2$ and $\theta = 4\beta$; the solution is

$$\tan\left(\frac{\theta(t)}{2}\right) = e^{-\gamma t}\tan\left(\frac{\theta_0}{2}\right), \qquad (10)$$

where θ_0 is the initial value.

If we choose $odd(x) = 1/x$, corresponding to rigorous perturbation theory, we obtain the eq.

$$\frac{d\psi_1}{dt} = \frac{1}{2} f \psi_2 \tan 2(\beta - \alpha), \tag{11}$$

in which λ disappears; putting, as before, $\alpha = 0$, $\theta = 2\beta$ we have the solution:

$$\sin\theta(t) = e^{-ft} \sin\theta_0 . \tag{12}$$

as a third example we put $odd(x) = sign(x)$ obtaining the eq.

$$\frac{d\psi_1}{dt} = f\psi_2\left(\lambda - \frac{1}{2}\right)\sin 2(\beta - \alpha) \tag{13}$$

with solution

$$\tan\left(\frac{\theta(t)}{dt}\right) = e^{-\rho t} \tan\left(\frac{\theta_0}{2}\right), \tag{14}$$

where $\theta = 2\beta$ and $\rho = 2f(\lambda - 1/2)$.

Comparing the three cases we see that the approach to the asymptotic value is always exponential, but the time constants are different: if we assume $1/2 < \lambda < 1$ we see that $\gamma < \rho < f$ showing that the most rapid approach is given by perturbation theory.

If the driving matrix, which in this example is represented by the angle α, fluctuates around its average value, the angle β also fluctuates around the time trajectory given by the preceding formulae with an amplitude proportional to α fluctuations.

In this simple example it is easy to understand what I mean by scenario stability: in order to prevent large variations of ψ_1, namely "jumps" on ψ_2 and viceversa, α fluctuations must be small with respect to $\pi/4$.

6. CONCLUSIONS

I have shown that the precise requirements of the first version of PNN can be released without compromising its behaviour and, more important, its capability to reproduce synaptic weights in many (hopefully all) perceptive systems. Only the stability and the variety of stimulating scenarios cannot be released: the last condition is necessary in order that the correlation matrix G has non degenerate eigenvalues: if this is not the case it is impossible for the corresponding eigenvectors to reach a definite asymptotic value, since they continue to jump within the degenerate space.

This circumstance occurs not only when the stimulating vectors occupy a space with less than n dimensions, but also when they are completely random: the first case corresponds to a strongly limited stimulation, highly ordered, the second one to a completely disordered one. It turns out that synaptic development is favoured by situations in which the entropy is neither too high nor too low.

I can conclude this communication by saying that a perceptive system, capable of extracting in the most complete and efficient way the available information, is the result of an interaction between the system itself and the stimulating environment. A careful analysis of a variety of natural environments together with a deep neurophysiological exploration about synaptic organization of living beings could confirm this idea, which suggests that living beings and their environment must be considered as unseparable parts of a unique system.

REFERENCES

Camiz, P., 1998a, A model for synaptic development based on a self organizing neural network: General treatment. In *First Italian Conference on Systemics* (G. Minati, ed.), Apogeo, Milano, pp. 149-159.

Camiz, P., 1998b, A model for synaptic development based on a self organizing neural network: Three applications to visual system. In *First Italian Conference on Systemics* (G. Minati, ed.), Apogeo, Milano, pp. 161-171.

Camiz, P., and Lo Meo, S., 1999, A perturbative neural network as a universal tool for information codification. In *4th Systems Science European Congress Systemic* (L. Ferrer *et al.*, eds.), Valencia, pp. 481-490.

Camiz, P., 2000, A neural network approach to synaptic development of labyrinth. In *International Conference on Space Cognition - Cognitive Processing* (M. Olivetti Belardinelli, ed.), Roma, p. 83.

Theoretical Issues in Systemics

Emergence and Ergodicity: A Line of Research

GIANFRANCO MINATI
AIRS - Associazione Italiana per le Ricerca sui Sistemi – E-mail: gianfranco.minati@iol.it

1. INTRODUCTION

This working paper is a contribution to the discussion on emergence based on research topics debated in the international scientific community and introduced in the Italian Systems Society by Pessa and Minati with the notes presented during meetings and available in the AIRS's web page (Pessa, 2001).

Some phenomena of collective behavior, like the emergence of flocks, studied in ethology (Echelle, 1984; Darling, 2000; Parrish, 1998), gave arise to different approaches when studied in Artificial Life (Bonabeau, 1994), (Cariani, 1991). The evolution of beings having artificial life, living by computational cycles, like computer viruses, is studied by Synthetic Biology. Many behavioral models have been introduced trying to carry out software simulation programs. Flocks of Boids, i.e. birds considered in Artificial Life, are considered, for instance, in the two following approaches:

a) the one used by C. Reynolds introducing flocks of Boids programs simulation in order to carry out cartoons animated by computer (Reynolds, 1987);
   - separation rules: individuals must manage, drive their proceedings in order to avoid the crowding of locally adjacent components;
   - alignment rules: individuals must manage, drive their proceeding towards the average direction of locally adjacent components;
   - cohesion rules: individuals must manage, drive their proceeding towards the average position of locally adjacent components;

Emergence in Complex, Cognitive, Social, and Biological Systems, Edited by Minati and Pessa, Kluwer Academic/Plenum Publishers, New York 2002

b) another one has been proposed in (Pessa, 2000):
   - the performing of each single cognitive system requires the processing of information coming from external stimuli. The information to be processed may be assumed to come from adjacent individuals (locale inhibition).
   - such an information processing requires a significant amount of energy. It may be considered equivalent to the effect produced on individuals by an opposite force, opposing single movements (medium range activation).
   - such a force may be considered dependent only on directly adjacent individuals (long range inhibition).

   Such a schema is a generalization of the one proposed by (Gierer, 1972).

In the same way different simulation models have been proposed for swarms (Beckers, 1990; Belic, 1986; Bonabeau, 2000; Bonabeau, 1996; Millonas, 1993), ants (Aron, 1990) and fish schools (Anderson, 1980; Aoki, 1984; Breder, 1954; Vabo, 1997) have been studied, by using sophisticated mathematical models. In this contribution a different conceptual framework is introduced taking in count some hypotheses proposed in approaches formulated for studying gases properties and behavior, by focusing on micro behavior, at particles level. In this paper it is proposed to explore the possible use of the concept of *ergodicity*, introduced in statistical mechanics to model the behavior of gases particles, to emergence processes. In both cases the role of the observer is crucial.

The suitability of ergodicity to study emergence processes may be demonstrated by its usage in modeling and simulating.

2. SET THEORY AND SYSTEMS THEORY

Most of the definitions referring to the concept of ergodicity, presented in par. 2, have been introduced with reference to the concept of set. The possibility to extend the results developed in set theory to systems theory must be not just assumed but discussed, generally and case by case. In literature the theoretical problem is discussed, for instance, after the introduction of the very well known and applied theory of fuzzy sets by Lofti A. Zadeh (1996). This theory has been initiated and introduced for the first time in (Zadeh, 1965). Making reference to the classical set theory the axiom of extension (two sets are equal if and only if they have the same elements), the concept of empty set, of reflexivity, of union and intersection, of belonging, are examples of concepts used in set theory and non immediately applicable to systems. Another example is given by the two concepts:

- in sets theory, the whole is greater than one of its part (even if it is not applicable to infinite sets like it happens for the correspondence $n \rightarrow n^2$);
- in systems theory, the whole is greater than the sum of its parts, making reference to emergence processes (i. e. flocks from sets of birds, books from sets of words, music from sets of notes).

The main difference between the concept of set and system may be referred to the difference between the concepts of *relation* and *interaction*.

1. Examples of *relations* between elements of a set may be given by ordering with reference to any kind of parameters, by timing with reference, f. i., to synchronicity, by exclusiveness with reference, f. i., to the concept of exclusive or. Relations between elements of a set have almost a common aspect with properties of elements of a set (f. i. oddness, color, weight, duration in time): by considering the abstraction of set, they are assumed permanent, independent from other elements or relations. Relations among elements may be considered sets characteristics .
2. *Interactions* between elements are assumed to take place when one's behavior affects the other's behavior. Classic example in physics is given when particles collide; in social systems when people compete or cooperate. Interactions may take place when some circumstances occur: i.e. crossing trajectories, people meet, chemical reaction. Elements of a set may not interact for any reason even if they can.

A set of interacting elements is said to be a system when emergent properties are established: in such a case the interest is for the emergent properties more than for the elements: a system may be stable, homeostatic, goal-seeking, living (Minati, 2001).

The distinction just introduced between relations (referred to sets) and interactions (referred to systems) may bring to distinguish between structural and behavioral properties, being the first ones referred to the relation among elements of a set and the second ones referred to the behavior emergent from the interactions among elements of a system. Making reference to an observer, elements must have relationships to interact. *Structural properties are used by systems elements in interacting*. That's one of the bridges between sets properties and systemic emergence.

In literature interaction among components is assumed necessary condition for establishing emergence processes.

The difference between *sets of agents* and *systems of agents* may be helpful to realize that elements of a set able to give rise to a system may be considered for their structural properties as well for their behavioral properties able to produce emergent behavior. As we will see the concept of ergodicity has originally been introduced in physics for set of interacting particles and this concept will be considered in discussing systems emergence.

3. DEFINITIONS OF ERGODICITY

In this paragraph some introductory definitions of the concept of ergodicity introduced in different fields are presented. The concept has been originally introduced in physics, in statistical mechanics, studying the movement of gas components. In some definitions quoted from literature the distinction between set and system is not carefully considered.

From the various definitions, ergodicity emerges as a process of identification of correspondences between the behavior of subset of elements when *ensemble average is equal to time average* (par. 3.2.1).

3.1 Ergodic

In the *Webster's Dictionary* we find the follow definitions:

1. "of or relating to a process in which a sequence or sizable sample is equally representative of the whole (as in regard to a statistical parameter);
2. "involving or relating to the probability that any state will recur, especially having zero probability that any state will never recur."

In (Krippendorff 2001a) we find the following definitions:

3. "A collection of systems forms an ergodic ensemble if the modes of behavior found in any one system from time to time resemble its behavior at other temporal periods and if the behavior of any other system when chosen at random also is like the one system. We do not require identical performance, only quite similar time averages and number averages. (If you cannot tell one youth from another or one adult from another, they belong to an ergodic ensemble.) In an ergodic population, any single individual is representative of the entire population. The salient "characteristics of this individual are essentially identical with any other member of the group" (Iberall, 1978).
4. "Attribute of a behavior that involves only *equilibrium states* and whose transition *probabilities* either are unvarying or follow a definite cycle. In statistics, ergodicity is called stationariness and tested by comparing the transition probabilities or different parts of a longer sequence of events.

Ashby's "theory of incessant transmission" refers to the analysis of *information* flows in systems whose transition probabilities are unvarying and hence ascertainable for the analysis. All systems eventually converge toward ergodic behavior." See also (Iberall, 1978).

3.2 Ergodicity

The concept has been originally developed in Physics, in statistical Mechanics and applied to behavior of particles. The word comes from the Greek: literally, *The road or pathway of energy*.

The adjective "ergodic" comes from German: "*ergodenhypothese*" (hypothesis on the path of energy), from Greek: *erg* + *hodos* (energy path).

3.2.1 Physics

The theory behind statistical mechanics is based squarely on the *Ergodic Hypothesis* and Boltzmann statistics. These principles guide the analysis of the mechanical behavior of large sets of particles. Ludwig Boltzmann (1844-1906) introduced the "Ergodic Hypothesis". The study of transformations and their long-term behavior is ubiquitous in mathematics and the sciences. They arise in applications to the real world the name "ergodic" was introduced by Boltzmann, who intended to use his "Ergodic Hypothesis" as a justification of thermodynamics[1].

A citation from (Uchii, 1991) may help to identify the concept:

"The Ergodic Hypothesis ... states that a system of molecules will assume, in the long run, all conceivable" micro-states that are compatible with the conservation of energy. And if we can assume this hypothesis," then there is a unique time-average for any macro-state; indeed, we can calculate its value within the Newtonian mechanics. To be more specific, remember that (1) the time-average was a mechanical concept, and that (2) Boltzmann defined probability in terms of this mechanical concept. This means that the mechanical concept of time-average *has the same mathematical structure* as that of probability. Thus on the assumption of Ergodic Hypothesis, we can calculate the value of probability, within the framework of mechanics ... [*Actually, the concept of ergodicity has raised many difficult problems; but for simplicity we will ignore them. The philosophical import of the Ergodic Hypothesis can be grasped without getting involved with such problems. If you have a difficulty with the preceding formulation, the following version may help: -A system of molecules will go, in the long run, arbitrarily close to every conceivable micro-state.-*]".

[1] http://www.ams.org/meetings/suminst.html

The theoretical physicist J. W. Gibbs, (1839-1903) introduced the following postulate about *Time Evolution and Ergodicity* (Gibbs Postulate, Averaging and Ergodicity). The postulate is based on considering two ways of *averaging*:

1. by observing a system for a long time (*time averaging*);
2. by taking many systems from the observed ensemble (i.e. with same energy) and by observing them all in one time (*ensemble average*).

The Ergodicity hypothesis says that *ensemble average is equal to time average*.

The same system can be *both* ergodic and non-ergodic depending on the assumed time scale, i.e. polymers[2].

3.2.2 Theoretical geomorphology

In *theoretical geomorphology* (Anderson, 1988, Kirkby, 1994) the following definitions (Thorn, 1988, page 47) has been introduced:

- "The mean of observations of an individual made over time is equal to the mean of observations made of many individuals at a single moment in time over an area."
  OR
- "If 15% of the population is in a particular state at any moment in time, can assume that each individual in the population spends 15% of time in that state"[3].

3.2.3 Economics

The concept of ergodicity is applied in economics (Landon-Lane, 2000) and specifically to prices dynamics study (Domowitz, 1997).

A simple model of ergodicity has been introduced in Economics in (Landon-Lane, 2000). Let consider a C (finite number) of classifications. The classification in a socioeconomic system may be assumed as income classes. Income classes may be described as follow

Class k	1	2	$n + 1$
	$M_1 <$ Income $< M_2$	$M_2 <$ Income $< M_3$	$M_n <$ Income $< M_{n+1}$

Let π_{kt} be the portion of the total population G in the class k at the time t.

The vector $\pi_t = (\pi_{1t}, \pi_{2t}, \dots, \pi_{ct})$ defines the status of the world at the time t.

[2] http://www.plmsc.psu.edu/~www/matsc597c-1997/introduction/Lecture2/node3.html

[3] http://geog-ain.lancs.ac.uk/courses/pt1mod2/lecture10.2.html

Let consider now the probability P of transition of the vector π_t from a distribution i in classes of income at the time $t-1$ to a different distribution in classes of income j at the time t:

$P\ (\pi_t = j \mid \pi_{t-1} = i\ ) = p_{ij}$

The assumption of the first order Markov chain is that the status of the world π_t depends only on π_0:

$$P\ (\pi_t \mid \pi_{t-1}, \pi_{t-2}, \ldots, \pi_{t-s}\ ) = P\ (\ \pi_t \mid \pi_{t-1}), \quad \forall\ s = 2, 3, \ldots$$

The transition Markov's matrix is $[P_{ij}]$. Then it is $\pi'_t = \pi'_{t-1}\ P$.

It will be $\pi'_t = \pi'_0\ P$ (i.e. P applied t times).

The distribution π' is unique if there is only one eigenvalue of P *mod* =1.

For further mathematical details the reader may refer to (Landon-Lane, 2000).

3.2.4 Collective Behavior

By applying the approach used in the previous economic problem to the study of geometrical and systemic properties of collective behaviors like in flocks and swarms we may have the following conceptual framework.

Let consider a population G of elements X_i and classes of distances instead of income. The elements X_i may be clustered in a finite number C of classes characterized by the fact that in the class k ($k < C$) at the time t (time and space are assumed discrete as for Cellular Automata: Cellular automata are mathematical idealizations of physical systems in which space and time are discrete, and physical quantities take on a finite set of discrete values) the distance d between two elements belonging to this class is $< k_{max}$ and $> k_{min}$:

Class k	1	2	$n+1$
	$M_1 < D(X_i,X_r) < M_2$	$M_2 < D(X_i,X_r) < M_3$	$M_n < D(X_i,X_r) < M_{n+1}$

One element X_i may

- to belong to a one single class.
- to simultaneously belong to J ($J < C$) classes;
- to change in time the classes of distance to which it belongs.

The elements X_i may be represented in n-dimensional ($n > 2$) space.

An example of distribution of two classes corresponding to

Class	1	2
	$M_1 < d(X_i,X_r) < M_2$	$M_2 < d(X_i,X_r) < M_3$
Example	• •	• •

in a 2D space in represented in fig. 1.

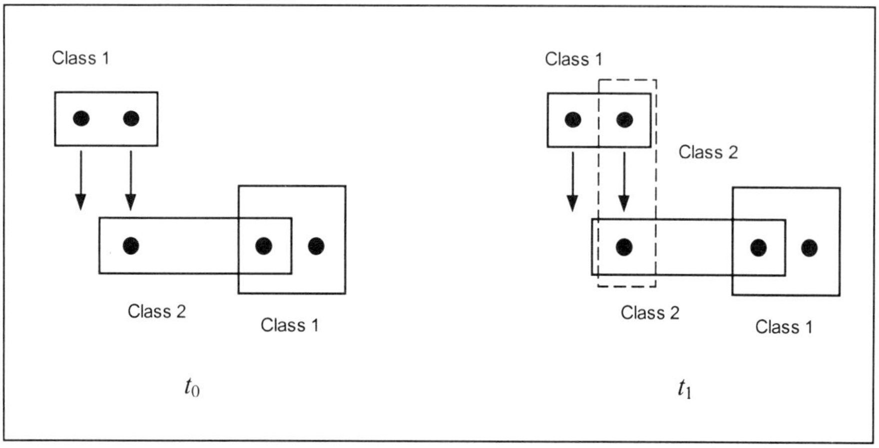

Figure 1. By considering only two classes of boids we have an example of spatial distribution at the time *t*.

Elements can migrate from a class to another one because they move in space inside the limits and finding themselves to belong to new classes.

Distribution is dynamic in time:

- elements move in such a way to have different: speed (acceleration different from zero) and direction, not one from the other, but from the previous step;
- elements may belong to different classes in time;
- one element may any time stop to belong to population G and then to a class because of internal (cognitive system) or external reasons.

Let's now discuss the ergodicity of the distribution π:

Let π_{kt} be the portion of the total population G in the class k at the time t.

The vector $\pi_t = (\pi_{1t}, \pi_{2t}, \ldots, \pi_{ct})$ defines the status of the distribution in classes at the time t.

Let consider now the probability P of transition of the vector π_t from a distribution i in classes of income at the time $t-1$ to a different distribution in classes of income j at the time t:

$P\,(\pi_t = j \mid \pi_{t-1} = i) = p_{ij}$

The first order Markov assumption implies that the status of the world πt depends only on π_0:

$P\,(\pi_t \mid \pi_{t-1}, \pi_{t-2}, \ldots, \pi_{t-s}) = P\,(\pi_t \mid \pi_{t-1}), \quad \forall\, s = 2,3, \ldots$

The transition Markov's matrix is $[P_{ij}]$.

Then it is $\pi'_t = \pi'_{t-1} P$.

It will be $\pi'_t = \pi'_0 P$ (i.e. P applied t times).

$\pi' = \pi' P$ ergodic.

The distribution π' is unique if there is only one eigenvalue of P $mod = 1$.

Ergodicity manifested in Collective Behaviors emerging from agents supposed provided with cognitive system (Anderson, 1985) may be supposed kept thanks to the relationship with neighbors (Minati, 1997). Such a Collective Behaviors may be supposed to take place when there are limited variations assumed by some critical parameters such as speed, acceleration, altitude, direction (Pessa, 2000).

3.3 Non-Ergodic

In (Krippendorff, 2001b) the following definition has been introduced:

"Attribute of a behavior that is in certain crucial respects incomprehensible through observation either for lack of repetition, e.g., by involving only transient states which are unique, or for lack of stability, e.g., when transition probabilities (see probabilities) are so variable that there are not enough observations available to ascertain them. evolution and social processes involving structural changes are inherently non-ergodic. To understand non-ergodic behavior requires either reference to the underlying organization of the system exhibiting it or the study of a large sample of systems of the same kind."

4. PROCESSES OF EMERGENCE

With reference to an observer, to its models, to its cognitive system (Benjafield, 1992) processes of emergence take place when interacting agents establish such a configuration having behavior different from the one of the composing agents.

From the definitions introduced above it is possible to assume that any kind of system has an ergodic behavior in time. The focus of this paper is on considering ergodicity as an indicator (see par. 5) of the kind of emergence:

- The interaction among agents may take place in an (1) *organized* way, for instance when components are structured in functions often corresponding to subsystems. Examples are departments in companies, schools, hospitals; devices accomplishing complex tasks in electronic equipment, software, biological systems; roles in natural systems like in behaves, anthills, termitery, nests, families. In this case the ergodicity of the system is hypothesized to be strong and stable.

- Other processes of emergence, i. e. self-organization, are hypothesized to take place when agents interacting at (2) *different dynamic levels* of ergodicity establish emergent systems. In this case levels of ergodicity are hypothesized oscillating in a confidence interval, that's adopting various values within that interval, changing smoothly or even by jumps. Different confidence intervals correspond to different kinds of emerging systems.

The distinction is not rigid because, as mentioned before, the same system may be ergodic (strong ergodicity) and non ergodic (very weak ergodicity) at the same time, having organized and/or non organized behavior as it is made evident by considering how much organization is required to simulate an ergodic behavior.

The differences between these two cases must be taken in count when considering the opportunity to manage, keep, start, modulate, control and replicate processes of emergence. This aspect will be discussed in par.4 when some proposals about an index of ergodicity will be introduced.

Processes of emergence take place when organized (case 1.) or non-organized (case 2.) interactions among agents give arise to a new form of reality having specific properties, i. e. *systems*.

Processes of emergence take place when an *observer* may detect that:

- New models are required to understand properties and behavior of the emerging system: the models used by the observer to understand agents behavior are not sufficient to explain the system's behavior;
- Agents deal with situations that they cannot cause or manage individually;
- Agents have a behavior organized in roles in the system or have an ergodic behavior.

Emergent properties arise from configurations of agents manifesting behavior based on an observer point of view:

- on rules locally applied, i.e. without processing information explicitly regarding effects in time and space referring to the global system. The global effects are not linearly produced but reached through micro behaviors able to give arise to emergent processes. The emergent system may be assumed by the observer, thanks to experiments, to have or not awareness of the global effects in time and space. Agents having micro behaviors impacting on environmental modifications and population dynamics may be assumed to have not awareness by an observer having different time and space scale. Swarms, flocks, herds, crown may be assumed to have awareness by an observer when defending from a predator, or using their large number of components. In this case the conceptual schema of Cellular Automata may be very suitable for simulation. *In such a kind of emergence processes an agent is indistinguishable by another by the detecting observer.*

– on specific functional roles of agents and subsystems. The ability of a configuration of agents to do something that single agents can't may be considered emergent when is based on specific roles assumed by agents or subsystems (like in an assembly line, a living being, a company, a beehive). The organization is explicitly designed for the purpose to reach a goal, to establish a behavior. The system may be aware of the global effects in time and space depending on the scale adopted by the user. *In such a kind of emergence processes, based on organization, an agent is distinguishable by another because of its role.*

The two aspects just introduced may be not exclusive inside the same system and inside the systems to which agents, even simultaneously, belong.

Examples of the two cases emerging from the same agents is given by swarms vs. beehives, crown vs. line. Emergent properties are assumed to have some stability in reference to the scale adopted by the observer.

When agents are supposed provided with cognitive model they may manage interactions, having the ability to decide, depending on the sophistication of the cognitive model, if and how to interact (Pessa, 2000).

The observer detects organization depending on its (cognitive) models. The concept of "functioning", based on the role of the "observer" may be helpful in making evident differences between organization and self-organization. The distinction between detecting emergence coming from organized, structured systems and from self-organization processes, reminds the difference between symbolic, procedural, algorithmic and sub-symbolic processing. In the first case emergence is related to designing devices having symbolically and explicitly knowledgeable functioning. In the second case it is still possible to represent functioning, but not symbolically explicit, as it happens with Neural Networks. The distinction refers to the cognitive models of the observer.

5. MEASUREMENT AND INDEXES OF ERGODICITY. ERGODICITY AND EMERGENCE

It may be of great interest to be able to have a *measurement* or indication of ergodicity, having the possibility to describe its evolution in time and to use it in order to compare two sets or processes displaying ergodicity. We may try to introduce a conceptual framework in which the availability of an indicator of ergodicity may describe the evolution of the ergodicity of a process. Making reference to the definitions of ergodicity introduced in 3.2.2 (Theoretical geomorphology) and considering x as the number of elements per single moment and y as the amount of time per single element, is possible to reformulate the definitions:

a) "If x % of the population is in a particular state S at any moment in time, can assume that each individual in the population spends y % of time in that state", is considered equivalent to the other formulation:
b) "The mean of observations of an individual made for a time y is equal to the mean of observations made of x individuals at a single moment in time over an area".

Ergodicity may be considered weak and strong in different disciplines. As introduced in theoretical geomorphology (Schumm, 1965) 'Weak ergodicity' allows geomorphologists to extend limited observational time-scales. By considering weak and strong ergodicity the focus is on the models of the observer, that's on the detectable effects of the ergodicity in the time and space scale. The concept of weak and (by extrapolation) strong ergodicity may be intended to make reference to the values of x and y, in (a) as percentage and in (b) with special reference to the time-space scale of the observer. *"Weak and strong" are referred to the models and the time scale of the observer.* A very weak level of ergodicity is difficult to be detected by the observer (its models must be very sensitive and working on other time scales) while a very strong level of ergodicity takes place when components of the considered set or process tends to have all the same behavior, to assume always the same configuration in time and space. The indicator of ergodicity we are looking for must detect the level of ergodicity of a process or set in the considered time-space, making the observer able to make correspondent sets, systems or processes of very different kind having the same ergodicity or having similar ergodic evolution in time and to detect the dynamics of different dynamic levels of ergodicity assumed by interacting agents establishing emergent systems.

Considering now the process of changing of the level of ergodicity in time for a set or system, depending on the behavior of components and on the learning process of the observer (possible because of similar time-space scale or because of assumptions about relationships between different time-scale phenomena like it happens in geology and astronomy), it is interesting to remind what has been introduced in par. 3.1, definition (4): "All systems eventually converge toward ergodic behavior." This not only refers to the dynamics of the system (if any: there is no dynamics in a set of non interacting, formal elements considering for instance statistical properties only), but specially refers to the observer, to its learning capabilities, its memory, its cognitive system, to the observation time and number of observations.

The interest in studying the process of establishing ergodicity in a set or in a system, its path from very weak levels of ergodicity towards stronger ones, is related to how much it may point out the establishing of processes of emergence. The interest is on having indicators able to point out that process of emergence are taking place. As we will see in the following the proposed indicator is the dynamics of the levels of ergodicity.

At the beginning of a process of emergence it may be assumed, by following the definition just introduced, that the ergodicity is very weak. At the other end the ergodicity may be assumed be at the maximum level and that means the set, the process is always equal to itself, that's identity in time.

The idea is that processes of emergence may be detected and managed (for instance slowed or speeded up) by considering the dynamics of their ergodicity. The idea is that when in a set, system or process showing high dynamics, the indicator of ergodicity tends to assume values in a confidence interval (changing smoothly or even by jumps), then it is possible to assume that there is the emergence of a new organization, new structure.

Making reference to the definitions just introduced and to the concept of weak ergodicity introduced in (Schumm, 1965):

a) "If x % of the population is in a particular state S at any moment in time, can assume that each individual in the population spends y % of time in that state".

The request about spending y % of the time in a particular state is very strong. We can try to introduce a different and more general formulation such as:

"If x % of a population is in a particular state S at *Any* moment in time, can assume that *All* subpopulations spends y % of time in that state".

With reference to the original strong definition introduced in theoretical geomorphology (Anderson, 1988; Kirkby, 1994; Thorn, 1988) and in[4] instead of "each individual" it is used the expression "all subpopulations": the request (to spend y % of time in a particular state) is not for the behavior of *Each* single agent, but for *All* "subpopulations" of them allowing them to have not constant but statistically significant properties.

Definitions of subpopulation come from the interest of the observer, the same observer who is expected to detect emergence.

Moreover it is conceptually more balanced to consider:

- percentage of a population; vs. all subpopulations, rather than
- percentage of a population vs. each individual.

[4] http://geog-ain.lancs.ac.uk/courses/pt1mod2/lecture10.2.html

For instance: "If x % of a population watches TV in *any* moment on time, can assume that *all* subpopulations (i.e. all families of the social system) spends y % of time watching TV". In this case there are not explicit rigid prescriptions for single components of the family: it is sufficient that it is the family to spend y % of time watching TV and not the single components. With respect to the question of subpopulations versus populations, it is possible to also consider the following example. Some individuals may be excluded from taking part in certain activities in a system by principle, for instance men giving birth to children, but nevertheless the mixed society of men and women may be ergodic, with reference to certain activities.

The reformulation of the concept introduced in theoretical geomorphology seems to be more appropriate for generalizations to different kinds of systems.

Assuming x / y as index of ergodicity, when

- $x / y \rightarrow 1$, there is ergodicity;
- x and $y \rightarrow 0$, there is weak ergodicity;
- x and $y \rightarrow \infty$, there is strong ergodicity.

In other cases there is no ergodicity.

When not only there is weak ergodicity, but, for instance, x and y assume in time unstable and limited values (min $< x <$ max and min $< y <$ max), in a confidence interval, then a process of emergence may be assumed to be going on.

b) "The mean of observations of an individual made for a time y is equal to the mean of observations made of x individuals at a single moment in time over an area".

Tacking in count the fundamental role of the observer, the models applied in its cognitive processing and saying

- O_x the observations made by x individuals at a single moment in time over an area (a picture of an area considered by x observer);
- O_y the observations made by an individual for a time y (a movie examined by a single observer for a time y);
- F the similarity function between observations with reference to an observer such as $F(w,z) = u$, with $0 < u < 1$ being 0 corresponding to total dissimilarity and 1 maximal similarity between w and z,

it is possible to assume the value of $F(O_x, O_y)$ as index of ergodicity.

When:

- $F(O_x, O_y) \rightarrow 1$, there is ergodicity;
- x and $y \rightarrow 0$, there is weak ergodicity;
- x and $y \rightarrow \infty$, there is strong ergodicity.

In other cases there is no ergodicity.

When $F(O_x, O_y)$ signals ergodicity and when not only there is weak ergodicity, but x and y assume in time limited values ($min < x < max$ and $min < y < max$) in a confidence interval (changing smoothly or even by jumps) then a process of emergence may be assumed to be going on.

The hypothesis introduced is that the continuous changing of values of the indexes of ergodicity are, in case of emergence processes, in a confidence interval. The values of the indexes of ergodicity assume constant values when the system emerges from a structure, an organization. The confidence interval and the dynamics of changing of the assumed values may possibly correspond to the emergence of some different dynamic systems from collective behaviors, such as flocks, swarms, herds.

c) let consider the eigenvalues of the transition matrix $[P_{ij}]$. The more they are and greater is the ergodicity. When the eigenvalue is the identity ergodicity is zero.

In the example regarding economics the elements have the possibility to change class without influencing each other. In the example referred to collective behavior elements may belong to different classes simultaneously and they interact by their position (one element may change class just because the move of another one).

The hypothesis that a Collective Behavior may emerge from a set of interacting Boids because of their particulat ergodic behavior is based on another important point. The subset of element behaving ergodically at the time t must have non empty intersection with the subset of elements behaving ergodically at the time t+1, and so on. In such a way the fundamental behavioral aspects are reproduced in time, diluted in the continuous non empty intersections between subsets having ergodic behavior.

6. CONCLUSIONS

In this working paper the topic related to using set theory principles in systems theory has been introduced.

Different definitions of ergodicity has been introduced making reference to different contexts, such as Physics, Geomorphology, Economics and Collective Behavior emerging from agents provided with cognitive system.

Considerations about the relationship between emergence processes and ergodicity have also been introduced focusing on the necessary role of the observer. The interest specially focuses on systems having components, i.e. agents, provided with cognitive system. That's because such a kind of agents may do not interact just following formal rules as properties of physic or biologic components. They are able to modulate, to change, to design their way to interact having levels of awareness.

The concept of ergodicity has been proposed as basis to model and control emergence processes like the establishment of flocks of birds (Boids) without using explicit symbolic rules as in other approaches, based f. i. on geometrical properties, speed, acceleration, direction. The interest to have theories and tools about ergodicity is for modeling, controlling, inducing emergence in systems. And this is specially for social systems where interacting agents are supposed provided with cognitive model, making them able to manage interactions, having the ability to decide, depending on the sophistication of the cognitive model, if and how to interact (Pessa, 2000).

From the definitions introduced above it is possible to assume that any kind of system has an ergodic behavior in time. The focus of this paper has been on considering ergodicity as an indicator (par. 5) of the kind of emergence:

- The interaction among agents may take place in an (1) organized way, for instance when components are structured in functions often corresponding to subsystems. Examples are departments in companies, schools, hospitals; devices accomplishing complex tasks in electronic equipment, software, biological systems; roles in natural systems like in behaves, anthills, termitery, nests, families. In this case the ergodicity of the system is hypothesized to be strong and stable.
- Other processes of emergence, i. e. self-organization, are hypothesized to take place when agents interacting at (2) different dynamic levels of ergodicity establish emergent systems. In this case levels of ergodicity are hypothesized oscillating (that's adopting various values within that interval, changing smoothly or even by jumps) in a confidence interval. Different confidence intervals correspond to different kinds of emerging systems.

The hypothesis introduced is that the continuous changing of values of the indexes of ergodicity are, in case of emergence processes, in a confidence interval. The values of the indexes of ergodicity assume constant values when the system emerges from a structure, an organization. The confidence interval and the dynamics of changing of the assumed values may possibly correspond to the emergence of some different dynamic systems from collective behaviors, such as flocks, swarms, herds.

Some hypotheses, possible lines of research and applications have been also presented. The application of the concept of ergodicity to systems must allow modeling, simulating and controlling in processes of emergence based on the introduced considerations.

REFERENCES

Allen, R. B., 1990, User model: theory, methods, and practice. *International Journal of man-machine Studies* **32**: 511-543.

Anderson, J., 1985, *Cognitive Psychology and its Implication*. 2nd Edition, W. H. Freeman and Co, New York, N Y.

Anderson, J. J., 1980, A stochastic model for the size of fish schools, *Fish Bull.* **79**: 315-323.

Anderson, M. G., (ed.), 1988, *Modeling geomorphologic systems*. Wiley, New York, p. 458.

Aoki I., 1984, Internal dynamics of fish schools in relation to inter-fish distance. *Bulletin Japanese Society Scientific Fishing* **48**: 1081-1088.

Aron, S., Deneubourg, J. L., Goss, S., and Pastels, J. M., 1990, Functional self-organization illustrated by inter-nest traffic in ants : the case of the Argentine ants. In *Biological Motion* (W. Alt and G. Hoffmann, eds.), Springer-Verlag, pp. 533-547.

Beckers, R., Deneubourg, J. L., Goss, S., and Pasteels, J. M., 1990, Collective decision making through food recruitment', insectes Sociaux 37, 258-267.

Belic, M.R., Skarka, V., Deneubourg J. L., and Lax, M., 1986, Mathematical model of honeycomb construction. *J. of mathematical Biology* **24**: 437-449.

Benjafield, J. G., 1992, *Cognition*. Prentice-Hall, Englewood Cliffs, NJ, USA.

Bonabeau, E., Dorigo, M., and Theraulaz, G., 2000, *Swarm Intelligence: from natural to artificial systems*. Oxford University Press.

Bonabeau, E.W., and Theraulaz, G., 1994, Why Do We Need Artificial Life? *Artificial Life* **1**: 303-325.

Bonabeau, E., Theraulaz, G., and Deneubourg, J. L., 1996, Quantitative study of the fixed threshold model for the regulation of division of labor in insect societies. *Proc. Roy. Soc.* London Ser. B **263**: 1565-1569.

Breder, C. M., 1954, Equations descriptive of fish schools and other animal aggregations. *Ecology* **35**: 361-370.

Cariani, P., 1991, Emergence and artificial life. In *Artificial Life II* (C. Langton, D. Farmer and S. Rasmussen, eds), Addison-Wesley, Redwood City, CA, pp 775-797.

Cruchtfield, J.P., 1994. The Calculi of Emergence: Computation, Dynamics and Induction. *Physica D* **75**: 11-54.

Darling, F., Fraser, 2000, *Bird Flocks and the Breeding Cycle*. The Blackburn Press; ISBN: 1930665164.

Domowitz, and El-Gamal, M., 1997, *Financial market Structures and the Ergodicity of Prices*. Social Systems Research Institute, University of Wisconsin.

Echelle, A. A., and Kornfield, I., (eds), 1984, *Evolution of Fish Species Flocks*. Univ. of Maine Pr; ISBN: 0891010572.

Forrest, S., (ed.), 1990, *Emergent Computation*. North Holland, Amsterdam, The Netherlands.

Gierer, A., and Meinhardt, H., 1972, A theory of Biological Pattern formation. *Kybernetik* **12**: 30-39.

Iberall, A. S., and Soodak, H., 1978, Physical basis for complex systems - some propositions relating levels of organization. *Collective Phenomena* **3**: 9-24.

Kirkby, M. J., (ed.), 1994, *Process models and theoretical geomorphology.*, Wiley & Sons, New York, p. 417.

Krippendorff, K., 2001a, Ergodic. In *Principia Cybernetica Web* (F. Heylighen, C. Joslyn and V. Turchin, eds), Principia Cybernetica, Brussels, (Modified Apr. 12, 2001), http://pespmc1.vub.ac.be/ASC/ERGODIC.html.

Krippendorff, K., 2001b, Non-ergodic. In *Principia Cybernetica Web* (F. Heylighen, C. Joslyn and V. Turchin, eds.), Principia Cybernetica, Brussels, (Modified Apr. 12, 2001), http://pespmc1.vub.ac.be/ASC/NON-ERGODIC.html.

Landon-Lane, J. S., and Quinn, J. A., 2000, *Growth and Ergodicity: Has the world converged*? The University of New South Wales, Sydney 2052, Australia.
Millonas, M. M., 1993, *Swarms, Phase Transitions, and Collective Intelligence*. Complex Systems Group, Theoretical Division, and Center for Non linear studies and Santa Fe Institute.
Minati, G., 2001a, Esseri Collettivi, Apogeo, Milano, Italy (Collective Beings, 2001, in progress)
Minati, G., 2001b, Experimenting with the DYnamic uSAge of Models (DYSAM) approach: the cases of corporate communication and education. In *Proceedings or the 45th Conference of the International Society for the Systems Sciences (ISSS)*, USA.
Minati, G., Penna, M. P., and Pessa, E., 1997, A conceptual framework for self-organization and merging processes in social systems. In *Systems For Sustainability*, Plenum Press, New York.
Mingers, J., and Gill, A., (eds), 1998, *Multi methodology: Towards Theory and Practice and Mixing and Matching Methodologies*. Wiley & Sons, U.K.
Parrish, J. K., and Hamner, W. M., (eds), 1998, *Animal Groups in Three Dimensions*. Cambridge Univ. Press.
Pessa, E., and Minati, G., 2001. *Systems community should focus on emergence in complex cognitive and biological systems*. http://www.airs.it
Pessa, E., Penna, M. P., and Minati, G., 2000, Collective Phenomena in Living Systems and in Social organization. *Chaos & Complexity Letters*, Italy.
Reynolds, C., 1987, *Computer Graphics* **21**: 25.
Schumm, S., and Lichty, R., 1965, Time, space and causality in geomorphology. *American Journal of Science* **263**: 110-19.
Thorn, C. E., (ed.), 1982, *Space and time in geomorphology*. Allen & Unwin, Boston, p 379.
Thorn, C. E., 1988, *An introduction to theoretical geomorphology*. Unwin Hyman, Boston, p 247.
Thornes, J. B., and Brunsden, D., 1977, *Geomorphology and time*. Wiley, New York, p 208.
Uchii, Soshichi, 1991, *Theory reduction: The Case of the Kinetic Theory of Gases*. Kyoto University, Japan, http://www.bun.kyoto-u.ac.jp/~suchii/reduction7.html.
Vabo, R., and Nottestad, L., 1997, An individual based model of fish school reactions: predicting anti predator behavior as observed in nature. *Fisheries Oceanography* **6**: 155-171.
Zadeh, L. A., et al, 1996, *Fuzzy Sets, Fuzzy Logic, and Fuzzy Systems: Selected Papers by Lotfi A. Zadeh*. (Advances in Fuzzy Systems - Applications and Theory , Vol. 6)
Zadeh, L. A, 1965, Fuzzy sets. *Information and Control* **8**: 338-353.

Decomposition of Systems

MARIO R. ABRAM
CESI S.p.A. - Centro Elettrotecnico Sperimentale Italiano
Via Rubattino, 54, I-20134 Milano, Italy – E-mail: abram@cesi.it
AIRS - Associazione Italiana per la Ricerca sui Sistemi

1. INTRODUCTION

The complexity of processes with which we interact and the variety of the relationships that exist between the elements or the components of a system are constantly increasing. In studying a system it is convenient to find specific methodologies that combine the analytical approach, useful for going deeper in the details of knowledge of the elements that constitute the system, with a position that gives the possibility to operate from a more general point of view.

The traditional analytical approach, in which the knowledge of critical information about a process is grasped decomposing the system into its simpler elements and analysing the properties of each element, is useful but it is inadequate to evaluate in a deeper way the systemic aspects of a problem. Typically we consider "systemic" all the aspect of a problem that involve the concepts of relation and interaction between the elements of a system (Minati, 1998, 2001). The understanding of the systemic aspect in the analysis of processes is indispensable to manage the increasing complexity we are facing with.

Scientific departments developed sophisticated methods to inquire and collect information in their own field of interest. The mathematical models, very fine and sophisticate, were developed for studying many processes and phenomena; they are useful, and sometime indispensable, to understand the behaviour of processes and to design their use in technological applications

Emergence in Complex, Cognitive, Social, and Biological Systems, Edited by
Minati and Pessa, Kluwer Academic/Plenum Publishers, New York 2002

of increasing complexity. In general the mathematical models describe the behaviour of a system and, in particular, express and show in detail, often completely, the relations between the components or the elements of the described system. But, when we attempt to build a model of the interactions between processes or phenomena evolving in time with very different time scales, even the simple coupling of mathematical models is difficult. In these cases the need to operate with great detail does not help to build a global vision of the problem, of subsystem involved and of the active relations or interactions between them.

With reference to a very general definition of system (Klir, 1991), we will investigate the possible decompositions of a system into subsystems and some basic properties of the relations between them. To do so we will keep a very general position and we will not consider the technical details normally used into mathematical modelling. In particular we will attempt to build a methodology that can help us to decompose a system into subsystems and to put in evidence the possible relations that can exist between the various subsystems.

In general the decomposition of a system into subsystems is not unique and is strictly connected with the motivations and the goals for which we start to study a particular system. Nevertheless, to keep a very general level of abstraction and to find the more useful decomposition into subsystems, gives one the position to see and to perceive the context of the problems. This point of view sufficiently general is very useful to drive in details the design and analysis processes.

After a recall of the definitions of system and relation (chapter 2) we will list some general properties useful in building the methodology. Then we will define the elementary decomposition (chapter 3), and we will recall the composition of systems (chapter 4). With this elements we will describe briefly the methodology to decompose a system into subsystems. Furthermore we will consider (chapter 5) some possible representations of a decomposition. In particular we will examine the visual representation and the topologic matrix of a decomposition. As an example (chapter 6) we will apply the decomposition methodology to the preliminary study of the control for a generic industrial plant. We will attempt to put in evidence how the constitutive properties of the system and the operating hypothesis produce very different decomposition paths. Many open problems need more investigation (chapter 7) nevertheless the use of this methodology suggests some general considerations (chapter 8).

2. SYSTEMS AND SUBSYSTEMS

We give some general definition.

Definition: A system $\boldsymbol{S}$ is the couple $\boldsymbol{S} = (T, R)$, where T is the set of the elements of the system $\boldsymbol{S}$ and R is the set of relations on the set T: $R \subset T \times T$.

Definition: A subsystem $\boldsymbol{S}_n$ of the system $\boldsymbol{S}$ is the couple $\boldsymbol{S}_n = (T_n, R_n)$ where: $T_n, T_n \subset T$ is the set of the elements of the system $\boldsymbol{S}_n$ and R_n is the set of relations on the set $T_n : R_n \subset T_n \times T_n$.

Definition: Given two systems $\boldsymbol{S}_h = (T_h, R_h)$ and $\boldsymbol{S}_k = (T_k, R_k)$, the set of relations R_{hk} between the systems is $R_{hk} \subset (T_h \times T_k) \cup (T_k \times T_h)$.

The relations are symmetric so $R_{hk} = R_{kh}$. If we want to specify a direction or an order between the systems $\boldsymbol{S}_h$ and $\boldsymbol{S}_k$, it is necessary to suppose that $R_{hk} \neq R_{kh}$ and we introduce the interactions specified as action and reaction. In particular the relations R_{hh} are the inerr relations of the system $\boldsymbol{S}_h$.

Definition: The action is the set $I_{hk} \subset R_{hk}$ with $h > k$; the reaction is the set $I_{kh} \subset R_{kh}$ with $h < k$. In general $I_{hk} \neq I_{kh}$.

If $h = k$ the actions and reactions are considered inner to the systems and are the same as the inner relations.

3. DECOMPOSITION OF SYSTEMS

A system $\boldsymbol{S}$ can be decomposed into two subsystems $\boldsymbol{P}$ and $\boldsymbol{N}$ where the subsystem $\boldsymbol{P}$ owns the property (P) while the subsystem $\boldsymbol{N}$ does not owns the property P (we can write (–P)). In general we can set the following:

Definition: The elementary decomposition $\boldsymbol{D} = [\boldsymbol{P}, \boldsymbol{N}]$ of a system $\boldsymbol{S}$ is the couple of systems $\boldsymbol{P}$ and $\boldsymbol{N}$, where $\boldsymbol{P}$ is the system defined by means of the property (P) and $\boldsymbol{N}$ is the system that is defined by means of the property (–P).

With reference to the definition of system and taking into account the relations of the new subsystems it is possible to write:

$$S = (T, R) = [\boldsymbol{P}, \boldsymbol{N}] = ((P, R_P),(N, R_N), R)$$

where R_P is the set of the relations between the elements of the system $\boldsymbol{P}$ and R_N is the set of the relations between the elements of the system $\boldsymbol{N}$.

Then it is possible to have the following hierarchy of decompositions, in which any level is associated with a decomposition $\boldsymbol{D}_n$:

$$\begin{aligned}
\boldsymbol{D}_0 &= \boldsymbol{S} = (T, R) \\
\boldsymbol{D}_1 &= [\boldsymbol{P}_1, \boldsymbol{N}_1] = ((P_1, R_{P1}),(N_1, R_{N1}), R) \\
\boldsymbol{D}_2 &= [[\boldsymbol{P}_2, \boldsymbol{N}_2], \boldsymbol{N}_1] = (((((P_2, R_{P2}),(N_2, R_{N2})), R_{P1}),(N_1, R_{N1}), R) \\
&\dots \qquad \dots \qquad \dots \\
\boldsymbol{D}_i &= [[\dots[\boldsymbol{P}_i, \boldsymbol{N}_i]\dots]\dots] = ((\dots((P_i, R_{Pi}),(N_i, R_{Ni})),\dots)), R)
\end{aligned}$$

Taking into account the relations, the following formulas are useful to visualize some properties of the structure of the decomposition:

$$\begin{aligned}
\boldsymbol{R} &\supset \boldsymbol{R}_P \cup \boldsymbol{R}_N \\
&\supset \boldsymbol{R}_{P1} \cup \boldsymbol{R}_{N1} \cup \boldsymbol{R}_N = \boldsymbol{R}_N \cup \boldsymbol{R}_{P1} \cup \boldsymbol{R}_{N1} \\
&\supset \boldsymbol{R}_{P1} \cup \boldsymbol{R}_{P2} \cup \boldsymbol{R}_{N2} \cup \boldsymbol{R}_N = \boldsymbol{R}_N \cup \boldsymbol{R}_{P1} \cup \boldsymbol{R}_{P2} \cup \boldsymbol{R}_{N2} \\
&\supset \boldsymbol{R}_{P1} \cup \boldsymbol{R}_{P2} \cup \boldsymbol{R}_{N2} \cup \boldsymbol{R}_N \cup \dots \cup \boldsymbol{R}_{Pi} \cup \boldsymbol{R}_{Ni} \cup \dots = \\
&\qquad \boldsymbol{R}_N \cup \boldsymbol{R}_{P1} \cup \boldsymbol{R}_{P2} \cup \boldsymbol{R}_{N2} \cup \dots \cup \boldsymbol{R}_{Pi} \cup \boldsymbol{R}_{Ni} \cup \dots \\
\boldsymbol{R} &\supset \boldsymbol{R}_N \cup \boldsymbol{R}_{P1} \cup \boldsymbol{R}_{P2} \cup \boldsymbol{R}_{N2} \cup \dots \cup \boldsymbol{R}_{Pi} \cup \boldsymbol{R}_{Ni} \cup \dots
\end{aligned}$$

In general, for symmetry reasons, in a decomposition of a system into n_s subsystems, the maximum number N_R of available independent relations R_{hk} is:

$$N_R = (n_s^2 - n_s)/2 = (n_s(n_s - 1))/2 \ ,$$

instead the maximum number N_I of available interactions I_{hk} and I_{kh}, $(I_{hk} \cup I_{kh})$ between the subsystems is:

$$N_I = (n_s^2 - n_s) = n_s(n_s - 1) \, .$$

4. COMPOSITION OF SYSTEMS

The elementary composition of two interacting subsystems can be seen as the inverse procedure of the decomposition. If in a decomposition the system

$\boldsymbol{P_n}$ has the property (P) and the system $\boldsymbol{N_n}$ does not have the property (P), have the same relations with the other subsystems, the two systems can be composed in the system $\boldsymbol{C_n}$. Then the doubled relations of the two subsystems can be generically considered a single relations.

The composition process can involve a greater number of subsystems in order to simplify the decompositions and to take in evidence only the subsystems really important.

5. METHODOLOGY

The definition and properties just listed help us to build an elementary logic methodology for the construction of the decompositions of systems and to find the particular decomposition that is more adequate to our goals.

Given a system we can operate in accordance with the following steps:

1. To apply the elementary decomposition to the system putting in evidence the relations that it is possible to derive from the properties characterising each subsystem.
2. To apply the elementary decomposition process more and more to the subsystems until the detail of the decomposition is adequate to visualise the relations involved by the problem.
3. With each elementary decomposition the previous existing relations with the other subsystems are doubled and are considered true for the new subsystems of the last decomposition step.
4. To reduce the doubled relations of the last decomposition eliminating the relations that are not consistent with the properties of the subsystems.
5. If more subsystems are not useful to go deeper in the analysis they can be composed in one subsystem; the existing relations are grouped again.

The composition of systems can be seen as the inverse process of decomposition. When the required detail level of information can be reduced this give the possibility to modulate the use of the methodology (decomposition/composition) for each subsystem and for specific investigation goals

6. REPRESENTATION OF DECOMPOSITIONS

The decomposition of a system into subsystems can be represented in various ways. Many types of diagrams and pictures were developed in the fields of science and technology.

Many methods are usually adopted in technical practice. We can remember: tree schema useful to formalize the structure of properties; block diagrams, used in control engineering to visualize the interconnections of subsystems and the flux of the variables; flow charts used in programming and planning, etc.. All these specific representation were developed in specialised fields and for well defined applications.

Now we want to build a graphical representation of the decompositions of systems that is simple, immediate and helpful to gain and to maintain a global vision of the process during the application of the methodology.

In the development of applications when the decomposition must be formalised using specific mathematical models the decomposition can be translated into alternative visualization forms.

6.1 Visual representation

For our goal we will represent the decomposition into subsystem drafting on a surface the blocks that identify the subsystems.

We will visualise the relations between the subsystems with arrows that identify the interactions. The identification of the interaction between two subsystems ask for the definition of the direction in the relation and then to choice between two possibilities. We will draw the interactions i_{hk} with a right arrow " $\rightarrow$ " if $h > k$ and with a left arrow " $\leftarrow$ " if $h < k$.

In figure 1 a subsystem $\boldsymbol{S}_i$ is visualized in relation with other generic subsystems. Applying the decomposition methodology the subsystem $\boldsymbol{S}_i$ is decomposed into two subsystems $\boldsymbol{P}_i$ and $\boldsymbol{N}_i$ using a specific property (P_i). A generic relation valued for $\boldsymbol{S}_i$ is supposed valued also for the subsystems $\boldsymbol{P}_i$ and $\boldsymbol{N}_i$, Then the consistency of each relation is verified and the void relations are deleted.

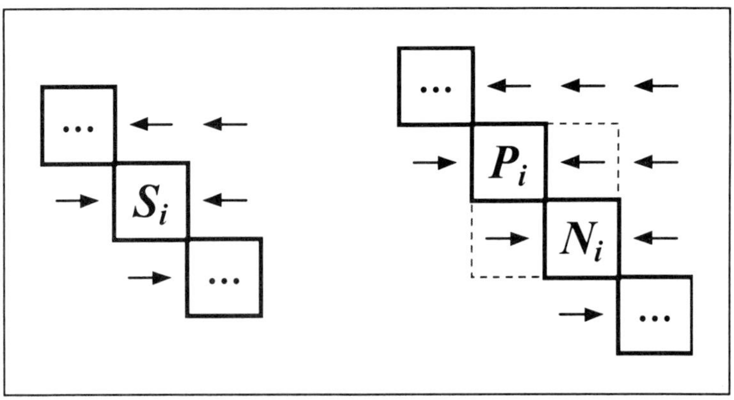

Figure 1. Representation of the decomposition of a system into two subsystems.

6.2 Topologic matrix

The visualization of the decompositions as shown in figure 1 suggests to use a matrix formalism. In particular the visualization of the decomposition by means of topological matrices is interesting because it depicts in a synthetic way the structure of the relationships between the subsystems.

So, each decomposition of a system into n subsystems can be visualized by a $n \times n$ square matrix $\mathbf{T}$ filled with the values:

$T_{hk} = 1$ if I_{hk} exist, $T_{hk} = 0$ elsewhere. We consider $T_{hh} = 1$.

The topologic matrix can be decomposed in the sum of three square matrices:

$$\mathbf{T} = \mathbf{I} + \mathbf{A} + \mathbf{R}$$

where $\mathbf{I}$ is the matrix of internal relations (diagonal), $\mathbf{A}$ is the matrix of actions and $\mathbf{R}$ is the matrix of reactions.

It is interesting to note that the topological matrix $\mathbf{T}$ is symmetric if we limit our analyses to the relations; the matrix $\mathbf{T}$ is not symmetric if we deal wit interactions (actions and reactions).

7. EXAMPLE OF APPLICATION TO CONTROL

The preliminary study of a control system for the managing of an industrial plant must take into account the specific features of the involved processes and of the interactions between the subsystems in order to reach the prefixed goals.

Furthermore the criteria used to select and reach the goals and the objectives of a plant must undergo the restrictions that guarantee the operation in safety conditions and with an acceptable level of interaction with the environment. These restrictions are often imposed by regulations. Furthermore the quality standards are becoming more restrictive and are increasingly important due to the rising sensitivity about the environmental impact and the search of the health care for the involved people. The operation procedures and the restrictions for safety operation are realized with the help of control systems. They are specific apparatuses that help the human operator in managing the plant.

Intuitively it is easy to identify the main subsystems that constitute the plant as, for example, the physico-chemical processes, the environment, the control functions, the human operator, etc. The exploration and the definition of the existing relation between the various subsystems is usually a difficult and complex task.

Let consider a system constituted by an industrial plant, as for example an electric power generation facility. We will attempt to apply the decomposition methodology; in particular we propose to identify the main subsystems of the power plant and to evidence the relation between them. We propose to understand how the relations between the subsystems are imposed by design strategies or derive from the specific characteristics of the plant.

The decomposition process can evolve in the following steps:

1. A generic system (S) for power generation can be decomposed into two subsystems: *Plant* and *Environment*. The Plant (I) is a structure in witch the specific processes for power generation take place. Environment (E) is the context in witch the plant is positioned and operate (all that is not Plant). Note that this first decomposition coincides with the definition of the system.
2. The plant (I) can be decomposed into the subsystems *Process* and *Control*. Process (P) is a physical, chemical, energetic transformations necessary to power generation. Control (C) is formed by apparatuses and functions that interact with the process implementing the procedure for driving and managing the generation process.
3. The control (C) can be decomposed into the subsystems: *Operator* and *Apparatus*. Operator (O) is the human agent that set the goal for managing the plant. Apparatus (A) is composed by hardware and software that realize the control functions.
4. The apparatus (A) can be decomposed into the subsystems *Interface* and *Functions*. Interface (I) is the portion of the apparatus that interact with the human operator and does not interact with the plant. Functions (C) is the part of the control apparatus that interact with the plant and does not interact with the operator.
5. The control apparatus (C) can be decomposed into two subsystems: *Modulating Control* and *Safety Control*. Modulating Control (M) manages the plant in normal conditions in order to reach the goals settled by the operator. Safety Control (S) assures that, for any operating condition, the plant operations evolve within safety constraints.

We can continue, but now we stop the decompositions to six subsystems as visualised in figure 2; for each decomposition the maximum numbers of relations and interactions are reported.

The detailed steps of the decompositions are reported in table 1.

The definition of the relation between the various subsystems must be fixed in each step of the decomposition; they must take into account the features of each subsystem. Each high level decomposition is strictly connected with the relations defined in accordance with operating hypothesis. Nevertheless the developing of the detailed decomposition analysis sometime give light to particular information about critical points. In some case it is necessary to define again the relations previously defined at higher level. It is then an iterative process that is refined step by step in the various phases of the analysis.

Decomposition	System	Environment				
		Plant	Process			
			Control	Operator		
				Apparatus	Interface	
					System	Modulating
						Safety

Decomposition	1	2	3	4	5	6
Relations	0	1	3	6	10	15
Interactions	0	2	6	12	20	30

Figure 2. Example. Decomposition of the system into subsystems.

For example, more detailed analysis can put in evidence that the interactions between the plant and the environment, considered null in a first approximation, become important for that particular subsystems. It is the case when we realize that it is necessary to meet additional safety constraints.

Then this give a different number of available relations in the same decomposition. We built the decomposition of the system concentrating on the main components of the control of the plant; if we want to analyse with greater detail interactions of the plant with the environment we must decompose the environmental aspects of the problem focusing the attentions, for example, on the subsystem involved with the user, the public services, the health care of people, the protection systems, the rescue systems and so on.

Table 1. Steps of the decomposition of system into subsystems and relations

n	SUBSYSTEMS	DECOMPOSITION
1	S System	
2	I Plant E Environment	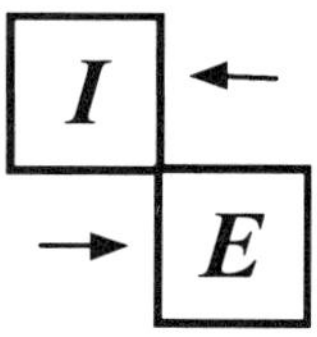
3	C Control P Process E Environment	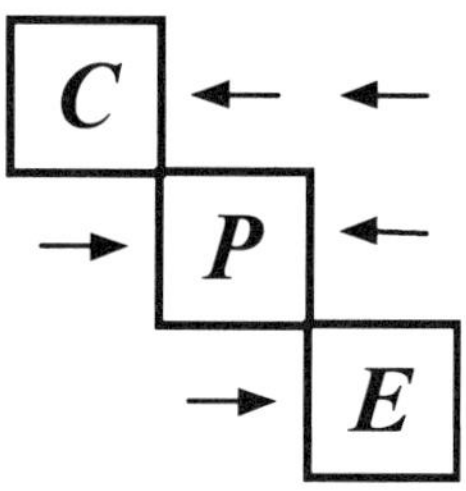
4	O Operator A Apparatus P Process E Environment	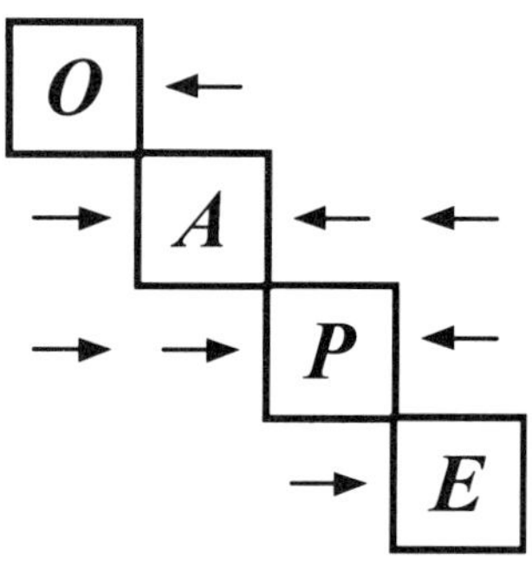

Table 1. (cont.) Steps of the decomposition of system into subsystems and relations

n	SUBSYSTEMS	DECOMPOSITION
5	O Operator I Interface C Control Functions P Process E Environment	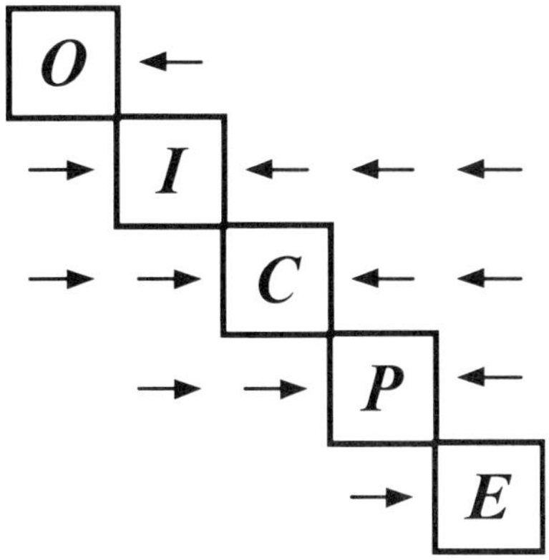
6	O Operator I Interface S Safety Control M Modulating Control P Process E Environment	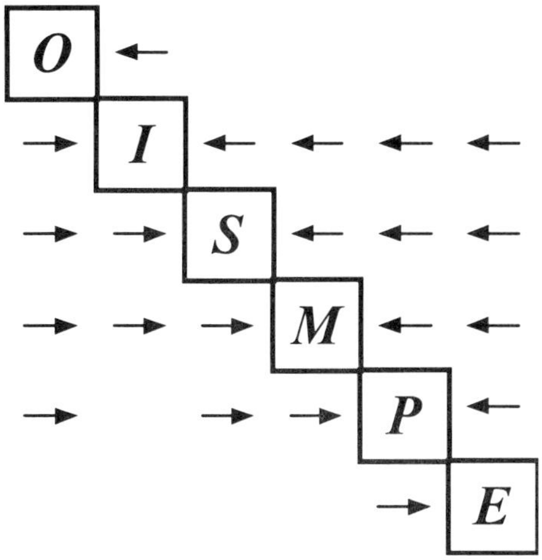

In a similar way if we want investigate the economical and organisation aspects of the system, a detailed decomposition must be developed for the organisation and people involved in the company (operator, chief operator, plant manager, operation manager, operation planning, etc.).

Now let we consider as example the decomposition in five subsystems (see table 1)

Safety considerations and design constraints can force us to consider additional relations between specific subsystems or to impose the existence of defined and specific relations between two or more subsystems. So, if the starting hypothesis change, the additional relations dotted in figure 3 must be investigated.

If it is necessary to concentrate on the environment subsystem (E) it is necessary to build more decompositions involving the various properties of the environment.

In this case it is possible to reduce the number of subsystems. May be useful to reduce the decomposition complexity reducing the number of involved subsystems eliminating the subsystems not necessary for the detailed analysis. In figure 3, for example, if it is sufficient to have only one subsystem for the plant, the subsystems (I, C, and P) can be composed or grouped into the unique subsystem (W). So he dimension of the decomposition gains a more manageable dimension and we can start de decomposition of the subsystem (E).

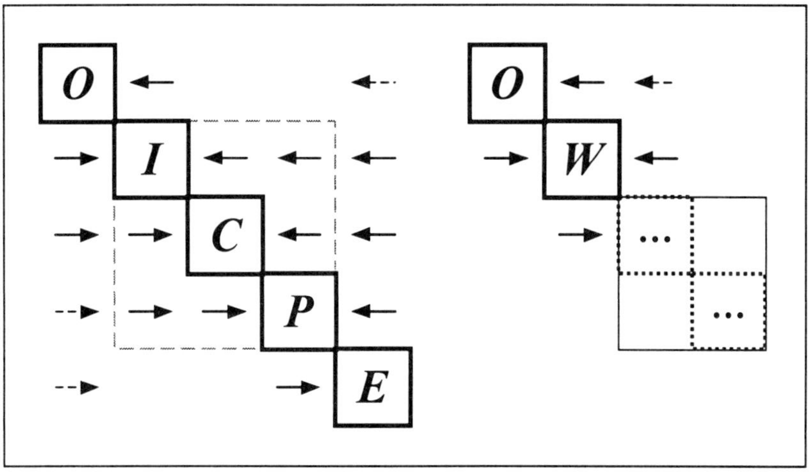

Figure 3. Composition and decomposition.

For the initial configuration of the example of figure 3 (see table 1) the topologic matrix has the form:

$$\mathbf{T} = \begin{bmatrix} 1 & 1 & 0 & 0 & 0 \\ 1 & 1 & 1 & 1 & 1 \\ 1 & 1 & 1 & 1 & 1 \\ 0 & 1 & 1 & 1 & 1 \\ 0 & 0 & 0 & 0 & 1 \end{bmatrix}$$

The components of topologic matrix are:

$$\mathbf{I} = \begin{bmatrix} 1 & 0 & 0 & 0 & 0 \\ 0 & 1 & 0 & 0 & 0 \\ 0 & 0 & 1 & 0 & 0 \\ 0 & 0 & 0 & 1 & 0 \\ 0 & 0 & 0 & 0 & 1 \end{bmatrix} \quad \mathbf{A} = \begin{bmatrix} 0 & 0 & 0 & 0 & 0 \\ 1 & 0 & 0 & 0 & 0 \\ 1 & 1 & 0 & 0 & 0 \\ 0 & 1 & 1 & 0 & 0 \\ 0 & 0 & 0 & 1 & 0 \end{bmatrix} \quad \mathbf{R} = \begin{bmatrix} 0 & 1 & 0 & 0 & 0 \\ 0 & 0 & 1 & 1 & 1 \\ 0 & 0 & 0 & 1 & 1 \\ 0 & 0 & 0 & 0 & 1 \\ 0 & 0 & 0 & 0 & 0 \end{bmatrix}$$

Structural properties of the system can emerge from the structure of the topologic matrices. They evidence in a synthetic way the dependence of the patterns that emerge from the decomposition process.

8. OPEN PROBLEMS

The methodology here presented arise some questions and open many interesting problems. The attempt to answer the following questions is useful to understand and to evaluate the effective utility of the methodology.

- Can the decomposition add information to the knowledge of the system?
- Can the methodology give more information about the structure of the system?
- Can the structure of topological matrix give information about the hierarchic structure emerging from the decomposition?
- There is a natural decomposition for the system between all the possible decompositions?
- Can the topologic matrices give information about the structural stability of the system?
- Can the eventual information on the stability of the system from topologic matrix be useful to evaluate the stability of the mathematical model in some canonical form?
- Can the building of system models be connected with the specifications of the relations and of the interactions between the subsystems?

To investigate these aspects will be useful to understand and to refine the steps of the methodology. In particular it is interesting to verify the effective utility of this approach and to set the goals and the key points for future investigations and improvements.

9. CONCLUSION

The use of methodology suggests some general operating considerations while open problems connected with the formalism of the decomposition process are emerging and suggest some possible and useful developments for future investigations.

The main advantage using this methodology is to gain a global point of view. From this position it is possible to evaluate the complete context for the relations of the system; so in the analysis of the possible relations one must chose to give meaning to the existence and not existence of the relations, without forgetting none.

A decomposition developed with this methodology can be detailed defining the relations and interactions and giving them the detailed formulation in mathematical models. Many formalisms can be use and, in this case, the topologic matrix of the decomposition may evidence the topological structure of the mathematical model.

The possibility to chose de decomposition more adequate to the study of a particular system means "to find" the decomposition more "interesting" and "economic" in connection with the problem. In other words we interact and manage only what we need or what we are involved with in our problem.

REFERENCES

Ilic, M., and Skantze, P., 2000, Electric power systems operation by decision and control. *IEEE Control Systems Mag.* **20**(4): 25-39.

Klir, G. J., 1991, *Facets of Systems Science*. Plenum Press, New York.

Mesarovic, M. D., and Takahara, Y., 1989, *Abstract system theory*. Springer-Verlag, Berlin.

Minati, G., 1998, *Sistemica*. Apogeo, Milano.

Minati, G., 2001, *Esseri collettivi*. Apogeo, Milano.

Voronov, A. A., (ed.), 1986, *Management and control in large systems*. Mir, Moscow.

Generic Properties of Complex Networks

ROBERTO SERRA
Centro Ricerche Ambientali Montecatini, Via Ciro Menotti 48, I-48023 Marina di Ravenna
E-mail: rserra@cramont.it

1. INTRODUCTION

The most interesting topics of systems science concern the discovery of common properties in seemingly very different systems, and the introduction of theories which apply to a variety of situations, an outstanding examples being the concept of order parameter (Haken, 1977; Serra *et al.*, 1986). Some recent researches have pointed out the possible existence of rather general structural features in systems as diverse as the nervous system, the Internet, cellular metabolism, social groups, electrical power distribution, and others (Watts, 1999; Reka and Barabasi, 2001; Wagner and Fell, 2000; Jeong *et al.*, 2000).

These features become apparent when one uses a graph representation, where the system variables are associated with nodes and the influence of one variable upon another is represented by drawing a link between the two corresponding nodes.

This sort of representation is of course very general, and can be applied to very different systems. In some cases (like e.g. power transmission, telephone systems, neural networks) the links have a direct physical counterpart (transmission lines, telephone cables, axons and dendrites), while in other cases they indicate the existence of a relationship which may be mediated by social contacts (friendship networks), by chemicals (genetic/metabolic networks), by meaning (semantic networks) a.s.o.

The graph representation performs a strong compression of the information. For example, in a neural network the nodes may be linked by connections which have very different values of their synaptic weights; in a military hierarchy the downward connections convey different kinds of informations (and orders!) than those which are directed upwards; in a genetic network some links represent proteins which activate genes and others may represent chemicals which alter the conformation of a protein, etc. All these differences are lost in a graph, which provides just a stylized representation of one feature of the system, namely its interconnectedness. In spite of this information compression, looking at the structure of the graph may prove very useful and effective.

The most familiar models of networks are either regular graphs (like e.g. a tree, or a feedforward neural networks) or random graphs. The former models are well suited for designed systems (like a military hierarchy) while the latter models are more appropriate to describe phenomena which grow in a largely spontaneous way (the Internet, the cellular metabolic network, etc.). Every such system has its own network structure, which can be described by a graph. How can we compare the graphs corresponding to systems which may differ in the nature of the nodes, their number, the number and topology of links? It turns out that some aggregate variables are particularly interesting in this respect:

- the sparseness of the network, which measures the role of direct connections among nodes with respect to indirect interactions
- the characteristic path length, which measures how many nodes one must pass, in order to go from a node to another
- the clustering coefficient, which measures how the network is locally structured, i.e. how likely it is that two nodes which are connected to the same node are also connected between themselves.

These quantities are precisely defined in section 2.

The random graph (RG) model is introduced in section 3, and its properties are described. It turns out that sparse random graphs have the remarkable property of displaying very short path lengths, in the sense that the average distance between nodes scales with the logarithm of the number of nodes (and logarithmic growth is of course a very slow growth). On the other hand, the random graph has no structure, a property which is reflected in a very small clustering coefficient.

In studying many real world networks, scholars have found out that there are several cases where short path lengths are indeed the case: the most popular example is the "six degree of separation", based upon a study from psychologist Steven Milgram who asked people in Kansas to send letters to a man in Boston, unknown to them, whose address was also unknown. The letters should have reached the target by passing through direct acquaintances: for example, if one knows a friend in Boston he may try to send the letter to him, who has greater chances to know the target, a.s.o. To the researcher's surprise, the number of steps required turned out to be indeed small (see e.g. Watts, 1999).

Although this study has been recently criticized on the ground that only a fraction of the letter recipients have actually responded to Milgram's requests (Gewolb, 2001), the finding that the separation is surprisingly small in many social networks have been confirmed in different settings, including actors participating to the same movie and researchers co-authoring scientific papers. Moreover, small path lengths have been identified in systems like the Internet, the World Wide Web, metabolic networks, neural networks, and others. A short discussion is presented in section 3.

The "small worldness" would be not surprising, if the graph were structureless. Indeed, let us suppose that I know personally about 1000 persons; if each of them knows 1000 persons, etc. then in just three jumps I should be able to reach 10^9 persons, if there were no correlation between our acquaintances. But this is certainly not the case: it is much more likely that two friends of mine know each other than they know someone in Himalaya or Palm Beach. The random graph model accounts for the short path lengths, but requires this lack of structure, while many interesting networks couple short pathways with relevant structure (i.e. relevant clustering). Two classes of models have been proposed to describe this kind of network, the Watts-Strogatz (WS) and the scale-free (SF) models: they are summarized in sections 4 and 5.

In section 6 some comments will be presented concerning the application of these different model to some real world network.

2. GLOBAL PROPERTIES OF COMPLEX NETWORKS

A graph is defined by a set of labelled nodes $P = \{v \mid v = 1 \ldots N\}$ and a set of edges $E = \{(a,b) \mid a = 1\ldots N,\ b = 1\ldots N,\ b \neq a\}$, where an edge is defined as a pair of nodes (a,b). If the pair is ordered, then the graph is ordered. In the following we will consider non ordered graphs, most generalizations to the case of ordered graphs being straightforward. An edge (a,b) is said to connect the nodes a and b. Synonims are often used, i.e "vertex" instead of node and "link" instead of edge.

Different graphs can be compared: two graphs are equal if it is possible to define a one-to-one mapping between the two sets of nodes in such a way that any two nodes connected in one graph are also connected in the other one, and vice versa. However, the case where two graphs describing the network structure of different systems are equal is rather exceptional and therefore, in order to compare different networks, which may also have a different number of nodes, it is important to define some global properties. In this way, a macroscopic description of a network will be given in terms of a few aggregate variables, instead of requiring the detailed pattern of connections (see also Wagner and Fell, 2000; Reka and Barabasi, 2001).

The degree of connectivity k_v of a node v in a graph is the ratio between the number of links which connect it (to other nodes in the graph) and the total number of other nodes in the graph, $N-1$. The degree of connectivity $k_v \in [0,1]$ is a property of a given vertex. If we average over all the nodes of the graph we obtain a property of the network, the average degree of connectivity

$$< k > = (1/N) \sum_v k_v .$$

Sparse networks are those where each node is directly connected (on the average) to a few other nodes, i.e. $< k > << 1$.

A path between node a and b is defined as an ordered set of links which start from a and end in b (or viceversa) such that any node is trasversed only once, i.e. it belongs only to a pair of successive links: $\{(a,c)(c,d)(d,e)(e,b)\}$ is a path, $\{(a,c)(c,d)(d,e)(e,c)(c,b)\}$ is not. The length of a given path is the number of links it contains. The distance between a pair of nodes a and b, $l(a,b)$, is defined as the length of the shortest path between the two nodes (it is easy to verify that this quantity satisfies the requirements for a distance).

If we take the average of $l(a,b)$ over all the pairs of nodes in the network $(a \neq b)$ we obtain a global property which is defined as the characteristic path length l of the network

$$l = <l(a,b)> = (2/N(N-1)) \sum_{a,b\,(a \neq b)} l(a,b)$$

where the coefficient 2 comes from the fact that the distinct pairs of different links is $N(N-1)/2$.

Let us consider now a third property of a network, which measures its tendency to cluster together, i.e. the likelihood that two nodes which are connected to the same node are also directly linked. For each vertex v, we define the clustering coefficient of that node as the ratio between the number of links which exist among its k_v neighbours, $n(v)$, and the maximum possible number of these links (corresponding to full connectedness among the neighbours):

$$C(v) = n(v) / n_{max}(v)$$
$$n_{max}(v) = k_v (k_v - 1) / 2$$

By averaging over all the nodes, one obtains the clustering coefficient of the graph

$$C = < C(v) >$$

Let us consider the values of the global parameters l and C for two different graphs: a regular graph and a random graph.

For the former case, let us consider for definiteness a one-dimensional regular ring, where the N nodes are aligned on a ring and where each node is connected to its k neirest neighbours (k is an even number, and there are $k/2$ neighbours on each site). In this case one easily finds that $C(v)$ is the same for every node, and $C = 3(k-2)/[4(k-1)]$, so that $C \to 3/4$ in the high k limit. On the other hand, the characteristic path length scales linearly with number of nodes: $l \approx N/2k$. Summarizing, in the regular lattice case we have a high clustering coefficient (actually, independent of N) and a fairly long characteristic path length, which grow linearly in the number of nodes.

A different model is that of the random graph (shortly, RG), where one supposes that any two nodes are connected by a link with a given probability p. For a thorough discussion, (see e.g. Reka and Barabasi, 2001); it turns out that the distribution of the number of connections per node is approximately Poissonian (supposing that $p << 1$) and that, in this case, it can be demonstrated that $l = ln(N)/ln(<k>)$, and $C \cong p = k/N$. Therefore in the random graph case one finds short path lengths, which grow in a logarithmic way with the number of nodes, together with a small clustering coefficient.

The two best known theoretical models of complex networks, i.e. the regular lattice and the random graph, therefore display opposite behaviours concerning the two variables of interest. But what happens in real networks?

3. SOME REAL NETWORKS

It is very difficult to provide a general answer to this question, however many networks which have been studied with a quantitative approach display a behaviour which differs from both models.

Examples of networks where the clustering coefficient and the characteristic path lengths have been investigated include (see e.g. Reka and Barabasi, 2001):

- social networks, like e.g friendship networks, where humans are nodes and a link exists between two of them if they know each other, or co-authorship networks, where scientists are associated to nodes, and a link exists if two scientists have co-authored at least one paper, or actor networks, where nodes represent actors, and a link exists if two actors have taken part to the same movie;

- the Internet, where the nodes are computers and routers and the links are physical links
- the World Wide Web, where the nodes are the documents and two documents are connected if there is at least a hyperlink between them
- metabolic networks, where the nodes represent chemicals which are present in a cell, and two of them are linked if there exists at least a reaction to which both participate
- electric power networks, where the nodes represent physical production and distribution plants and the links represent physical wires.

In these case, a number of authors have found a rather surprising result (reviewed in Reka and Barabasi, 2001), namely that, although these networks are sparse (i.e. the number of links per node is small compared to the network size), they have high clustering coefficients and, at the same time, short characteristic path lengths.

It seems therefore that many real netowrks, in many different domains, display a combination of properties which is different both from the regular lattice and from the random graph models. It is therefore interesting to investigate models which also display this so called small-world character.

4. THE WATTS-STROGATZ MODEL

Let us first of all notice that it is not difficult to devise a class of networks with the required properties. Let us consider a regular tree structure, where each node in layer m is connected to k nodes in the following layer. There is one node in layer 0, k nodes in layer 1, k_2 nodes in layer 2, a.s.o. Let us also modify this classical hierarchical scheme by connecting all the nodes which have the same parent node (a topology which might well represent the actual communication structure of classical organization theory). In this case the clustering coefficient is high (since all the neighbours but the parent node are fully connected) and the path length is short (it grows with the number of layers, i.e. with the logarithm of N), so this network has indeed a small-world character.

However, this example is highly artificial, while many of the networks which have been shown to be of the small-world type show a higher degree of randomness.

The first model which couples randomness and small-worldness is due to Watts and Strogatz (WS), who started from the one dimensional regular ring considered in section 2, and modified it by rewiring some connections at random. The WS recipe is the following (Watts and Strogatz, 1998): choose at random a node i, choose at random one of its connections, and modify it by connecting node i to another one, also chosen at random. Double connections are prohibited.

In this way it is possible to generate a family of graphs, with increasing randomness, by increasing the fraction f of rewired nodes. It is apparent that if $f = 0$, the graph is a regular lattice (with high l and high C), while $f = 1$ corresponds to a fully random graph (with small C and small l). If one plots the variables l and C as a function of f (while keeping the number of nodes and the number of connections constant), one therefore observes that, as f grows from 0 to 1, both quantities decrease. However, computer simulations show that the decrease rate is different for the two variables, so that there exists a region of f values where C is still high while l has dropped to low values, i.e. where the network is of the small-world type.

Therefore the WS model shows how the introduction at random of some long range connections in a regular lattice may lead to this behaviour. The introduction of this model has raised much interest in the topology of real networks, and it has also led to the introduction of yet another class of models which provide both randomness and small worldness, the scale-free models.

5. THE SCALE-FREE MODEL

The distribution of the number of connections per node, $p(k)$, follows a Poisson (exponential) distribution in both RG and WS models. However, it has been observed that in some of the cases discussed in section 3 the shape of the probability distribution is best represented, in a wide range, by a power law of the kind

$$p(k) = Ak^{-g}$$

A possible mechanism for the appearance of such laws in the case of complex networks has been suggested by Barabasi and co-workers (see Reka and Barabasi, 2001 and further references quoted therein), who have pointed out that the genesis of many networks is different from the way how the previous models are built. On the one hand, networks like the Internet, the WWW, social networks, metabolic networks are not likely to derive from perturbations of a regular lattice, as in the WS model. On the other hand, it is also unlikely that they are completely random. Instead, a mechanism of preferential attachment is likely to be at work: a site which has many visitors is likely to have even more visitors, i.e. to sharpen its edge with respect to less popular ones; a man with many friends is likely to know ever more people (his friends' friends), a.s.o.

In order to capture this phenomenon, Barabasi *et al.* have introduced a simple model which exhibits preferential attachment, which they named the scale-free (SF) model. The network is built in successive time steps: at $t = 0$ a number m_0 of nodes is created; at each successive time step a new node is created, and it is linked with m ($\leq m_0$) links to m (different) existing nodes. The probability $p(i)$ that one of the existing nodes, say the i-th, is chosen for linking to the new one is in turn a function of its connectivity k_i: $p(i) = k_i / \Sigma_j k_i$. In this way a node which has already many links is likely to increase its advantage even more: a mechanism of increasing returns is at work.

It can be shown that, after $T >> m_0$ steps, the probability distribution is indeed a power law, i.e. $p(k) = Ak^{-g}$. Moreover, the characteristic path length scales with the logarithm of the number T of nodes, and the clustering coefficient is higher than in the RG case, since the process of network growth introduces correlations among the connection degrees of different nodes. However, in contrast to RG and WS, the clustering coefficient vanishes as $T \rightarrow \infty$.

So also the SF model exhibits the small world phenomenon.

6. DISCUSSION

Different real world networks are of the small world type, as they combine short path lengths with high clustering. In some cases the pattern of connections follows a power law like that of the SF model, while in other cases an exponential distribution is observed. Even in the power law case, an exponential cutoff is often found. Therefore, the different models which have been developed (and their numerous variants, which have not been described here, see Reka and Barabasi, 2001 for a recent review) may be suited to different cases; a very interesting comparison of the distribution of connections per node for different small world networks can be found in (Amaral *et al.*, 2000).

The fact that the paths are short is not infrequent. Indeed, some artificial network structures display a logarithmic scaling, like the modified tree of section 3 Also regular lattices, e.g. hypercubes, whose path length scales as $N^{1/D}$, may have very slow growth if the number of dimensions is sufficiently high.

Random graphs have short path lengths, too. So the fact that networks, whose connections embed randomness, have short pathways is perhaps not so surprising. The interesting fact is that they couple this feature with a certain degree of structure, which manifests itself in the high level of the clustering coefficient. In the WS case, the result is achieved by introducing at random some long range connections in a regular lattice, in the SF case by developing some highly connected nodes (hubs) which are linked to many nodes.

The choice of the most adequate model is thus a matter of comparison with experimental data, or it can be inferred from the network growth process (e.g., preferential attachment).

The study of the topological properties of complex networks is likely to shed new light on some important features: for example, global properties of SF networks (like their characteristic path length) have been shown to be highly robust with respect to damages of the network, provided that the hubs are not affected (Albert *et al.*, 2000).

A very interesting and open issue concerns the influence of these different topologies upon the dynamical behaviours of the networks, like e.g. genetic or neural networks. The question which should be addressed is whether there are any reasonable general properties which describe the influence of these topological properties on the dynamical landscape.

Finally, it should be observed that the literature on this subject is very interesting and fast growing. The references provided here are not exhaustive, but they represent a good starting point to deal with this topic.

REFERENCES

Albert, R., Jeong, H., and Barabasi, A. L., 2000, Error and attack tolerance of complex networks. *Nature* **406**: 378-382.

Amaral, L. A. N., Scala, A., Barthelemy, M, and Stanley, H. E., 2000, Classes of small world networks. *PNAS 97*, 11149-11152.

Gewolb, J., 2001, Six degrees of uncertainty. *Science* **294**: 777.

Haken, H., 1977, *Synergetics*. Springer Verlag, Berlin.

Jeong, H., *et al.*, 2000, The large scale organization of metabolic networks. *Nature* **407**: 651-654.

Reka, A., and Barabasi, A. L., 2001, *Statistical mechanics of complex networks*. arXiv:cond-mat/0106096v1.

Serra, R., Zanarini, G., Andretta, M., and Compiani, M., 1986, *Introduction to the physics of complex systems*. Pergamon Press, Oxford.

Strogatz, S. H., 2001, Exploring complex networks. *Nature* **410**: 268-276.

Wagner, A., and Fell, D., 2000, *The small world inside large metabolic networks*. Tech.Rep. 00-07-041, Santa Fe Institute.

Watts, D. J., 1999, *Small worlds: the dynamics of networks between order and randomness*. Princeton University Press, Princeton, New Jersey.

Watts, D. J., Strogatz, S. H., 1998, Collective dynamics of small world networks. *Nature* **393**: 440.

The Role of Stability Theory in the Great Theories of the XX Century

UMBERTO DI CAPRIO
Stability Analysis, Via A. Doria, 48 /A - 20124 Milano, Italy
E-mail: umbertodicaprio@tiscalinet.it

1. INTRODUCTION

Three great theories dominated the first half of the twentieth century: Special Relativity, General Relativity and Quantum Mechanics (QM). Exceptional practical applications (radio, television, telecommunications) were made possible by a fourth theory: Electromagnetism. These theories revolutionized classical physics throwing new light on what we knew so far. However, since the death of Einstein in the second half of the twentieth century, the initial impetus has slowed down and developments of some importance are hard to be found, despite the frantic search for a "unifying theory" and the enormous effort to widen the application field of the existing theories.

Meanwhile the rapid advance of experimental science put together an impressive amount of data, but the existing theories do not make it possible to group all such data in one mathematical framework.

As is well known Einstein acknowledged the inability of General Relativity to unify gravitation and electromagnetism and he denied theoretical value to Quantum Theory. A lot of physicists share this point of view: Gerard't Hooft, 1999 Nobel prize for particle physics, has recently reasserted he regards quantum mechanics as a transitory mean.

Richard Feynman, the great physicist Nobel Prize in 1968, thought Quantum Mechanics to be precarious and contradictory. In his book "The Character of Physical Law" he wrote: "I think I can safely say that nobody

Emergence in Complex, Cognitive, Social, and Biological Systems, Edited by Minati and Pessa, Kluwer Academic/Plenum Publishers, New York 2002

understands quantum mechanics .... Do not keep saying to yourself, if you can possibly avoid it, 'But how can it be like that?' because you will go "down the drain" into a blind alley from which nobody has yet escaped. Nobody knows how it can be like that".

This paper shows the advantages of an approach to physical problems based on stability theory, particularly according to a modern formulation. Furthermore it attempt to put in evidence the negative consequences of the fact that this theory was mislead. Namely some specific theoretical models were privileged, thus avoiding to analyse alternatives which could have bee profitable.

2. THE NEED FOR A NEW THEORY

We need a total different approach. Stability Theory provides an affective and satisfactory answer. With regard to this, we must point out that the science of Automatic Control, Computer Science and Telematics marked the second half of the twentieth century, making possible to reach the moon and to begin exploring the cosmos.

The Theory of Automatic is the carrying structure of these sciences and stability is the lintel of such theory. The book "Stability of Motion" was the first and fundamental contribution, published in 1892 by the Russian mathematician M. A. Lyapunov. Over the following 60 years the analysis of stability remained prerogative of the russian school, with hundreds of scientific contributions and with decisive impact, both scientific and technological, upon the soviet world. In 1957 only, year of the launch of the first Sputnik, the western world became aware of the enormous delay that had been accumulated. President Eisenhower and J. F. Kennedy boosted an impressive plan of research, theoretical also, directed to recovering a leadership. Subsequent developments, from the landing on the moon to "star wars", are history of our epoch. In 1959 the first translation, in German, of Lyapunov's work was published in Western World; in 1963 the first version in English language was put forward: "The Direct Method of Lyapunov" by the Austrian mathematician Hahn. From that date the applications start, with increasing intensity up to nowadays explosion with Fractals, Strange Attractors, Deterministic Chaos. This explosion takes place in the context of an impressive development of Systems Theory (Kalman *et al.*, 1969), which includes Stability, Optimization, Identification, Automatic Control, Probability Calculus, Decision Theory, Dynamics of Dissipative Systems, Game Theory, Forecasting. The following conceptual connections bring out the fundamental role of Stability Theory in understanding the physical world.

3. STABILITY AND NON-LINEARITY

Stability and Non-Linearity: stability theory is essentially a theory of non linear systems. On the contrary contemporary Physics is essentially a science of linear phenomena, with the important exception of the contribution brought by Prigogine who asserts: "life better than any other physical phenomenon expresses some essential laws of nature. Life is the kingdom of non-linearities, life is the kingdom of the autonomy of time, it is the kingdom of multiplicity of structure".

4. STABILITY AND ENERGY

Stability and Energy: it is shown that Stability is a local property around an Equilibrium point. It can be led again to the property that Energy had a minimum value and the Dissipated Power had a maximum value. The concept of Maximum Dissipated Power is deeply innovative and represents one of the keys for understanding reality. However, "official "physics doesn't make due allowances for this. Already at the end of 19th century Wilhelm Ostwald, founder of the "energetic school" intended to show not only that the discovery and the setting up of the concept of energy is the height of science; not only that the ensemble of natural processes can be taken back to the transfer and the mutual conversion of different kinds of energy, but, in addition, the history of civilisation itself can be placed under the sign of energy: history of the more and more economical and efficient utilization of raw energy to be found in nature. Energists thought that energy represents an "invariant" for understandability of change, or rather for understanding change thanks to what, through it, remains unchanged. This approach continues to be up to date: we are to decide weather the sciences we are concerned with are sciences of becoming or sciences of being. In some cases the attitude of Quantum Mechanics give rise to a neglect of dynamics by viewing reality as a sequence of static states. When QM is used in this way, it appears as poorly suitable to explain the becoming.

5. STABILITY AND DYNAMICS OF SYSTEMS

As is known the Bohr atomic model implied and postulated an absurdity from the point of view of classical mechanics: the electron, charged particle in motion around a charged nucleus, does not radiate; otherwise it would loses energy steadily and should slow down and fall into the nucleus.

Heisenberg, one of the fathers of Quantum Mechanics, thought that the problem could be solved by eliminating "tout court" the concept of trajectory and reducing reality to a sequence stationary states, in which the different levels of energy show themselves without mutual interaction. In reality, in Bohr's time, Classical Mechanics was not expressed in enough general terms. We can say to-day that limits of Classical Mechanics appear in right evidence, and can be brilliantly overcome, in the light of Modern Stability Theory (as it has developed from 1963 on). Note in fact that Classical Mechanics was based upon the Lagrange's formalism and upon that one, more abstract, of Hamilton. These are physically significant only for a very restricted class of systems which are the conservative systems (or systems which retain energy). Such systems are the only ones integrable: they keep memory of the past and turn out to be completely determined by the initial condition.

On the contrary, Modern Stability Theory (second half of 20th century) using Lyapunov Method, proposed a Mechanics of Non-Conservative Systems and pseudo-dissipative systems, like, e.g., the atom.

The elementary concept of initial condition is then replaced by that of Region of stability, i.e. the ensemble of initial conditions, starting from which the system turns out to be asymptotically stable.

Energy no longer is an invariant of the motion "and therefore it no longer is a "limit imposed to freedom" (as Poincarè said).

Also, changes are irreversible: if, after an evolution from given "initial conditions" the system "turns back", it no more has the original energy. However, a component of the energy is kept unchanged: it is called the Conservative part of the energy and determines the system behaviour as to. stability. The variation of the conservative energy in a unit of time determines the Dissipated Power. Unlike for conservative systems, the Dissipated Power is different from zero. Another important feature is that the systems which are the objects of Stability Theory are nonlinear, non-symmetrical and multivariable.

These aspects are not present in Quantum Theory. With reference to the orbital motion of the electron around the nucleus of an atom, QM discriminates between quantities which are "observable", as the energy levels, and theoretical quantities which are not directly observable, as the position and the velocity. These latter are a-priori excluded, thus creating the illusion that one overcomes the intrinsic limits of classical mechanics. In particular, it is considered illusory to describe the motion, at each instant, regardless from the fact that it is "observed" or not.

Coherently with this, the classic approach of Hamilton is referred to new mathematical (not physical) quantities, called operators. The hamiltonian itself is an operator: it is not directly observable but, however, turns out to be directly connected to Energy, i.e. to the preminently observable quantity in QM. In the same way as, according to classical mechanics, the hamiltonian gives the value of the energy and, at one time, determines the evolution of the system, so, according to QM, the Hamiltonian determines the evolution of the wave-function that yields, at any instant, the probability of different results of measurements that are made on the system. The evolution of the wave-function of the system is given by the Schröedinger equation.

In a sense, QM "tout court" suppresses dynamics non-linearities, dissymmetries, pseudo-dissipations and the multivariable behaviour of the physical world.

6. STABILITY AND IRREVERSIBILITY

Galileian motions are reversible, or, they are characterized by the equivalence, at any instant, between identifiable cause as producing an effect and the effect itself. Therefore it is possible to restore the initial condition by inverting the arrow of time. According to the Galilei formulation, only changes of height (in a gravitational field), are pertinent, while it is indifferent whether they have been carried out after a route formed by russian mountains or after a vertical fall. Such formulation, followed by Lagrange, Hamilton, Jacobi, leaves out the majority of real physical processes, which are non-conservative processes and thus are irreversible. Quantum Mechanics avoids the problem of the connection cause-effect and thus eludes the problem of irreversibility, which, on the contrary has an extreme importance in Thermodynamics. At present time the only theory able of affording the problem, and thus to give an interlocutory judgement on thermodynamics is the Theory of stability of non-linear and non-conservative systems.

7. STABILITY AND LOCALIZATION

Since in general a non-linear system has a multiplicity of equilibrium points, stability essentially is a local property, namely it is bonded to a specific equilibrium point. In particular the Conservative Energy is a state function which depends on the coordinates of the equilibrium point under consideration. In other words, contrary to the results given by Lagrange, Hamilton, Jacobi, we cannot define in a global way the concept of Energy. A typical example is given by the atom: stable orbits are those with radii $n^2 r_{Bhor}$, n odd integer, while the orbits with n even integer, are unstable. Consequently the stability of the fundamental orbit of Bohr is only a local property. Around each stable orbit we can identify a region of stability in the large which gives us the correspondent of the Orbitals postulated by quantum mechanics.

8. STABILITY AND SYMMETRY

A fundamental property of conservative system is the connection between Stability and Symmetry (or, the symmetry of convenient mathematical structures bonded to the differential equations of the motion). As is well known, a sufficient condition for the integrability of the differential equations is that the matrix of the partial derivatives of the positional forces be symmetric. Clearly this is a strong condition which leaves out the great majority of physical systems. The theory of stability of non-conservative systems leads to a fundamental generalization as follows: if a system is stable then it must be either symmetric or symmetrizable.

Symmetrizability means that there exists a suitable transformation of coordinates so that the matrix of the partial derivatives becomes symmetric.

A non-conservative system is stable only if it is symmetrizable. Symmetrizability is a local property: i.e. with regard to any stable equilibrium point one defines a transformation of coordinates so that the given system is mathematically equivalent to a conservative system. The breaking of symmetry is a frequent property of physical systems (e.g. biological) but is not in contrast with stability. We can say, in talkative language, that the requirement of stability sets a limit on the maximum allowed dissymetry.

Consider a dynamical system represented by the vector differential equation

$$\mathbf{M}\ddot{\mathbf{x}} = \mathbf{f}(\mathbf{x}) \tag{1}$$

$\mathbf{x}$ $n \times 1$ vector whose elements represent positions

$\mathbf{f}(\mathbf{x})$ $n \times 1$ vector of continuous and differentiable functions $-\mathbf{f}(\mathbf{x})$ that represent forces

$\mathbf{M}$ $n \times n$ positive-definite matrix whose elements represent masses

$\dot{\mathbf{x}}$ $n \times 1$ vector whose elements represent speeds

$\ddot{\mathbf{x}}$ $n \times 1$ vector whose elements represent accelerations

$\begin{bmatrix} \mathbf{x} \\ \dot{\mathbf{x}} \end{bmatrix}$ $2n \times 1$ state vector

System (1) is said to be conservative if

$$\frac{\partial f_i}{\partial x_j} = \frac{\partial f_j}{\partial x_i} \qquad \text{for any couple } i, j \tag{2}$$

Otherwise system (1) is said to be non-conservative:

$$\frac{\partial f_i}{\partial x_j} \neq \frac{\partial f_j}{\partial x_i} \quad \text{for at least one couple } i, j \text{ with } i \# j \tag{3}$$

Now consider an equilibrium point ($\mathbf{x}=\mathbf{x}_0$; $\dot{\mathbf{x}}=\mathbf{0}$) with $\mathbf{f}(\mathbf{x}_0)=\mathbf{0}$, and the linearized system equation

$$\ddot{\mathbf{x}} = \mathbf{A}(\mathbf{x} - \mathbf{x}_0) \tag{4}$$

$$\mathbf{A} = -\left[\frac{\partial f_i}{\partial x_j}\right]_{\mathbf{x}=\mathbf{x}_0} \qquad n \times n \text{ matrix} \tag{5}$$

If system (1) is conservative $\mathbf{A}$ is a symmetric matrix

$$\mathbf{A}^{\mathrm{T}} = \mathbf{A} \tag{6}$$

On the contrary, if system (1) is non-conservative, then

$$\mathbf{A}^{\mathrm{T}} \neq \mathbf{A} \tag{7}$$

Viceversa, if condition (6) is satisfied, system (1) is said *locally conservative* in $\mathbf{x}_0$. Note that a system can be locally conservative without being conservative. This property is fundamental in view of the analysis of the stability of nonlinear systems. Further note:

1. System (4) is stable if and only if the eigenvalues of $\mathbf{A}$ are real and negative. In fact the $2n$ eigenvalues of system (4) $l_1, l_2, \ldots, l_{2n}$ are equal to the root squares of the n eigenvalues of $\mathbf{A}$, $\lambda_1, \lambda_2, \ldots, \lambda_n$.
2. If $\mathbf{S}$ is any positive-definite matrix, then premultiplying both sides of equation (4) by $\mathbf{S}$ we obtain an equivalent equation

$$(\mathbf{SM}) = (\mathbf{SA})(\mathbf{x} - \mathbf{x}_0) \tag{8}$$

Equation (8) represents a conservative system if, for some $\mathbf{S}$,

$$(\mathbf{SA})^{\mathrm{T}} = \mathbf{SA} \tag{9}$$

Therefore equation (4) represents a conservative system if either $\mathbf{A}$ or $\mathbf{SA}$ is symmetric.

3. Matrix $\mathbf{A}$ is said to be *symmetrizzable* (Taussky, 1968, 1969) if it satisfies equation (9). A necessary condition for $\mathbf{A}$ to be symmetrizzable is that its eigenvalues be real and the eigenvectors linearly independent (Huseyin, 1978).
4. Above condition is sufficient too (Di Caprio, 2001). In fact, if $\mathbf{A}$ has n real eigenvalues and n independent real eigenvectors, the matrix

$$\mathbf{S} = \mathbf{G}^{-1\mathrm{T}}\mathbf{\Theta}\mathbf{G}^{-1} \tag{10}$$

with $\mathbf{G}$ the matrix of the eigenvectors, $\mathbf{\Theta}$ an arbitrary diagonal and positive-definite matrix, is a simmetrizing matrix.

Proof. As $\mathbf{\Theta} > \mathbf{0}$, then, for well known relationships, $\mathbf{S} > \mathbf{0}$. Moreover, as $\mathbf{A}$ satisfies the equation

$$\mathbf{A} = \mathbf{G}\mathbf{\Lambda}\mathbf{G}^{-1} \tag{11}$$

with $\mathbf{\Lambda}$ diagonal matrix of the eigenvalues, it is

$$\mathbf{SA} = \mathbf{G}^{-1\mathrm{T}}\mathbf{\Theta}\mathbf{G}^{-1}\mathbf{G}\mathbf{\Lambda}\mathbf{G}^{-1} = \mathbf{G}^{-1T}(\mathbf{\Theta}\mathbf{\Lambda})\mathbf{G}^{-1} \tag{12}$$

As $(\mathbf{\Theta}\mathbf{\Lambda})$ is diagonal, we get

$$\left[\mathbf{G}^{-1\mathrm{T}}\mathbf{\Theta}\mathbf{\Lambda}\mathbf{G}^{-1}\right]^{\mathrm{T}}\mathbf{G}^{-1\mathrm{T}}\mathbf{\Theta}\mathbf{\Lambda}\mathbf{G}^{-1} \tag{13}$$

which implicates

$$(\mathbf{SA})^{\mathrm{T}} = \mathbf{SA} \tag{14}$$

Consequently $\mathbf{S}$ in (10) is a symmetrizing matrix.

Theorem 1: (Conditions for symmetrizzability). A real $n\times n$ matrix is symmetrizzable if and only if it has real eigenvalues and n linearly independent eigenvectors. If $\mathbf{A}$ is symmetrizzable there exist infinite matrices that symmetrize $\mathbf{A}$: they are given by equation (10).

Corollary: (Selection of matrix **S**). If **A** is symmetrizzable, a particular matrix $\boldsymbol{\Theta}$ in (10) is

$$\boldsymbol{\Theta} = \mathbf{I} \tag{15}$$

With this choice, it is **S=I** if **A** is already symmetric.

Theorem 2 (Conditions for symmetrizzability of $\mathbf{M}^{-1}\mathbf{K}$). With reference to equation (1) and to the Equilibrium Point ($\mathbf{x}_0$,**0**) the linearized system equation is

$$\ddot{\mathbf{x}} = \mathbf{A}(\mathbf{x} - \mathbf{x}_0) \tag{16}$$

$$\mathbf{A} = -\mathbf{M}^{-1}\mathbf{K} \tag{17}$$

$$\mathbf{K} = -\left[\frac{\partial \mathbf{f}}{\partial \mathbf{x}}\right]_{\mathbf{x}_0} \tag{18}$$

the following condition is necessary and sufficient for $\mathbf{M}^{-1}\mathbf{K}$ to be symmetrizzable: $\mathbf{M}^{-1}\mathbf{K}$ has real eigenvectors and linearly independent eigenvalues. In particular a sufficient condition is that $\mathbf{M}^{-1}\mathbf{K}$ has *n* real and distinct eigenvalues.

A particular symmetrizing matrix is

$$\mathbf{S} = \mathbf{G}^{-1\mathrm{T}}\left[\mathrm{diag}\,\mathbf{G}^{\mathrm{T}}\left(\mathbf{M}^{-1}\mathbf{K}\right)\mathbf{G}\right]\mathbf{G}^{-1} \tag{19}$$

Such matrix satisfies

$$\mathbf{S} = \mathbf{I} \qquad \text{if} \qquad \mathbf{K}^{\mathrm{T}} = \mathbf{K} \tag{20}$$

which implicates that, if system (1) is "locally conservative in $\mathbf{x}_0$", **S** equals the Identity matrix.

Proof. The first part directly derives from the general Theorem 1. As regards equation (19), under the given assumptions

$$\mathbf{M}^{-1}\mathbf{K} = \mathbf{G}\boldsymbol{\Lambda}\mathbf{G}^{-1}$$

with $\mathbf{\Lambda}$ real diagonal matrix. Consequently the matrix $\mathbf{S}$ defined by (19), not only is by construction positive definite but satisfies the equation

$$\begin{aligned}\mathbf{S}\left(\mathbf{M}^{-1}\mathbf{K}\right) &= \mathbf{G}^{-1\mathrm{T}}\left[\mathrm{diag}\,\mathbf{G}^{\mathrm{T}}\mathbf{M}\mathbf{G}\right]\mathbf{G}^{-1}\mathbf{G}\mathbf{\Lambda}\mathbf{G}^{-1} \\ &= \mathbf{G}^{-1\mathrm{T}}[\mathrm{diag}\,\mathbf{G}^{\mathrm{T}}\mathbf{M}\mathbf{G}]\mathbf{\Lambda}\mathbf{G}^{-1}\end{aligned}$$

which as the product $[\mathrm{diag}\ \mathbf{G}^{\mathrm{T}}\mathbf{M}\mathbf{G}]\mathbf{\Lambda}$ is diagonal, brings about that

$$\mathbf{S}\left(\mathbf{M}^{-1}\mathbf{K}\right) = \left[\mathbf{S}\left(\mathbf{M}^{-1}\mathbf{K}\right)\right]^{\mathrm{T}}$$

Theorem 3 (Relationship between stability and symmetrizzability).

a) If matrix $\mathbf{M}^{-1}\mathbf{K}$ is symmetrizzable and $\mathbf{S},\mathbf{M},\mathbf{K}$ satisfy the condition

$$\mathbf{S}\mathbf{M}^{-1}\mathbf{K} > 0 \qquad \text{(positive-definite)}, \tag{21}$$

the system represented by the linear equation

$$\mathbf{M}\,\ddot{\mathbf{x}} = -\mathbf{k}(\mathbf{x} - \mathbf{x}_0) \tag{22}$$

is stable.

b) If $\mathbf{M}^{-1}\mathbf{K}$ has n real positive eigenvalues and *n* independent eigenvectors the linear system represented by (22) is stable and equation (21) comes out verified with

$$\mathbf{S} = \mathbf{G}^{-1\mathrm{T}}\mathrm{diag}\left[\mathbf{G}^{\mathrm{T}}\mathbf{M}\mathbf{G}\right]\mathbf{G}^{-1} \tag{23}$$

c) If $\mathbf{M}^{-1}\mathbf{K}$ real positive eigenvalues and linearly independent eigenvectors, in particular if $\mathbf{M}^{-1}\mathbf{K}$ has real positive and distinct eigenvalues, then equation (22) is equivalent to

$$(\mathbf{\Gamma}\mathbf{M})\ddot{\mathbf{x}} = -(\mathbf{\Gamma}\mathbf{K})(\mathbf{x} - \mathbf{x}_0) \tag{24}$$

with

$$\mathbf{\Gamma} = \mathbf{G}^{-1\mathrm{T}}\mathrm{diag}\left[\mathbf{G}^{\mathrm{T}}\mathbf{M}\mathbf{G}\right]\mathbf{G}^{-1}\mathbf{M}^{-1} \tag{25}$$

It is

$$(\mathbf{\Gamma}\mathbf{M}) = \mathbf{G}^{-1\mathrm{T}}\mathrm{diag}\left[\mathbf{G}^{\mathrm{T}}\mathbf{M}\mathbf{G}\right]\mathbf{G}^{-1}\mathbf{M}^{-1} \tag{26}$$

$$
\begin{aligned}
(\mathbf{\Gamma K}) &= \mathbf{G}^{-1\mathrm{T}}\mathrm{diag}\left[\mathbf{G}^{\mathrm{T}}\mathbf{MG}\right]\mathbf{G}^{-1}\mathbf{M}^{-1}\mathbf{K} \\
&= \mathbf{G}^{-1\mathrm{T}}\mathrm{diag}\left[\mathbf{G}^{\mathrm{T}}\mathbf{MG}\right]\mathbf{G}^{-1}\mathbf{G\Lambda G}^{-1} \\
&= \mathbf{G}^{-1\mathrm{T}}\mathrm{diag}\left\{\left[\mathbf{G}^{\mathrm{T}}\mathbf{MG}\right]\mathbf{\Lambda}\right\}\mathbf{G}^{-1} > 0
\end{aligned}
\tag{27}
$$

d) Under the same assumptions, the non-linear equation (1) is equivalent to

$$
(\mathbf{\Gamma M})\ddot{\mathbf{x}} = -\mathbf{\Gamma K}\,\mathbf{f}(\mathbf{x}) \tag{28}
$$

Equation (28) represents a system with masses $(\mathbf{\Gamma M})$ subjected to "locally conservative" positional forces $(\mathbf{\Gamma Kf(x)})$. Such system is "locally stable" in $\mathbf{x} = \mathbf{x}_0$.
Also (Di Caprio, 1987), the function

$$
V(\mathbf{x},\dot{\mathbf{x}}) = \int_{\mathbf{x}_0}^{\mathbf{x}} \mathbf{\Gamma}\,\mathbf{f}(\mathbf{x})\,d\mathbf{x} + \frac{1}{2}\dot{\mathbf{x}}^{\mathrm{T}}\mathbf{\Gamma M}\,\dot{\mathbf{x}} \tag{29}
$$

is "locally positive-definite in $\mathbf{x}_0$". However its time derivative $dV(\mathbf{x},\dot{\mathbf{x}})/dt$ is indefinite, although its gradient and its Hessian equal zero in $\mathbf{x}_0$.

<u>Proof</u>

a) See reference (Huseyin, 1978).
b) Under the given assumptions $\mathbf{M}^{-1}\mathbf{K}=\mathbf{G\Lambda G}^{-1}$ with $\mathbf{\Lambda}$ diagonal and positive-definite. Then the matrix $\mathbf{A}=-\mathbf{M}^{-1}\mathbf{K}$ has real negative eigenvalues. But the $2n$ eigenvalues of system (4) are the roots square of the eigenvalues of $\mathbf{A}$ and then consist of n imaginary couples. This brings about that system (4) is weakly stable. Also the matrix $\mathbf{S}$ given by (23) is a symmetrizing matrix and moreover, as

   $$
   \mathrm{diag}\mathbf{G}^{\mathrm{T}}\mathbf{MG} > \mathbf{0} \quad \text{and} \quad \mathbf{\Lambda} > \mathbf{0}
   $$

   then

   $$
   \mathbf{SM}^{-1}\mathbf{K} = \mathbf{G}^{-1\mathrm{T}}\left[\mathrm{diag}\mathbf{G}^{\mathrm{T}}\mathbf{MG}\right]\cdot\mathbf{G}^{-1} > 0
   $$

c) Under the given assumptions the matrix $\mathbf{\Gamma}$ defined by equation (25) is real and nonsingular. Consequently

<u>Note 1</u> (Stability in the large). If $\mathbf{M}^{-1}\mathbf{K}$ has real positive and distinct eigenvalues, or real positive eigenvalues and linearly independent eigenvectors, system is "stable in the small" and the vector

$$-\mathbf{g}(\mathbf{x}) = -\mathbf{\Gamma f}(\mathbf{x}) \tag{30}$$

represents "locally conservative" positional forces. Then equation (28), (which is equivalent to (1)), can be written

$$(\mathbf{\Gamma M})\ddot{\mathbf{x}} = -\mathbf{g}(\mathbf{x}) \tag{31}$$

with $\mathbf{g}$ satisfying the condition

$$\left[\frac{\partial \boldsymbol{g}_i}{\partial \boldsymbol{x}_j}\right]_{\mathbf{x}_0} = \left[\frac{\partial \boldsymbol{g}_j}{\partial \boldsymbol{x}_i}\right]_{\mathbf{x}_0} \quad \text{for any couple } i, j \tag{32}$$

or, in equivalent terms

$$\left[\frac{\partial g}{\partial \mathbf{x}}\right]_{\mathbf{x}_0}^{\mathrm{T}} = \left[\frac{\partial g}{\partial \mathbf{x}}\right]_{\mathbf{x}_0} \tag{33}$$

We recall, without proof, the following result: there exists a functional transformation $\mathbf{T}$ such that

$$\mathbf{T}(\mathbf{g}(\mathbf{x})) = \mathbf{g}_1(\mathbf{x}) \tag{34}$$

and

$$\mathbf{g}(\mathbf{x}) = \mathbf{g}_1(\mathbf{x}) + \mathbf{g}_2(\mathbf{x}) \tag{35}$$

with $\mathbf{g}_1(\mathbf{x})$ and $\mathbf{g}_2(\mathbf{x})$ n x 1 vectors satisfying the conditions

$$\left[\frac{\partial \boldsymbol{g}_1}{\partial \mathbf{x}}\right]^{\mathrm{T}} = \left[\frac{\partial \boldsymbol{g}_1}{\partial \mathbf{x}}\right] \tag{36}$$

Equations (35) and (36) bring about that $-\mathbf{g}_1(\mathbf{x})$ represents conservative positional forces while $\mathbf{g}_2(\mathbf{x})$ represents non-conservative positional forces which in the neighbourhood of $\mathbf{x}_0$.

The following function is called "conservative part of the Energy"

$$E(\mathbf{x}, \dot{\mathbf{x}}) = \int_{\mathbf{x}_0}^{\mathbf{x}} \mathbf{g}_1^{\mathrm{T}}(\mathbf{x})\, d\mathbf{x} + \frac{1}{2}\dot{\mathbf{x}}^{\mathrm{T}}(\mathbf{\Gamma M})\,\dot{\mathbf{x}} \tag{37}$$

The function (37) is not a Lyapunov function (since its time derivative along the system trajectories is indefinite, but is, however, a quasi-Lyapunov function, since it is positive definite" in $\mathbf{x}_0$ and its time derivative $\dot{E}(\mathbf{x}, \dot{\mathbf{x}})$ has the gradient and the Hessian equal to zero in $\mathbf{x}_0$.

Note 2 (A necessary condition for the existence of a Lyapunov function). A Lyapunov function exists provided that we include in our dynamical system dissipative forces proportional to the speeds (Di Caprio, 2001, 2002):

$$\mathbf{M}\ddot{\mathbf{x}} = -\mathbf{f}(\mathbf{x}) - \mathbf{D}\dot{\mathbf{x}}$$

$\mathbf{D}$ $n \times n$ real and constant matrix, and provided that $\mathbf{M}^{-1}\mathbf{K}$ has real positive eigenvalues and linearly independent eigenvectors and, in addition $\mathbf{D}$ satisfies a convenient condition.

9. STABILITY AND AUTOMATIC CONTROL

The Theory of Automatic Control (born about 1940) sets stability at the center of its attention: a control system must perform assigned tasks keeping adequate stability margins (in the intuitive sense of the term) or, even must stabilize processes that by themselves would be unstable. Stabilization entails real-time control using feed-back techniques and on-line computing systems and telemetering plus telecommand facilities. The physical world is full of examples of spontaneous stabilization. E.g. in the atom the spin gives an automatic correction that asymptotically stabilize the orbit of a revolving electron.

10. STABILITY AND REPRESENTATION OF REALITY

The Lagrange, Hamilton, Jacobi leave out of consideration stability. Moreover they are prerelativistic, i.e. do not allow to account for the variation of mass with the speed, except for purely formal mechanisms, as the one appearing in many textbooks on classical mechanics. The Stability analysis of a special-relativistic system, as it requires formulation of mathematical models in canonic form, shows that even in the most elementary case (the plane motion of a particle under the action of a Central Force) the classical 2-nd order model is unacceptable, since the relativistic effects lead to a third order differential equation for each particle. Consequently the so called relativistic quantum mechanics is erroneous at its roots.

We underline that Stability theory has the same conceptual reach and the same rigours of the traditional concepts of limit, derivative, integral; also, it proofs to be multidisciplinar, unitary, adaptable to various realities. It is clear on the other hand that the observability of the physical world is strictly bonded to the existence of a stable organization of existing structures, e.g. the biological structures or the atomic structures. What is then the connection between stability and order?

The requirement of stability strongly limits the "possible realities" without, however, compel them to a deterministic mechanicism. As a matter of fact physical reality is made up by a variety of non-linear, non-symmetric, non-conservative structures which must be, and are, stable. An example for all: the Newton law gives a linear tie between force and mass, while Special Relativity gives a non linear bond. While there exists only one linear behaviour, there are infinite non linear behaviours, which, however, owing to stability requirements, reduce themselves to a restricted spectrum. If stability imposed linearity, conservation of the energy, symmetry, reversibility, we would have a brutal determinism.

REFERENCES

Di Caprio, U., 1987, Accounting for the transfer conductance effects in Lyapunov stability analysis of multimachine power systems. *Int. Journal of EPES*, July, 1987.

Di Caprio, U., 2001, Matrix properties relating to stability analysis. *Int. Journal of EPES* **23**: 229-235.

Di Caprio, U., 2002, The effect of friction forces upon stability in the large. *Int. Journal of EPES*, May 2002.

Di Caprio, U., and Saccomanno, F., 1970, Nonlinear stability analysis of multimachine power systems. *Ricerche di Automatica*, **1**.

Edelen, D. G. B., 1977, *Lagrangian mechanics of nonconservative nonholonomic systems.* Noordhoff Int. Publ., Leyden.

Gantmacher, F. R., 1959, *The Theory of matrices. Vol. 1 and 2.* Chelsea, New York.

Hhan, W., 1963, *Theory and application of Lyapunov's Direct Method.* Prentice-Hall, .

Hhan, W., 1967, *Stability of motion.* Springer, .

Huseyin, K., 1975, *Nonlinear theory of Elastic Stability.* Noordhoff Int. Publ., Leyden.

Huseyin, K., 1978, *Vibrations and stability of multiple parameter systems.* Sijthoff & Noordhoff, Leyden.

Kalman, R. E., Falb, P. L., and Arbib, M. A., 1969, *Topics in mathematical system theory.* McGraw-Hill, New York.

Lyapunov, A. M., 1992, The general problem of the stability of motion. (commemorative issue). *Int. J. Control*, Vol. **55**: 531-773.

Taussky, O., 1968, Positive definite matrices and their role in the study of characteristic roots of general matrices. *Advances in Mathematics* **2**: 175-186

Taussky, O., and Zassenhaus, H., 1969, On the similiarity transformation between a matrix and its transpose. *Pac. J. Math.* **9**: 893-896.

A Generalization of Compensation Mechanism in Variable-Structure Systems

ELIANO PESSA*, and GERMANO RESCONI#
*Dipartimento di Psicologia, Università di Pavia, Pavia, Italy
E-mail: eliano.pessa@unipv.it
#Dipartimento di Matematica e Fisica, Università Cattolica del Sacro Cuore, Brescia, Italy
E-mail: resconi@numerica.it

1. INTRODUCTION

Compensation mechanisms are mental tools widely used in building models of complex systems. These mechanisms enable the restoration of conservation laws and of invariances in systems where the inner interactions between the elements would seem to give rise, from the starting, to an unpredictable variability and to random behaviors. A typical example is given by the development of theoretical physics, from Newton's times to our times. As is well known, the mathematical scheme underlying the initial formulation of classical mechanics, based on second-order dynamical equations and on energy conservation principle, was so successful and convenient that, every time it appeared as impossible to fit new experimental data to this scheme, the latter was suitably generalized just through a compensation mechanism. This happened, for instance, when dealing with the domain of thermal phenomena, where the energy conservation principle was generalized through the First Principle of Thermodynamics. Here the compensation was consisting in the introduction of a new quantity – the heat – just compensating for the losses of mechanical energy associated to these phenomena.

There is a variety of different compensation mechanisms, but the most important – and the most extensively studied – have been the ones associated to *field-like* descriptions. Generally the term "field" denotes a suitable entity, whose character (mathematically specified, according to the context, in different ways: e.g. through a real number, a real vector, the form of an operator, and so on) depend on space-time coordinates (or on other independent variables analogous to spatio-temporal ones). This concept is of paramount importance in physics, where *field theories* gave the theoretical basis for the description of all kinds of interactions. When spatial coordinates are neglected and only the time variable is taken into consideration, we obtain, as a special subcase of field theories, the *dynamical systems models* which lie at the heart of Systemics.

Within field theories the main compensation mechanism is based both on the mathematical concept of *isomorphism* and on the so-called *gauge transformations* . Such a kind of mechanism experienced its triumph in the theoretical physics of Twentieth Century, owing to theories such as General Relativity, Nonabelian Gauge Theory of Electroweak Interactions, Superconductivity Theory, and many others. In this paper we will show, however, that this form of compensation is unsuitable to deal with complex systems, characterized by a structure varying with time. Therefore we will introduce a new generalization of it, allowing the description of a class of these systems, and we will explore some consequences of our proposal.

2. GAUGE TRANSFORMATIONS

In order to clarify what is the concrete nature of compensation mechanism within field theories, we must first recall some mathematical concepts. Let us start with the one of isomorphism. It applies to the case in which we have two different algebraic structures, each one associated to a set of objects and to a set of operations acting on these objects, and a law of biunivocal correspondence between these structures (as regards both the objects and the operations). The two structures are isomorphous when they behave in the same way, that is when the result of an operation performed on a given object within one of the two structures is just the image (in the biunivocal correspondence) of the result obtained in the other structure by applying the corresponding operation to the corresponding object. In more formal terms, let us condider the simplest case in which each structure is associated to only one operator, denoted by T in the first structure and by T' in the second structure. Besides, let us denote by x a generic element belonging to the first structure and by φ the law of biunivocal correspondence between the objects of the two structures. Then, the two structures are isomorphous if Tx is the object corresponding, through φ , just to $T'(\varphi x)$. Therefore, an isomorphism between the two structures exists if and only if the following relation is satisfied (see Figure 1):

$$\varphi(T x) = T'(\varphi x),$$

in turn implying:

$$T' = \varphi T \varphi^{-1}.$$

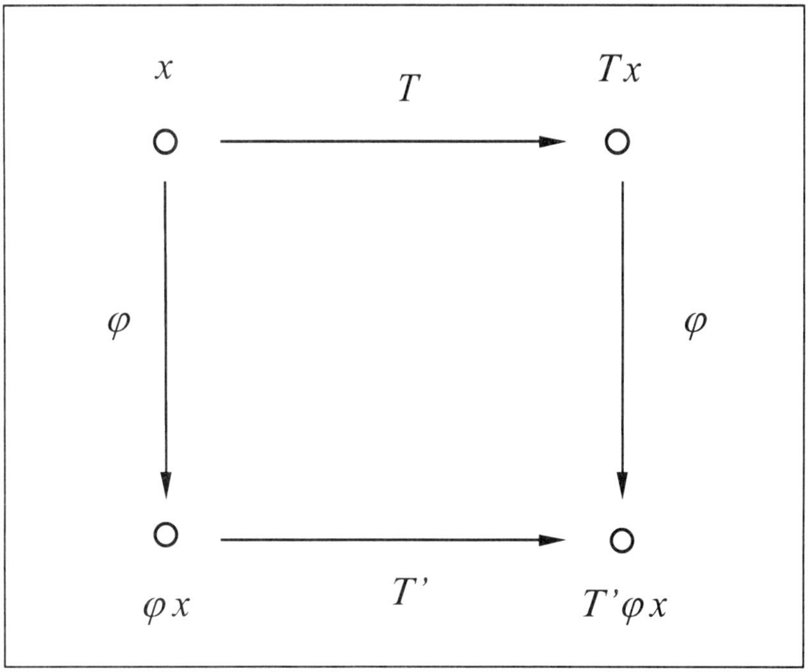

Figure 1. Graphical representation of an isomorphism.

As regards gauge transformations, the simplest (abelian) case assumes the form:

$$\psi'(x, t) = \psi(x, t) \exp(i \varphi)$$

where the symbols $\psi(x, t)$ and $\psi'(x, t)$ denote, respectively, a (scalar) field and its gauge-transformed version, whereas φ is a space-time dependent *phase* and i denotes the imaginary unit. The importance of the gauge transformations case stems from the fact that, if the phase is space-time dependent, the compensation mechanism imposing invariance (isomorphism) with respect to gauge transformations gives rise to the introduction of the concept of *force* and, most generally, of *interaction.* In other words, the presence of an interaction compensates for the changes induced by a gauge transformation.

A simple example is related to the behavior of a system in Quantum Mechanics, which is described through a wave function ψ, which is a solution for a dynamical equation prescribing its space-time evolution: the celebrated Schrödinger equation for a single particle ($h = 1$),

$$i\frac{\partial \psi}{\partial t} = \left(\frac{1}{2}m\right)(-i)^2 \sum_j \partial_j^2 \psi + U\psi ,$$

The wave function carries no physical meaning per se: only its square modulus is associated to a measurable quantity, interpreted as a probability. This implies that the wave function can be represented by a complex scalar, function of space-time coordinates, which has the form

$$\psi = \rho e^{i\theta},$$

where ρ and θ are, respectively, its modulus and its phase. As only ρ has a physical meaning, we expect that the quantum-mechanical description of phenomena be invariant with respect to phase transformations, leaving ρ unaffected, and defined by

$$\psi' = \psi e^{i\varphi}$$

What happens if we let φ depend on space-time coordinates? It is easy to see that we will obtain a new equation no longer invariant with respect to phase transformations, that is:

$$\begin{aligned} i\frac{\partial \psi}{\partial t} &= e\left(\frac{\partial \varphi}{\partial t}\right)\psi = \\ &= \left(\frac{1}{2}m\right)(-i)^2\Big\{\sum_j \partial_j^2 \psi - 2ie\sum_j (\partial_j \varphi)\partial_j \psi + \\ &- \Big[ie\sum_j \partial_j^2 \varphi + e^2 \sum_j (\partial_j \varphi)^2\Big]\Big\}\psi + U\psi \end{aligned}$$

However, the difference between the two equations will disappear when we deal with a particle interacting with an electromagnetic field. In this case the Schödinger equation takes the form

$$i\frac{\partial\psi'}{\partial t}=\left(\frac{1}{2}m\right)\sum_j D_j D_j\,\psi'+(e\phi'+U)\psi',$$

where:

$$D_j\,\psi'=-i\partial_j\,\psi'-eA'_j\,\psi'$$

in which A'_j is the usual 3-dimensional vector potential and ϕ is the scalar electric potential. If we define the transformed values of A'_j and ϕ by:

$$A_j = A'_j + (\partial_j\,\varphi), \quad \phi = \phi' - (\partial\varphi/\partial\, t),$$

(sometimes called *gauge transformations of second kind*) we will find that the Schödinger equation for the transformed wave function ψ will assume the form:

$$i\frac{\partial\psi'}{\partial t}=\left(\frac{1}{2}m\right)\sum_i D_j D_j\,\psi+(e\phi+U)\psi,$$

where:

$$D_j\,\psi=-i\,\partial_j\,\psi-eA_j\,\psi\,.$$

The generalized derivative D_j coincides with the old derivative of the classical physics plus a compensating term that is proportional to the electromagnetic potential. Associated to the local change of the phase, we have a new interpretation of the four-potential A_j . It appears as a *compensating term* whose role is that of keeping unchanged the previous formal Schrödinger structure. Moreover, the generalized derivative D_j is the analogous of the covariant derivative ∇_λ introduced in General Relativity, and is known as "*gauge-covariant derivative*".

Summarizing the previous arguments, we can conclude that the dynamical evolution equation (that is the Schrödinger equation) of a suitable basic field (substratum), described by a wave function ψ, is invariant in form with respect to local gauge transformations of the first kind if and only if it is subject to an electromagnetic interaction whose field potential transforms according to a gauge transformation of the second kind. In other words, this gauge transformation "compensates" for the effects of the gauge transformation of the first kind on the substratum itself. As this "compensating" transformation is possible only in presence of an electromagnetic field, we can say that this latter acts as a "compensating" field.

We thus have two equivalent representations of the dynamical evolutions of ψ. Alternatively, we can say that there is no electromagnetic field, but that the Schrödinger's equation is not invariant with respect to gauge transformations of the first kind, or we may say that Schrödinger's equation is invariant in form with respect to these latter, but that such an invariance is due to the presence of an electromagnetic field. Starting from the latter representation, it is possible to build a general theoretical apparatus for describing all physical interactions and for formulating *unified theories* of these interactions. As is well known, such an approach was very successful in the last forty years and opened the way to a number of new exciting discoveries within the domains of particle physics and of condensed matter physics (see, e.g., Frampton, 1987).

3. A GENERALIZED GAUGE TRANSFORMATION

Despite the successes obtained by such an approach, it needs a suitable generalization. Namely it can be applied only to fixed-structure systems, that is to systems in which the number of possible interactions is fixed in advance and the coupling parameters are to be considered as constants, determined once and for all, like the electron charge or the gravitational constant. On the other hand, in many physical and nonphysical complex systems we need a generalized description able to take into account the occurrence of a variable number of interactions and of space-time dependent coupling parameters.

To bear witness to the interest of physicists in this kind of complex systems, we remind that in recent times, such a description was introduced in Quantum Field Theory under the name "Third Quantization" (Maslov and Shvedov, 1998). Whereas the expression "Second Quantization" refers to a situation in which we allow a nonconservation of particle number, associated to processes of particle creation or destruction, the expression "Third Quantization" refers to a more general situation in which even the *number* and the *kind* of physical fields (that is of possible interactions) are allowed to vary.

The practical implementation, however, of a Third Quantization procedure is very difficult, owing to the fact that we still lack a description of its classical counterpart, that is of a classical situation in which the kinds of physical fields can vary as a consequence of suitable interactions. From a general point of view, this would require a mathematical approach based on functionals instead of ordinary functions; however, even if someone tried to explore such a possibility (cfr. Kataoka and Kaneko, 2000, 2001), it still appears as implying considerable technical difficulties. To this regard, the simplest way to describe the "kind" of a physical field is to focus on its coupling constant with the other fields (e.g. with the substratum field as in the previous section). If this constant were no longer a constant, but a space-time dependent quantity, we had a possibility of introducing a simple description of a kind-variable field.

Inspired by these developments and these considerations we then introduced a generalized gauge transformation:

$$\psi' = \psi e^{iq\varphi}.$$

where q is the (space-time dependent) "charge" of the gauge field (that is its "coupling constant") and φ is the gauge field itself.

In the particular case in which q is a function only of the "compensating field" A_μ we can develop the general compensating field in term of a Taylor series in A_μ, , thus obtaining, at the first order, a gauge derivative of the form:

$$D_\mu = \partial_\mu \sum_\rho \chi_{\mu,\rho} A_\rho$$

From the expression of such a gauge derivative we can obtain a generalized field equation describing the evolution of the compensating field A_μ. Such an equation is based on a number of logical steps (for the details see Mignani, Pessa & Resconi, 1999) which characterize all gauge theories which, in a sense, generalize Maxwell's theory of electromagnetic field. These steps can shortly be summarized as follows:

1. definition of a generalized gauge derivative and whence of the field potentials;
2. computation of the commutator of the generalized gauge derivatives defined in the previous step; the obtained result lets us find the relationship between the field strength and the field potentials;
3. the Jacobi identity applied to the commutator gives the dynamical equations satisfied by field strength;
4. the commutator between the generalized gauge derivative and the previous commutator (triple commutator) lets us define the field currents and find the connection between field strength and field currents.

When applying this procedure to our previous definition of generalized gauge derivative, we obtained, after a number of calculations, that the step 4) gave rise to the following complicated equation connecting the field potentials with the field currents:

$$\Box A_\nu = - J_\nu$$

where:

$$- J_\nu = [(\partial_\mu{}^\mu \chi_\nu{}^\delta) A_\delta - (\partial^\mu{}_\nu \chi_\mu{}^\gamma) A_\gamma] +$$
$$+ [\chi_\nu{}^\delta (\partial_\mu{}^\mu A_\delta) - \chi_\mu{}^\gamma (\partial^\mu{}_\nu A_\gamma)] +$$
$$+ [2 (\partial^\mu \chi_\nu{}^\delta) (\partial_\mu A_\delta) - (\partial_\nu \chi_\mu{}^\gamma) (\partial^\mu A_\gamma) - (\partial^\mu \chi_\mu{}^\gamma) (\partial_\nu A_\gamma)].$$

From such an equation it is possible to derive a number of consequences, the most important of which is that, in presence of space-time dependence of the "charge" (or "coupling constant") of the gauge field, the compensating field undergoes a spontaneous spatial decay in crossing the physical medium in which the field itself is embedded. Such a circumstance follows from the fact that the previous expression of the field current contains terms which are directly proportional to the field potentials themselves. This means that the dynamical equation ruling the behavior of field potentials has solutions decaying to zero.

Such an effect is analogous to the Meissner effect in superconductors and shows that in the case of variable-structure systems the variability of the interactions prevents from a complete transmission of the information (carried by the gauge field) within the system itself. In other terms, inner coherence opposes to the propagation of disturbances mediated by the interactions themselves and can preserve the stability of the system itself; related, in turn, just to the variability of the interactions.

4. CONCLUSION

The introduction of a new generalized gauge transformation led us deal in a more suitable way with complex systems characterized by a variable structure and whence by varying kinds of interactions. Needless to say, most complex systems of biological, economic, social or psychological nature belong to this category, so that our proposal could allow Systemics to face the problems connected to the study of these systems, thus overcoming the limitations of the traditional tools used by theoretical physics. A first application of our method led us discover a new general principle ruling the reaction of these systems to disturbances. Such a principle, stated in less formal and more intuitive terms, could sound as: *variability of the interactions could grant just for the survival of a highly complex system.* We hope this principle be the first of a new Systemics able to better deal with complexity.

REFERENCES

Frampton, P.H., 1987, *Gauge Field Theories*. Benjamin-Cummings, Menlo Park, CO.

Kataoka, N., and Kaneko, K., 2000, Functional Dynamics. I: Articulation Process, *Physica D* **138**: 225-250.

Kataoka, N., and Kaneko, K., 2001, Functional Dynamics. II: Syntactic Structure, *Physica D* **149**: 174-196.

Maslov, V. P., and Shvedov, O. Yu., 1998, *Large-N Expansion as Semiclassical Approximation to the Third-Quantized Theory*. Preprint hep-th/9805089.

Mignani, R., Pessa, E., and Resconi, G., 1999, Electromagnetic-Like Generation of Unified-Gauge Theories, *Physics Essays* **12**: 61-79.

Education

A Systemic Proposal on the Use of a New Technology as a Learning Tool in School Context

MARIA PIETRONILLA PENNA*, VERA STARA#, and NATALE BONFIGLIO*
*Dipartimento di Filosofia, Sezione di Psicologia, Università di Pavia
#Facoltà di Scienze della Formazione, Università di Cagliari

1. INTRODUCTION

The spreading within scientific world of *General Systems Theory* (GST), proposed by L. Von Bertalanffy since 1950, revolutionized the theoretical apparatus of several disciplines, including the humanistic ones.

In fact, Von Bertalanffy, with GST, suggested a new approach to the study of human subjects, viewed as open systems belonging to a network of integrated dimensions and related with the environment (Von Bertalanffy, 1968). In a system like this any modification influences the state of the whole system, and every educational action upsets the subject state, modifying the various dimensions related to it.

Such a circumstance led pedagogists 1) to give up a study of educational processes based on cause-effect principles, for promoting a new global approach interested in the structural complexity of the organism and in relationships between different phenomena, and 2) to revalue the variety of aspects of communicative and educational relation.

To this regard, several researchers (Milton, 2000; Lian, 2000) supposed that the utilization of hypermedial educational tools can produce relevant changes of student cognitive abilities and of their knowledge representation.

A responsible use of the computer could lead to a positive effect on the subject; related to a suitable re-orienting of his own emotions. The affective-relational attitudes could become easily oriented to the group cooperation

Emergence in Complex, Cognitive, Social, and Biological Systems, Edited by Minati and Pessa, Kluwer Academic/Plenum Publishers, New York 2002

(Isenhour *et al.*, 2000; Olguin *et al.*, 2000) so that the single human behaviour be dependent on the group and vice versa, through action and feedback phenomena amongst persons. Of course, the possibility of this action is strictly dependent on the tool perception (typically the computer).

Starting from these premises, this contribution tries to experimentally analyse the relation connecting tool's perception and its use, as regards subject's cognitive responses. In particular we investigated children and teacher perception of PC when the latter is used as a didactic tool.

We assume that *subject expectation* determines the tool *perception* and influences his *learning* process. The expression "subject expectation" denotes what users expect relatively to the benefits deriving from computer use; whereas "tool perception" denotes the degree of positivity the subject associates to an approach based on the PC. In case of positive perception of PC, we should find significant differences between traditional and computer-supported lessons; learning, attention, understanding ability, and so on, should sensibly improve as a result of computer use.

2. THE RESEARCH

We investigated both teachers and students of an elementary school of Monserrato (Cagliari), that includes two different educational units. The difference between the unit A and the unit B is that the former has a computer laboratory, whereas the latter lacks any computing facility.

The subject sample was constituted by 306 children (45% male and 55% female) and 26 teachers, all female. 51% of children was belonging to the unit A and 49% to the unit B. However, the subjects were in four different grades (II, III, IV, and V elementary grade).

We built two different questionnaires, one for the students and the other for the teachers; their formulation was based on a previous work by Russell (1999). The areas dealt with by the questionnaires were the following:

Students questionnaire:

1. Approaching the computer
2. Effective and possible computer use at school.

Teachers questionnaire:

1. Approaching the computer
2. Effective and possible computer use as a didactic tool
3. The opinion about some features of student mutual cooperation, and of the cooperation between students and teacher.

2.1 Results of student questionnaire

From an analysis of the answers to the items of the questionnaire we found that 68% of children made use of computer at least one time at week, and the majority of them used PC especially for playing (61%).

76,5% asserted that they were able to use a computer, while 47% declared they were highly skilful.

Moreover, we obtained that the use of computer at school was or should be pleasant and fun for 88% of students, while, between the subjects unable to use a PC, 70% of them would learn willingly to do it.

The use of computer for writing and attending to lessons was preferred by 80,7%; 80,1% would use it within all scholastic disciplines, and still 80,1% would make homeworks using the PC.

When searching for the existence of suitable relationships between some variables, in order to detect subject cognitive features, we obtained that children that *should like to use the computer at school* asserted they better understood the lessons in which the teacher used a computer ($\chi^2 = 25$; df = 4; $p \leq 0{,}001$), and they had a clearer vision of the treated topics ($\chi^2 = 25$; df = 4; $p \leq 0{,}001$).

The students *feeling they were able to use a computer* asserted they better understood the lessons when the teacher used a PC ($\chi^2 = 9{,}25$; df = 2; $p \leq 0{,}05$) and the explained arguments were more clear ($\chi^2 = 6$; df = 2; $p \leq 0{,}05$).

The subjects asserting *they were highly skilful in using a computer* argued too they better understood the lessons ($\chi^2 = 21$; df = 4; $p \leq 0{,}001$) and considered more clear the arguments with the use of PC ($\chi^2 = 14$; df = 4; $p \leq 0{,}001$); they asserted, moreover, they better remembered the lessons ($\chi^2 = 10$; df = 4; $p \leq 0{,}05$) and they were more attentive ($\chi^2 = 29$; df = 4; $p \leq 0{,}001$).

Finally, the student that *declared they knew a word processing* platform affirmed they better understood the lessons ($\chi^2 = 23{,}4$; df = 4; $p \leq 0{,}001$), they considered as more clear the arguments explained with the computer ($\chi^2 = 26{,}2$; df = 4; $p \leq 0{,}001$), they better remembered the lessons ($\chi^2 = 16$; df = 4; $p \leq 0{,}01$) and they were more attentive during explanations in which PC was used ($\chi^2 = 13{,}7$; df = 4; $p \leq 0{,}01$).

As regards the comparison between the two units A and B, it appears that the children belonging to the unit A (presence of a computing laboratory) better understood the lessons ($\chi^2 = 31$; df = 2; $p \leq 0{,}001$), considered as more clear the arguments explained with the aid of a PC ($\chi^2 = 19{,}7$; df = 2; $p \leq 0{,}001$), better remembered them ($\chi^2 = 23{,}9$; df = 2; $p \leq 0{,}001$), and devoted to them more attention ($\chi^2 = 19{,}3$; df = 2; $p \leq 0{,}001$).

3. DISCUSSION

The above results showed that there exists a positive expectation as to regard the computer use at school, as for students as for teachers, and we have not find significant negative expectation to confirm that this one could determine the refuse of the tool. On the contrary, the data allow us to assert that the subjects that cannot use the computer at home, at school or elsewhere, have developed positive expectations. These positive perceptions seemed to be more evident in children who used the computing laboratory, the ones that had more confidence with the PC. But we must interpret this result by taking into account the presence of a possible noise variable, that is the "tool novelty".

Why, in fact, we found an unanimous opinion in a positive direction? Why the use of this tool should produce enthusiasm between both students and teachers? Relatively to the children the association between computer and entertainment is intuitive, and so we explained the reason for the observed positive consensus. But this "association" should not be considered as negative in the case of a functional management: a good way to allow a child to learn to use a new tool is to let him play with it!

For teachers, the "tool novelty" variable could create an "halo" effect on the results; this is easily understandable if we take into account that the computer has several advantages with respect to the traditional learning tools, first of all the flexibility.

In fact, the data we collected seem to suggest that the use of a computer could simplify the lesson course by supporting the learning, the understanding and the memory recall of information. But we must be prudent with this interpretation too, because any new tool, or a tool that is sensibly different from traditional ones, could lead to an increase of motivation that derive from the use of the new tool, a circumstance which could explain a better understanding of arguments, an increase of activation condition and, consequently, of memory.

4. CONCLUSION

In spite of the presence of the noisy variable, we have demonstrate the presence of a positive perception of the tool analysed. Therefore, we can assume that the PC puts the users in a favourable condition, on the cognitive side, and such a circumstance could influence the learning, in conformity with the initial hypothesis.

As we have seen, some authors feel that this feature justifies an increase of the presence of this tool within educational setting, whereas others think that the insertion of PC within the school could produce a worsening of didactics. Our position is for a mediation of both opinions. We will assume that PC use could improve didactics but, in order to give rise to such an effect, it would be necessary to train in a suitable way the teachers in order to let them use this tool by exploiting its full potentiality.

Our hypothesis is that the introduction of computer within the school - with the goal of improving the didactics - will be effective only if the teacher will be able to use the computer in an appropriate and functional manner, and this will be possible only in the case of a good predisposition. A tool like the computer could be effective only if properly inserted within school context. Doing so, the full potentiality of the PC could be used to support the didactics. Therefore this new technology will ever improve the quality of life and work, but only if properly used; in the contrary case, its introduction will produce strong limitations in the educational activities.

REFERENCES

Benussi, I., and Voltolina, A., 1999, *Multimedialità E Processi Cognitivi: Una Nuova Teoria Dell'istruzione.* http://www.iwn.it/is.

Bringelson L. S., and Carey, T., 2000, Different (Key) Strokes For Different Folks: Designing Online Venues For Professional. *Educational Technology & Society* **3**(3).

Cornoldi, C., 1995, *Metacognizione E Apprendimento.* Il Mulino, Bologna.

Gasparetti, M., 1997, *Il Computer In Classe.* Garamond, Roma.

Hartley, R., 2000, On-Line Collaborative Learning Environment. *Educational Technology & Society* **3**(3).

Henrickson, L., 2000, Communication Technology And Personal Identify Formation. *Educational Technology & Society* **3**(3).

Hill, W. F., 2000, *L'apprendimento.* Bologna: Zanichelli.

Isenhour, P. L., Carroll, J. M., Neale, D. C., Rosson, B., and Dunlap, D. R., 2000, The Virtual School: An Integrated Collaborative Environment For The Classroom. *Educational Technology & Society* **3**(3).

Kreutz, R., Kiesow, S., and Spitzer, K., 2000, Netchat: Comunication And Collaboration Via Www. *Educational Technology & Society* **3**(3).

Le Louche, R., and Seguran, M., 2000, Can Abstraction Be Used As A Unifying Guideline To Design Intelligent Educational System? *Educational Technology & Society* **3**(3).

Lian, A., 2000, Knowledge Transfer And Technology In Education : Toward A Complete Learning Environment. *Educational Technology & Society* **3**(3).

Milton, J., 2000, Virlan: Collaborative Foreign Language Learning On The Internet For The Primary Age Children: Problems And A Solution. *Educational Technology & Society* **3**(3).

Olguin, C. J., Delgado, A. L. N., and Ricarte, I. L. M., 2000, An Agent Infrastructure To Set Collaborative Environment. *Educational Technology & Society* **3**(3).

Russel, M., 1999, *Testing On Computer: A Follow Up Study Comparing Performance On Computer And On Paper.* http://www.epaa.asu.edu/v7n20/.

Von Bertalanffy, L., 1968) *General system theory. Essay on its foundation and development.* Braziller, New York, (it. trans. *La teoria generale dei sistemi.* Mondadori, Milano).

A System Approach to the Educational Research: Reflection on an Experience

ADALBERTO CODETTA RAITERI, and RENZA CAMBINI
OPPI, Via Orseolo 1, 10144 Milano, Italy
E-mail: codetta@tin.it - renza.cambini@formain.it

1. THE RESEARCH ON ZERO AND COMMON SENSE

1.1 The reasons of the research

The research sprang from the idea that all students build up opinions, concepts, naive theories, meanings and explanations about the different aspects of their school experience and this forms the bases for interpreting, receiving and organising new knowledge. On this hypothesis we elaborated a research method to put *the same questions* to students from every level of school and to some adults, too. More precisely, we interviewed the students at the end of:

- *first cycle of primary school* (8 years-old)
- *second cycle of primary school* (11 years-old)
- secondary school (14 years-old)
- *biennium of secondary school* (16 years-old)
- *triennium of secondary school* (19 years-old)
- *some adults*.

Our study set out to analyse the psychological development of the relationship between some mathematical concepts and "common sense" notions in the different age groups. The set of questions encouraged answers referred not only to mathematical models but also to emotions and practical use. For this reason we produced a questionnaire open to a variety of responses expressed in various ways: pictures, cartoons, formal languages, stories .... In fact, in our opinion, "common sense" is connected not only to "the world of tables and chairs" but also to emotions. We want to emphasise that for pragmatic culture "common sense" has a generally positive meaning but for idealistic culture it has an ambiguous meaning connected with appearances ("common sense is deceptive!"). Both cultures, in any case, underestimate the importance of emotions in the construction of the common sense of a social community.

1.2 The questionnaire

The formulation of the questionnaire was based on the following underlying queries:

- how can we put the questions in a way that fits the theme of this meeting?
- how can we express the questions using a common language?
- how can we encourage the students to express themselves in the various ways mentioned above?
- how can we stimulate the students to use their internalised knowledge when answering the questionnaire?

The answers were suggested by research studies originating from cognitive sciences, post-Piaget cognitive psychology, constructivist and ecosystem epistemology. We put forward two solutions:

what does zero mean to you?

what difference would it make to you if there was no zero?

These questions were put in different contexts in order to evoke ideas and images linked to mathematical concepts. In fact we accept the metaphor of semantic net used in cognitve sciences. Memory is not a store or a library of our accumulated knowledge but involves the construction of a semantic net. We start from a concept which is the first node in a net linked to other concepts by the principle of affinity of types.

The general form of our questionnaire was suggested by Gardner's idea of a plurality of intelligence: only one form of which generally comes to the fore. Given the goals of our research we produced a questionnaire designed to stimulate the area of feelings and emotions of the interviewees, their consciousness about their own feelings and emotions, space and interpersonal intelligence. The language and structure of the questionnaire was suggested by the work of the Geneva post-Piaget school. Their studies on mental representations of structured scientific concepts show that, in order to verify the presence of a scientific concept in an individual's

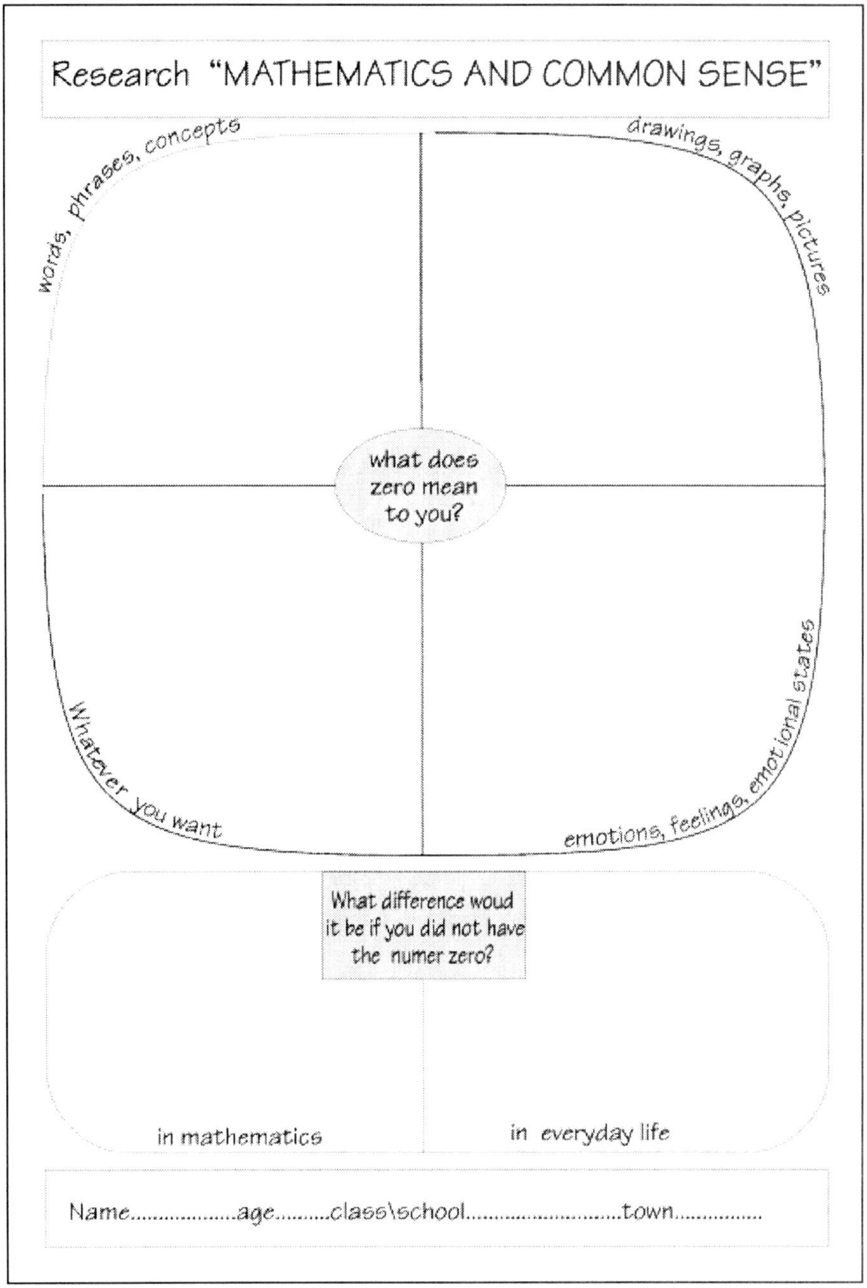

Figure 1.

semantic net, it is necessary to formulate a question using common sense language, related to his everyday experience. We solved this problem by formulating the questions in everyday language while not renouncing the

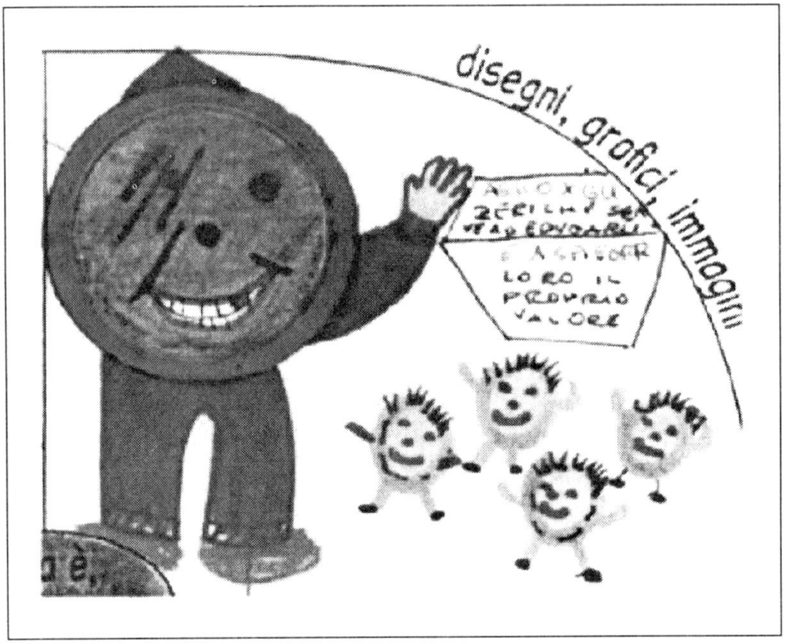

Figure 2. Nursery school for zeros which serves to teach and explain their value to them.

Figure 3. What does zero mean to you?

correct use of scientific terms related to the concept tested in the questionnaire. If the language used is too formal and specific there is a danger that the interviewees only give an automatic response which does not demonstrate the presence of the scientific concept in the semantic net. The questions about emotions were suggested by the work of epistemologists that analyse the emotional roots which form the basis of rational constructions.

1.3 Methodology and the difficulties of the research

Hundreds of teachers, from many regions of Italy, were involved in the research. They collaborated with OPPI, a teachers' associations involved in professional training. OPPI was interested in directly involving a large number of working groups and we chose not to adopt a formal statistical approach. There were two possible ways of administrating the questionnaire:

- decentralised supply and centralised analysis at the risk of systematic errors
- decentralised supply and decentralised analysis based on a common model with the risk of random errors.

We chose the second solution focusing on the series of operations shown in the figure 4. We collected 2463 forms from Aosta, Avellino, Bari, Lucca, Milano, Napoli, Novara, Palermo, Piacenza, Varese, Vercelli. We underestimated the wealth of ideas and inventiveness of students' answers. This made it difficul to produce coherent and non-subjective interpretation of answers . In fact the complexity of the questionnaire induces people to interpret the answers in different ways. The number zero was not only described with words and definitions more or less suitable to its mathematical significance but using metaphors, graphs and images which reveal the involvement of symbolic suggestion. The number zero was also linked with reality, particularly with its practical use, and with man's emotional sphere.

The answers have been surprising for the integration of the drawing and texts and for the knowledge and the richness of meanings in very young children like in the answer of figure 2, A. L., 8 years old. (Nursery school for zeros which serves to teach and explain their value to them),but also for the symbolic richness and the quality of the graphic elaboration, like in the answer of L. P., 19 years old. (figure 3). We had to decide how to take these consideration into account:

- *by classifying items generically with the risk of producing an ambiguous, imprecise classification,*
- *by classiyfing items too specifically with the risk of covering only a part of the range of answers.*

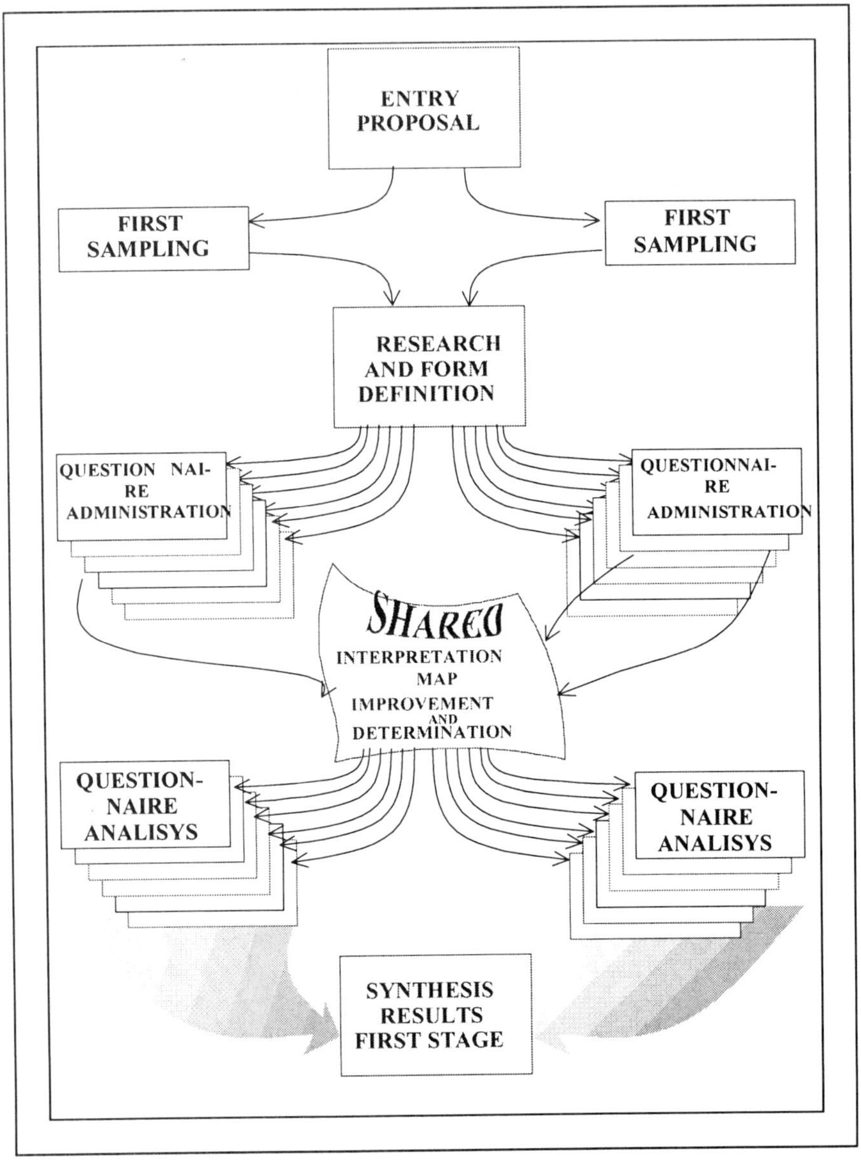

Figure 4.

We could have analysed the different languages or compared the mathematical meanings of zero with the subjective perception of the interviewees. The research group examined a sample of answers and decided to classify them into three categories which do not depend on the particular form of communication (words, images, graphs, ...) or on the particular question. The three macro-categories (mathematical models, emotions, practical use) were further subdivided into many subcategories. This let us

Table 1. Classification

A mathematical models	B emotions, suggestions, analogies	C every day life
A1 consciousness of numerical aspects; epistemological ideas	B1 deep emotions: death, sleep, peace, sadness, tedium	C1 zero referred to money
A2 cardinal number	B2 analogies referred to symbol's shape	C2 zero referred to competitions
A3 ordinal number	B3 suggestions from social language	C3 zero referred to time
A4 separator between R+ and R-	B4 emotions referred to school experiences	C4 zero referred to physical phenomenon
A5 origin of Cartesian axes		C5 zero referred to social considerations
A6 properties in the additions, multiplication and references with algebraic structures		
A7 attention to place value		
A8 references to infinitesimal and infinite		
A9 zero's other particular propriety		

compare conceptions of zero connected with mathematical meanings learned at school with pre-conceptions from the emotional sphere generally not completely controlled by the interviewees. Such pre-conceptions are the result of a cultural inheritance. With our choice of classification method, we attempted to investigate the complexity of a learner's cognitive net which is a product of rationality and emotivity and links between objectivity and subjectivity. We hoped to find some recurring motifs among the different age groups and people coming from different types of schools.

1.4 The quantitative analysis

With the quantitative method the frequency of every item was tallied for every school cycle and the research group compared the frequencies of the answers linked to emotional and practical meanings. Less than 50% of answers refer to mathematical concepts, about 40%, to emotions, about 10% to practical uses (figure 5). But the mathematical significance only emerges in the period in which the school syllabus provides for the teaching of it. Afterwards it gets confused with the others!

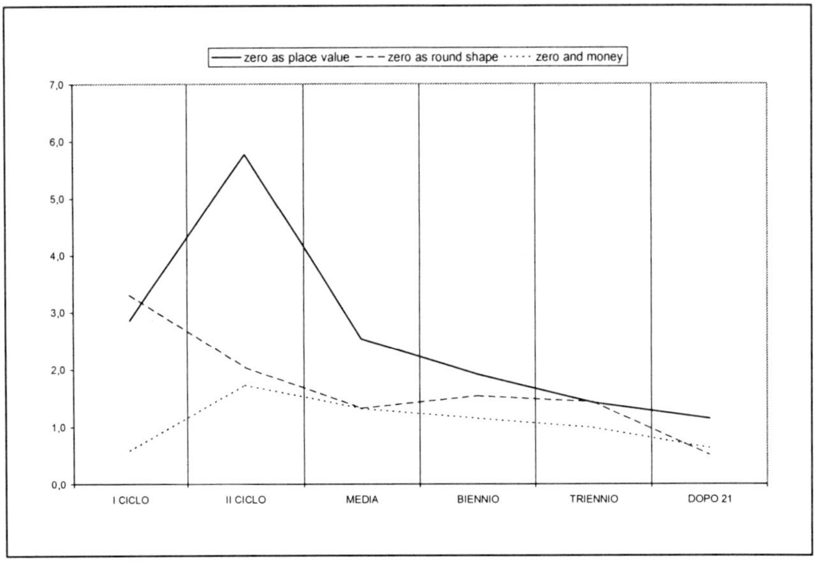

Figure 5.

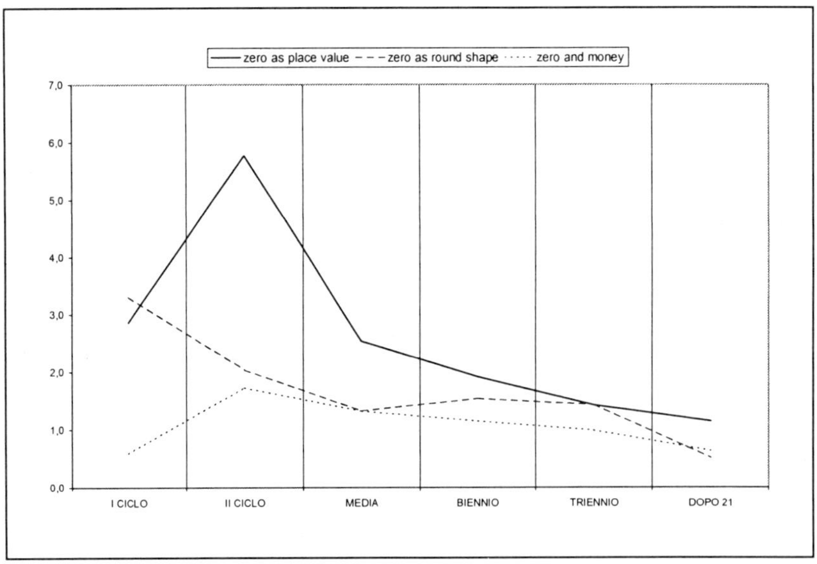

Figure 6.

The research proves that the principal mathematical meanings of the number zero, proposed by the school system (positional numeration, cardinality, ordinality, separation R^+ R^-, cartesian axes origin, additional neuter element, ...) live only temporarily with the more stable ones of emotional and practical origin. At the end only the emotional ones remain unchanged (emptiness, nothing, death, peace, delusion, round outline, self estimation.). It is very interesting to emphasize that a treasure trove of shapes, colours, and expressions present, demonstrate that the mathematical meanings proposed by the school system live only temporarily with the more stable ones of emotional and practical origin (figure 6). It is as if the mathematical concepts were in conflict with everything emotional and practical, as if emotions and practical life were an obstacle to disciplinary learning. The large number of answers which didn't concern mathematics were examined very carefully and it clearly emerged that the many drawings were answers expressed in graphic language: in many cases this was the main and sometimes the only means used.

1.5 Other methods of analysis of the answers

The research group formulated the hypothesis that the separation between the rational sphere and the emotional one, deeply rooted in the scholastic tradition is the cause of the temporariness of the learning emphasized from quantitative research. On this hypothesis the research group decided to utilise many other methods (quantitative analysis, comparison with different schools, graphical and linguistic interpretation, cognitive maps, analysis of classes with cognitive maps and quantitative method) which are explained in the report of the research. In this communication we will refer to the answer interpretations through the cognitive maps.

1.6 The analysis through cognitive maps

With this procedure the knowledge is represented by a net where the connecting lines are the attributes and the knots are the concepts. It means that every element is in relation with the others. In this representation of the knowledge a concept is defined by the whole of the relations which link it to the other concepts, none of which is primitive. The maps show that the students organise the disciplinary knowledge they learn about the zero during the school years according to the structure of their emotional spheres. There are many students who permanently link the zero to balance and harmony, while the concept of empty whole is forgotten and replaced by the positional system one.

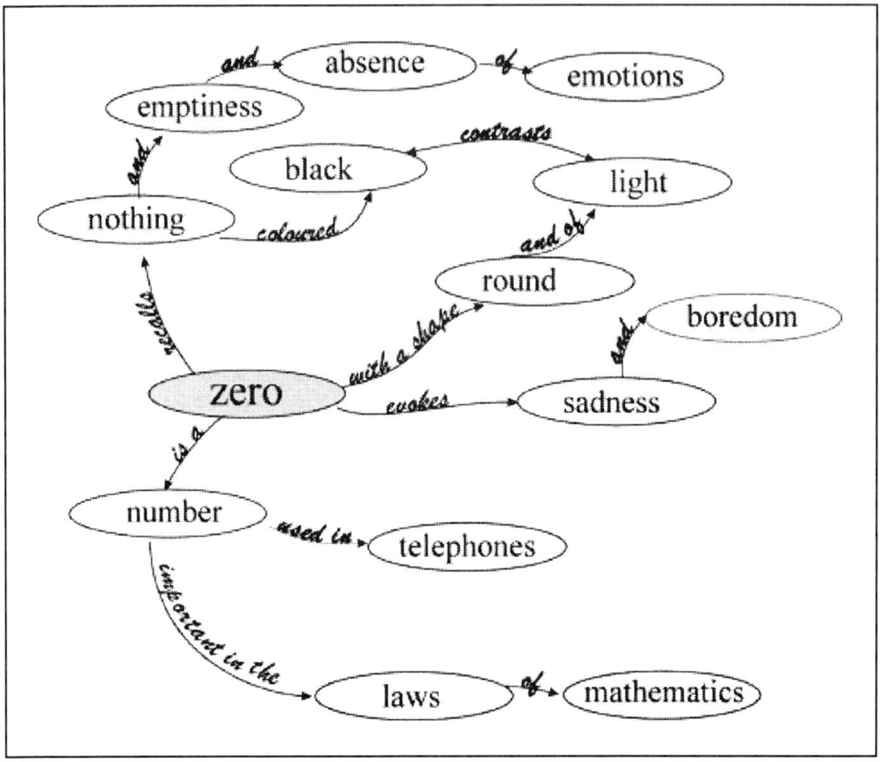

Figure 7. A cognitive map drawn from the answer of A. S., 15 years old.

For the students who link the number zero to sadness and melancholy, the concept of empty whole seems more permanent because linked to … sadness. In many maps of this kind of students we find that the links found in the 8 year old student maps are still present! We observe that the emotional sphere seems so strong as to persist continuously from primary school through to the school leaving examination, while the learning connected to the meanings of zero is not able to become a stable structure which grows, is enriched and reorganised.

None of the mathematical meanings learnt at school connects to others to assume stability that corresponds in some way to the structure of the discipline. From the analysis of the maps an incapacity of the school to build stable and organized knowledge which allows the pupils to have a conscious reflection on what they are learning emerges. The contents alternate, substituting one thing with another without activating cognitive processes which can lead to the construction of conceptual maps representative of the disciplines.

1.7 From the cognitive maps to the conceptual maps

The results of this method suggest refining the definition of cognitive maps:

A cognitive map is a net of meanings that represents the subjective organization of the ideas, concepts, emotions that everybody builds around one or more key-words.

But the teachers, specially those of the secondary school and in particular the mathematics teachers, are not interested in the "subjective organization of the ideas". Their teaching is directed at bringing the students to the learning of codified knowledge which is the temporary result of the research of the Scientific Communities; so they are interested in the building of conceptual maps of which the research group gives the next definition.

Conceptual map is a net of concepts shared by a Scientific Community or by a working community in which the meanings expressed in the knots and the links between the knots find a relation in one or more disciplinary fields.

The research 'zero and common sense' shows that it is not possible to build conceptual maps ignoring the cognitive ones because they are the beginning on which the conceptual ones have been based. The teachers, reflecting on the student maps, will work out the cognitive maps that show the knowledge organization in every school-class. The student's cognitive maps will be a reference to plan itineraries of teaching-learning. The subjective organization of the ideas, will reorganize itself through appropriate stimulus to the reflection about one's knowledge, the comparison with class-pupil's ones in appropriate group-work and by giving meaning to the codified forms of knowledge. In this way the students will lead a research somewhat similar to the Scientific Community ones which have built the same knowledge

1.8 From the research it's clear that:

1. The deeply-rooted separation between the rational and the emotional sphere in mathematics teaching is the cause of the temporariness of its learning. The pupils can give meaning and stability to the acquired structures of rational knowledge if they can connect them to their gushing " fountain "of emotions and experience.
2. A didactic strategy aimed at stimulating a conscious reflection in the pupils, on the continuous grouping and reorganising of knowledge in their minds, can lead to the construction of stable disciplinary learning.
3. Stable learning cannot happen only through the separation and distinction of concepts. Once a concept is acquired by separating and distinguishing it from others, it is necessary to dedicate just as much time on seeing how

it is connected to previous ones and, if necessary, reorganizing the structure.

4. Conceptual maps representing the disciplines can be very useful in controlling the process of learning from the viewpoint of curricular continuity.

2. REFLECTIONS ON EDUCATIONAL RESEARCH

2.1 A map of an educational system

In July 1995 this research was presented in Berlin at the 47 CIEAEM (International Commission for the study and Improvement of Mathematics Teaching) Congress. The following autumn the group concluded the work. The research had reached its objectives. Some small groups have contined this activity in the following years going more deeply into numerous other aspects, above all from the cognitive point of view. The most ardous work remained: to make known, legible and interpretable, that rich experience to a much wider public than the people directly involved: influence the processes of teaching-learning. That reflects a well-Known limit, typical of research-action. The participants live a "strong" professional experience which allows important revision and reorganisation of their convictions and behaviour. It is, however, very difficult to make a wider public participate in the conclusions of this experience. Anyone who is both the subject and the object of an investigation at the some time can't describe the process of transformation induced in him or herself by the investigation he or she is conducting. In our case it has taken five years of re-elaboration to make this experience legible for a wider public.

"*I remember a personal experience very well which goes back to the period of the national didactic centres on orthographical errors. It was a well done experimental research but with the result that the errors continued to proliferate. The teaching principle of orthographics has not been influenced by the research in this sector*" (Scurati 1993). In this way Professor Scurati efficiently summed up the central problem of educational systems: what effect does educational research have on the every-day teaching-learning which regards millions of students and teachers? Certainly having a deeply self-referenced object, educational research can't think of transferring the interactive models present between the scientific research and the productive world. The risks are clear that one runs in trying to normalise the "happy Babel" of models and languages which stop efficient interaction between researchers and teachers.

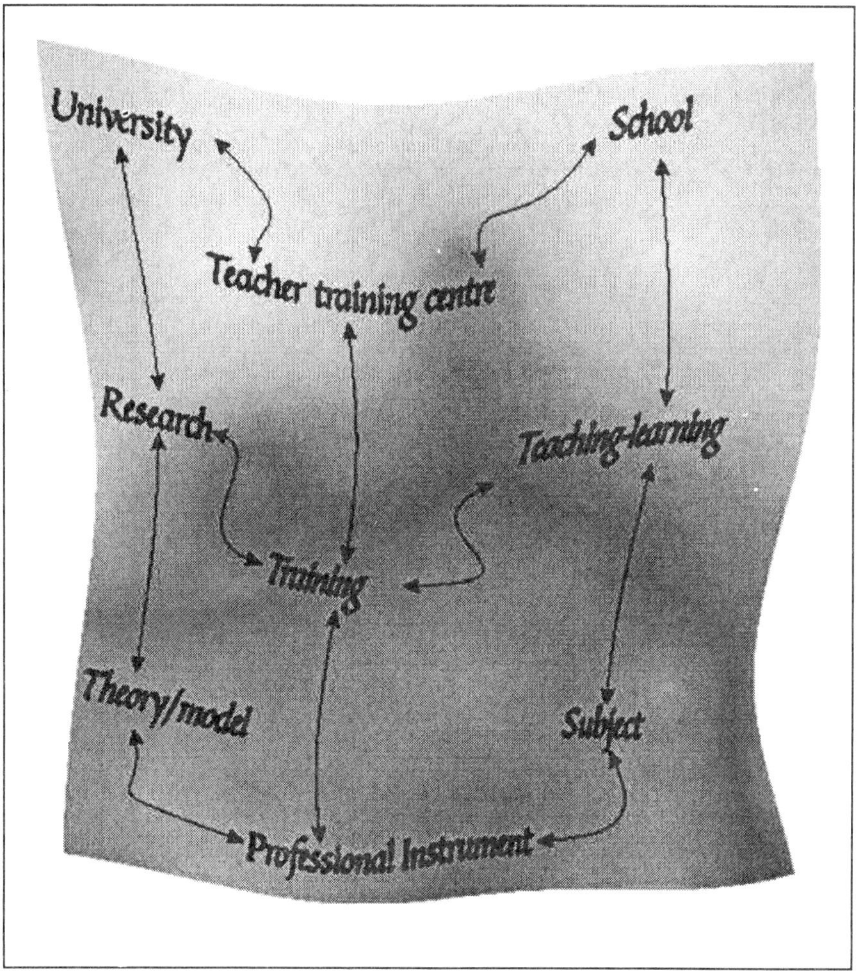

Figure 8.

Above all, educational systems need ample space which is not formalised where youngsters can test themselves with the most strategic factor, creativity. However the epistemological weakness of educational research is born from the fact that there is no agreement on the places and procedures for affirming the validity of an educational theory. Using a system approach, numerous agents can be identified which contribute to the construction of educational practice without forgetting the social context. It may be useful, only to focus some aspects, to examine each isolated component of the system described in figure 8.

UNIVERSITY ⇔ RESEARCH ⇔ THEORIES/MODELS

Let's take into considerations some examples of scientific theories which became known in the years 1920 and '30 and which have profoundly changed previous concepts.

K. Godel	Logic	Undecidable propositions of formal mathematical systems
W. Heisenberg	Physic	Uncertainty principle
L. Bertalanffy	Biology	System theory
J. Piaget	Psychology	Genetic epistemology
Prague Linguistic Circle	Linguistics	Structural Phonology

It's interesting to observe that these five scientific theories were elaborated in the same ten years but they influenced didactic activities in schools at very different times. Some of them have never arrived in educational institutions. This leads one to think that each discipline has its own time for the spreading of its canonical models. Those of mathematics are among the slowest: the number zero took half a millennium in Europe to overcome the Roman numeration.

An important work for the affirmation and diffusion of canonical models of the school disciplines can be developed, in an era of great cultural and scientific innovations by centres of formation and service for teachers. These have an interface role between university research and the teaching-learning processes carried out in schools.

RESEARCH ⇔ TRAINING ⇔ TEACHING-LEARNING

The need for an interface between research and teaching-learning comes out in the following considerations:

- The language of theoretical research has a mastery limited to a very narrow scientific community
- The formulation of models doesn't take into account the motivations of those who do not belong to the research community
- Cognitive psychology reminds us that learning doesn't take place without emotion and therefore without motivation
- The motivations of a scientific community are evidently different from those of the teachers and learning students

All this confirms the need for teacher training centres which know how to link scientific logic with affective motivations. Of course, the universities themselves, as already happens in many cases, can play a role in the teacher training centres; anyway, it's important to observe that in this case they play a completely different role from that of a research centre.

TEACHER TRAINING CENTRE ⇔ TRAINING ⇔ PROFESSIONAL INSTRUMENTS

A formation centre is interested in canonical forms of a large number of theoretical models which come from various disciplines.

It must use two models: a disciplinary theory and a learning theory. In reality teacher training centres are places of sincretism of numerous and different theories and of sequence operative definition (training) capable motivating teachers towards the study of disciplinary and learning theories. In this way, teachers are supplied with cultural, professional and organisational instruments to define re-elaborations which are generally very different (above all, for the language), from the original theoretical models.

SCHOLASTIC INSTITUTES ⇔ TEACHING-LEARNING ⇔ SUBJECTS

In scholastic institutes a large number of disciplinary models and theories of learning live side by side stratified one on another. The students' learning comes not only through the action of the teacher but above all, through the exchange between teacher and pupil. In this context the teacher must regulate the cognitive, affective and metacognitive processes. The belief is gaining credit that the process of teaching-learning must be controlled through a system of evaluation which does not only concern the students but also the teachers and the scholastic institute itself.

In the evaluation, attention will move from the results of learning (getting through tests and exams) to the process of teaching-learning which generated the learning.

2.2 Elements for evaluating educational research.

The impressive social, technological and epistemological transformation we are living through impose a much stronger interaction on all the components of formative systems than in the past. On the basis of this hypothesis, the quality of an educational research must be evaluated above all on the interaction developed between the various components of the formative system.

It's not the authors'intention to proceed to a self evaluation of their research, but it may still be useful to examine certain significant aspects in the light of the proposed model:

- This research was born through a proposal of an international educational research organisation (CIAEM) and on the strong impact of constructive theories formulated by H. Maturana
- It established formal agreements with the Directors of numerous Italian schools to start the research action, by the same means it conducted the research in an informal way through the teachers.
- More than one hundred teachers were involved in carrying it out and 35 in the analysis of the questionnaire

Figure 9.

Because at the time of the research this model of evaluation had not been refined and we were understandably worried about bringing an extremely complex activity to a conclusion, the documentation of the complicated network of relations which developed has been very much reduced because it was finalised prevalently to the production of a result. Vice versa the richness of the relations which mature during an educational research constitutes on its own a precious result which needs to be accurately documented because it represents the qualitative element. This work has brought the authors to the conclusion that every educational research must also be research-action. Naturally the risk of exclusively self referenced research exists in which the external product becomes insignificant in comparison with the attention given to the relations between the actors in the research. Self-reference, even if it can't be eliminated, must be controlled by a wide route through problems outside each of the actors involved. An educational research is like a journey between relations and problems at the end of which all the agents involved have been transformed.

As in Esher's picture "Gallery of printings" (figure 9) commented by D.R. Hofstader (1979) "*An extraordinarily beautiful illustration of Entangled Hierarchy. We can see a gallery in which a young man, standing, is looking at a picture which shows a ship in the port of a small city, perhaps Maltese, judging by the architecture with its little towers, domes every now and then and flat stone, roofs on one of which a boy is sitting resting from the heat while two floors below a woman, perhaps his mother, is standing at the window of her flat which is exactly above a gallery in which a young man, standing, is looking at a picture of a port in a small city, perhaps Maltese ...*". In plain terms an educational research is born through questions teachers, in relation with students, put in order to rule the learning processes which are becoming more and more complex and demanding in the new cultural and technological context. A good question can't always find an immediate answer but imposes a research route which involves - transforms a lot of actors in the educational system and makes them return with appropriate strength to transform the processes of teaching learning which take place in the same classes it started from.

REFERENCES

Albertazzi, P., Regis, A., and Roletto, E., 1993, Capire che cosa capiscono gli studenti: le mappe concettuali. *Didattica delle scienze* **164**, Febbraio 1993.

Barbieri, G., Bernardini, P., and Gagliardi, R., 1988, Analyse des conceptions sur le rapport homme environment des élève des licèes de la Lombardie. *Journées Internationales sur l'education scientifique*, Actes XIII.

Barbieri, G., Bernardini, P., and Gagliardi, R., 1988, Le rappresentations en ecologie: un mojen pour aborder l' interdisciplinarité dans l'education à l'environment. *Journées Internationales sur l'education scientifique*, Actes X.

Chevalier, J., and Gheerbrandt, A., 1982, *Dictionnair des symboles*. Juppiter, Paris.

CIEAEM 47, 1996, Mathematics (education) and common sense. In *CIEAEM 47 proccedings*, Freie Universitat, Berlin.

Codetta, A., and Peroni, F., Recherche-action pour amener les enseignants a l'utilisation des tests dans l'evolution des processes d'einsegnement-apprentissage.Considérations sur cinq ans d'activité: un modéle. *CIEAEM 50 proceedeings*.

Codetta, A, Cappucci, and G., Cazzaniga, G., 2001, *Lo zero e il senso comune*, Adelphi, Roma.

Damasio, A. R., 1995, *L'errore di Cartesio, emozione ragione e cervello umano*. Adelphi, Milano.

Freudental, H., 1980, The great problem of mathematic education. *Proceeding of ICMI 4*.

Gardner, H., 1983, *Frames of mind: the teory of multiple intelligence*. Basic Book, N.Y.

Gardner, H., 1982, *Il bambino come artista*. Anabasi, Milano.

Hofstader, D. R., 1984, *Godel, Escher, Bach*. Adelphi, Milano.

Ifrah, G., 1983, *Storia universale dei numeri*. Mondadori, Milano.

Liber, M., and Sesenna, G., 1988, *Ritrovare il sapere*. O.P.P.I. Milano.
Maturana, H., 1992, *Linguaggio ed emozioni*. In For, Milano.
Maturana, H., and Varela, F., 1987, *L'albero della conoscenza*. Garzanti, Milano.
Novak, J. D., and Gowin, D. B., 1983, *Learning how to learn*. Combridge University Press.
OPPIDOCUMENTI, 1995, Pensare e formare, epistemologie a confronto. *Strumenti per la formazione*. Anno XVIII, n. 68, Milano
OPPIDOCUMENTI, 1996, Il pensiero dov'è. *Atti del seminario con Humberto Maturana*. Anno XIX n.75-76, Milano
Pellerey, M., 1986, Psicologia cognitiva e attività didattica. *Orientamenti Pedagogici*. 6.
Pellerey, M., 1987, Psicologia cognitiva e attività didattica. *Orientamenti Pedagogici*. 1.
Rotman, R., 1988, *Semeiotica dello zero*. Spirali, Milano.
Scurati, C., 1993, *La ricerca azione*. Bollati Boringhieri, Torino.
Varela, F., 1989, Il circolo creativo: Abbozzo di una storia naturale della circolarità. In *La realtà inventata. Contributi al costruttivismo* (P. Watzlawick, ed.), Feltrinelli, Milano.

Evaluating an Educational Course

A Constructivist Approach

RENZA CAMBINI, and ADALBERTO CODETTA RAITERI
OPPI, Organizzazione per la Preparazione Professionale degli Insegnanti
Via Orseolo 1, 20144 Milano
E-mail: oppi@oppi.mi.it

1. INTRODUCTION

Educational systems are receiving special attention by governments and social designers in all the so called advanced countries. Education is more and more emerging as a strategic resource in the international economic scenario.

The need to professionally train the young generations in order to make them able to compete in the global market asks for having goal oriented, comparable and measurable educational activities.

English speaking countries particularly have long experiences in evaluating by using procedures based on tests. They also experienced procedures to evacuate single schools and educational systems. Numerical rankings produced by such educational testing procedures (also called "test machineries") have great influence in decision making activities because of their role of objective synthesis. This aspect reinforce the importance of such an approach.

The international usage of educational testing procedures may have a role in making generally accepted, if not standardized, educational systems approaches similarly to what is already established in the international business community (quality, industrial standards, information to the user, trade procedures, etc.). Quality certification in education has difficulties to take off: it is intended not as an opportunity, but as an unavoidable procedure, for instance applying for founding (quality certification is one of the items setting the global evaluation when submitting requests to the European Social Found). Besides the need to make students able to change course of studies and kind of school is a strong stimulus to validate different educational systems thanks to equivalence among different educational qualifications.

The adoption of evaluation systems from other cultural contexts without any adapting processing has the risk to weak or destroy cultural specific properties able to make different communities able to compete and cooperate in the global market. Therefore the attention to cultural identities is matter of effectiveness and quality for the evaluator.

The selection of questions and topics, the way of wording, unavoidably related to cultural, racial, religious, sexual issues privilege particular aspects of some of the many ways in which the human intelligence is expressed.

This consideration introduces the awareness that any kind of evaluation, even trough testing procedures, does not assume the characteristic of objective measurement. Moreover even natural sciences (physics first of all) are not based anymore on the objectivity principle only. In the case of education any evaluation has meaning in a particular context, locally defined by relational processes established by the process of evaluating.

1.1 The operations-based approach and education

Translation of concepts always had a very important role in sciences. The translation of the operation-based approach in physics (i.e. quantities are defined by the operations allowing to measure them) to psychology of learning is a good example of the effectiveness and of the risks combined with the process of translating concepts and approaches from one discipline to another. The success of the approach mentioned above in Physics convinced other disciplines to adopt similar approach. By linearly assuming this approach in psychology, Binet (1917) said at the beginning of the '900 that "intelligence is what is measured by my test".

In measuring students knowledge there are two classical approaches:

- The reduction process to a common scale focusing on balancing evaluation tools. The measurement of the performances is based on variables statistically identified;
- The evaluation of the scores in testing individual performance is based on considering he global result. Researchers focus on the global balancing of the different tests.

The second approach became the standard one in education.

1.2 The state of the art in Italy

In order to really understand the Italian situation we must make reference to the fact that "freedom to teach" is decreed by the Constitution. This makes the teacher a very peculiar worker, having individual characteristics and independence of the system in which s/he is working.

In Italy testing procedures are not very much practiced. In order to promote the usage of testing procedures in Italian school it has been established in 1997 the ADAS (Archivio Docimologico per l'Autovalutazione della Scuola)[1]. This Institute has the purpose to make available to schools information about this approach. Schools may then select the more appropriate tools and procedures.

The continuous process of maintaining and keeping up to date the ADAS is based on the collaboration of the CEDE (Centro Europeo dell'Educazione), Department of Electronic Engineering of the Rome University, the general management of the Minister of Education, supervisors, directors, deans, teachers.

In order to transform such a documentation in real practice in schools it is important to introduce and manage process of changing and to make available suitable tools and supports.

1.3 A training course proposed by OPPI

To deal with the just mentioned problem OPPI set up a well-tested and many times readjusted course, also focused on the usage of information systems solutions. This course has the purpose to make available supports to design, to give multiple answer tests, able to manage interactions generated by the evaluation process itself. Specifically, the course has the purpose to make the teacher able to adjust procedures based on testing with procedures used in evaluating curricula. Objectives of the course are:

- To introduce testing procedures in the framework of the general approach based on models for evaluation,
- To introduce and discuss techniques to design tests,
- To introduce and discuss techniques for evaluation, specifically information systems able to:
  - Get information from the whole set of the single answers, suggesting corrective actions or general and not only contextual reflections;
  - Evacuate the suitability of the educational approach used;
  - Evacuate the suitability of the adopted testing procedures;
  - Experiment the model in some classrooms.

[1] http://www.cede.it/adas

During the course both students and teacher use a software program to facilitate the tabulation of testing results. This software has been carried out by using standards tools and may be easily adjusted and modified. This software generates statistics about testing results. The software also automatically generates different kinds of graphics with reference to results regarding the classroom, the single student and the single item.

The software, carried out by a component of the OPPI's equip Umberto Perini is made available to the users at the end of the course to facilitate them in the test's practice.

2. EXPERIENCE EVALUATION

In the course held at the end of August 2001 an external evaluator has been required to attend, evaluating the consistency of purposes and activities. It has also been asked for collecting the evaluations expressed by the attendants.

- The course introduced some very interesting issues:
- The issue of globalization and the mirrored one of localization asking to focus on its own cultural identity in order to do not pervert its own roots.
- The crucial role of information systems in the adopted models, in order to give meaningful feedbacks to teachers, classrooms and single students.
- The recursive aspect due to the evaluation process of a course on evaluation, making an action object of the course itself.
- The concept of external observer on a topic (testing) having great influence in decision making because of its "objectivity".
- The name "machinery", introduced in literature, inducing reflections on the dialectics between mechanism/system, machine/ system. In education and specifically in evaluating such reflections focus on the usage of feedback as tool to design-redesign the process of teaching and not only to control it.

2.1 A proposal

In order to effectively cope with such a difficult task it is important to establish a framework. The Schema in table 1 has been introduced in literature (Peroni, 1997).

In the light of the synthesis introduced by table 1. it is possible to consider the course as a way to introduce and invite teachers to adopt the testing approach combining educational testing procedures with diagnostic aspects typical of the curricula evaluation.

Table 1. Evaluating in the Senior High School

	Tradition	Psychometric Testing	Programmed Instruction	Clinic Theory	Curricula Approach	Constructi-vism
REASONS (WHY)	to select	to careful measuring	to reach the goals	to express meanings	To appreciate a process	to reinforce global abilities
ACTORS (WHO)	teachers	technicians (also teachers)	processors of instructions packages	contracting parties	all the participants	all the negotiators in the context
TIMING (WHEN)	terminals	terminals	pre-established sequences	(before) during (after)	before during after	before during after
TOOLS (HOW)	marks and evaluations	statistical scores	successful detection by results	verbal expressions	integrated set of tools	integrated set of tools and subjects
OBJECT (WHAT)	results	knowledge	performance	the "being" the person	the learning process	the process of changing of subject to be studied
SYSTEM (WHICH BASIS)	model, deviation and range	testing	setting	conversation	criterion	the negotiated sharing
DIRECTION (TOWARDS WHERE)	acquiring codified knowledge	confidence in the acquired knowledge	functionality and effectiveness of behavior	awareness and self esteem	knowledge know how know how to be	autopoietic inclusion in the context

Italian teachers are still strongly adopting traditional approaches and use in different ways methodologies and tools interesting all the items of table 1.

The course focused on the columns dedicated on procedures for testing evaluation (No. 2) and curricula evaluation (No. 5).

The approach introduced, based on methods and tools, has the purpose to provide the attendants with designing, interpreting and analyzing test procedures (control feedback) abilities in order to allow teachers, the classroom, the single student to establish second level Creative (and not only controlling) loops (fig. 1).

The external evaluator decided to prepare tools and supports to carry out the approach indicated in the last column of table 1, the constructivist approach. The goal was to introduce new epistemological issues (fig. 2).

The idea is that each approach changes nature and borders if considered from the perspective of another one. The different approaches must be considered not alternative, but as a system of solutions to be used depending on contents, objectives, timing, contexts, in which the evaluation process takes place.

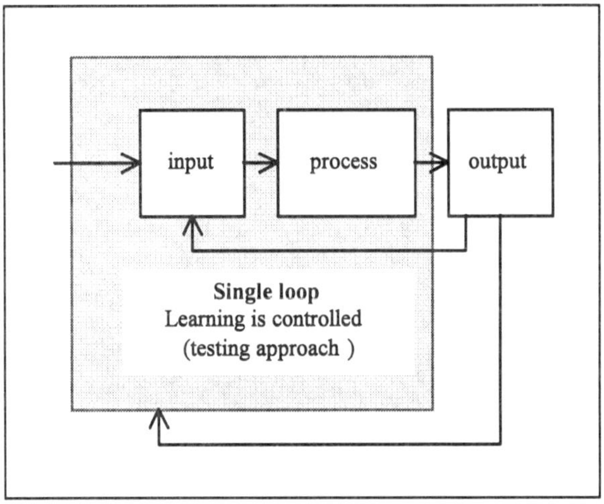

Figure 1. Dooble loop (Minati, 1998) System variables are changed, the learning/taching process to redesigned (curricola approach).

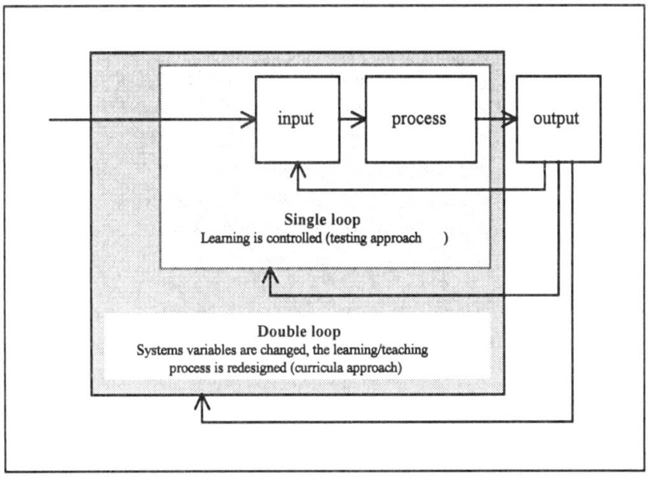

Figure 2. Higher loop. System variables are changed, le lesrning/teaching process is redesigned, renegotiating it in the interaction between different actor (constructivist approach).

2.2 Methodology and tools used to evaluate course's results.

The program of the course, articulated in two days, 8 hours per day, was based on elaborating information, on team work, on lab experiences by using information systems both to process test results and to use Internet searching programs looking for documentations and results on testing procedures.

Some of those activities gave to the evaluator the opportunity, keeping the role of observer, to find room to carry out feedbacks for single micro-evaluation (to be proposed in real time to the attendants).

Two tools have been used:

a) related to expectances (first day), in two events:
   1. at the beginning of the first day each attendant was asked to write an expectance related to the course going to start;
   2. at the end of the first day each attendant reread and comments his/her expectance
b) related to learning (second day)
   1. each attendant is expected to write three learning, meaningful aspects of the course.
   2. the realized learning are explicated, communicated and processed.

2.3 Analyzing expectances (A)

2.3.1 Collecting and organizing the expectances expressed by attendants.

In table 2 the expectances expressed by attendants have been listed and organized corresponding to doubts, perplexities and questions as presented by attendants in introducing themselves.

Table 2. The expectances

Form No.1 – listing expectances expressed by attendants	
Doubts, perplexities and questions	**Expectances**
It is just the latest fashion?	To correspond to an individual professional need
Can a structured trial evaluate competences? If it is so, which competences?	To correspond to a professional need, not only individual but institutional also
Because the testing approach is more suitable for adults, how to use it with children?	
Which gradualness between extreme subjectivity and extreme objectivity to be used in evaluating performances and competences applicable to social systems?	To learn how to carry out methods for evaluation and self evaluation.
	To look for and find out new evaluation criteria to share with colleagues
	To learn about evaluating/self-evaluating as teachers
	To realize limits of simple test
Who is expected to structure structured	To be able to evaluate processes

Form No.1 – listing expectances expressed by attendants	
Doubts, perplexities and questions	**Expectances**
It is just the latest fashion?	To correspond to an individual professional need
Can a structured trial evaluate competences? If it is so, which competences?	To correspond to a professional need, not only individual but institutional also
Because the testing approach is more suitable for adults, how to use it with children?	
Which gradualness between extreme subjectivity and extreme objectivity to be used in evaluating performances and competences applicable to social systems?	To learn how to carry out methods for evaluation and self evaluation.
trials?	(educational, institutional quality control)
How does the designer structure them?	To be able to evaluate products
What kind of product can they evaluate?	To be able to detect feedback both in processes and in activity of the teacher
Which significance?	To detect attention points, crucial issues
Are they able to detect competences or only knowledge?	
Is it possible to evaluate in very well focused actions, without disciplinary continuity, having limited observational time span?	To be able to find out a model, an evaluation effective approach to be used in different contexts.
Is it possible to find out evaluation criteria and tools in very non-homogeneous contexts?	To be able to properly used existent models affecting professional practice and process.
How to manage differences between quantitative and qualitative evaluation?	To overpass old approaches and to have available on going tools.
Is it possible to limit discretionary power?	To have tools to manage conflict.
What kind of evaluation content to the expression "S/he could do better"?	To induce a reflection on measuring.

2.3.2 Notes on expectances

Form No. 1 has been given back to attendants at the end of the first day with a new form, No. 2, to be filled into, as in table 3.

In this form attendants are invited to fill into keywords. In the left part they are invited to write keywords selected from words used in the form No. 1. These are the keywords proposed by attendants at the beginning of the course.

In the right part attendants are expected to write new keywords, generated after the first day of the course.

Room for observations has been reserved for deducing, in case, information useful for interpreting bare or contradictory results.

The lists have been reorganized in two different ways in order to classify and to verify possible misinterpretations and changes introduced by classification itself.

Table 3. Form No. 2

Form No. 2 – Notes and keywords	
Please reread problems and expectances in form No. 1 and write:	
At least one (maximum three) keywords selected *between the ones used in form No. 1* that you retain now fundamental	At least one (maximum three) keywords *not present in form No. 1* and that now you retain now fundamental
Which are your reactions rereading words written at the beginning and at the end of the day?	

The first listing resulted very rich, presenting some recurrences allowing grouping. For instance the entry Competence-Knowledge groups the two entries both if they have been written separately and together. The organizer did only this kind of actions.

The pie chart in fig. 3 shows the results of this first tabulation with reference to the problems and interests manifested by the attendants.

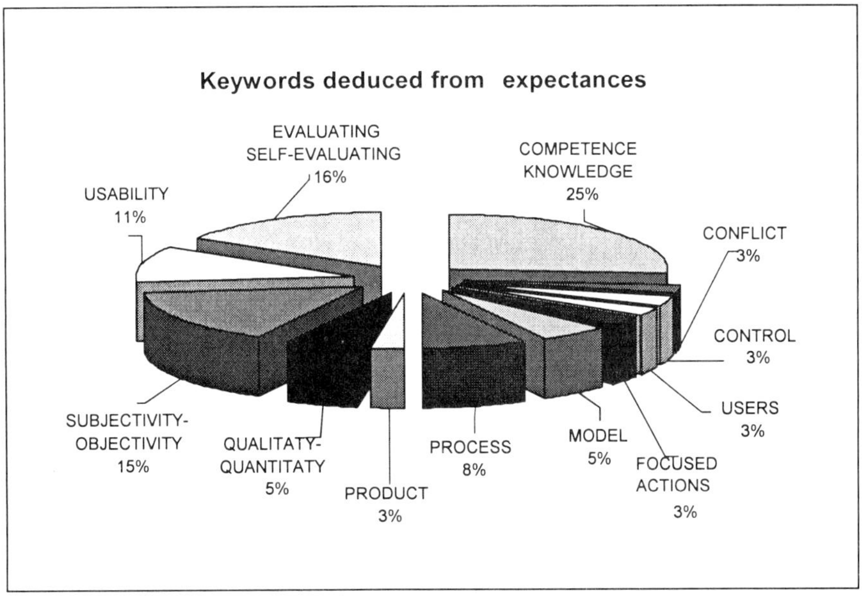

Figure 3. Organizing the results.

Words written in the right part of form No. 2 (keywords after the first day of the course) were very different one from the other, almost without repetition. This fact indicated that the educational activity had been successful stimulating personal reflections and processes.

A classification based on the indicated keywords was impossible.

The evaluator, tacking in count notes and observations (last field of form No. 2), introduced a new criteria based on three categories:

Table x.

Reference to designing	Words related to testing, its items, its design
Reference to the process	Words related to the educational process
Main points	Words related to risks, weakness of the testing procedure

Even if the data available are very limited (a single course has been considered at the moment, limited numbers of participants, only two days duration) we want to propose tentative ways to interpret the experience and key points to be considered in designing future activities. The results of this new classification are displayed in table 4.

Table 4.

Categories	Number of answers
Reference to designing	6
Reference to the process	11
Main points	7
Total number of keywords	24

The pie chart in fig. 4 displays the grouping of the keywords presented by the attendants after the first day of the course. It may be interpreted assuming that attendants after the first day of the course:

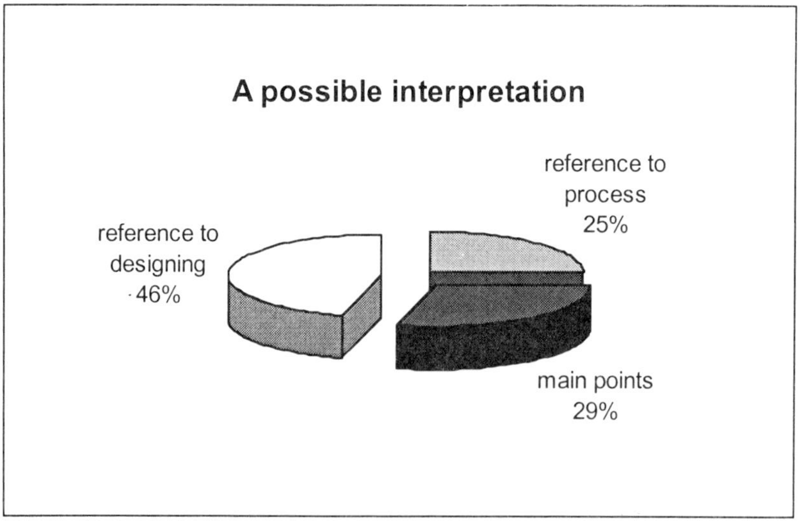

Figure 4. Results tabulated in table 4 displayed in pie chart.

- Are interested and available to design tests and to experience them in their classrooms (46%)
- Are interested in the elements of the educational process crucial for testing procedures (25%)
- Transformed perplexities in points of interest (29%).

2.4 Analyzing learning: results and elaboration (B)

In this activity the attendants have been asked to write one (max three) learning as result of the course.

Answers indicate the difficulty to consider what is in processing from a point of view different from the topics directly involved. That's avoiding explicit judgments (positive or negative) related to what has been presented. Some of the answers are tabulated in table 5.

Table 5. Learning results from the course

To be able to design an item
To be able to objectively and critically read the results of testing
To be able to use computer to tabulate
To be able to design a structured trial
How teachers may self-evaluate structured trials
How to concretely adapt evaluation procedures based on testing on their own educational context
How is possible to use testing for educational activities.
Confirmation of the inadequacy of evaluating only using structured trials
To evaluate/verify by using structured trials by using software systems helps to diffuse information and knowledge on evaluating in the educational system
How to designing a test based on reliable and effective elements
The ability to focus on what superficially may appear insignificant..
Graphics as a system to communicate results

Programming a long didactic process whit many events and settings, the teacher's trainer could project three steps: in the first the students are asked to describe their own learning; in the second the student compares each other the different learning; in the third step the result of the comparison produces the materials helpful for reflections about the expected learning. The purpose of such approach is to:

- have the purpose to improve general abilities;
- involve all the actors of the process teaching/learning;
- be useful in the knotty moments of the process,
- be integrated in different kinds of tools;
- be oriented to detect processes of changing.

3. CONCLUSION

We all experience how in the daily life results of testing procedures influence important events (to enter a school, to be hired , etc.). This kind of experiences, specially if negative, induce skepticism and even rejection of testing.

What H. von Foerster said

"… tests evaluate tests, evaluations evaluate evaluations (and not who are supposed to be "evaluated), Examinations evaluate examiners and not who is supposed to be examined. This "situation appears evident when we see students studying to pass an exam, totally different from "studying a discipline to know it. Then the question arises: how can we know what they know of a "discipline? Indeed, how can we know? …" may be interpreted as an authoritative opinion to confirm such a rejection.

By avoiding to excessively focus on the problem and by considering the problem in general, it is possible to interpret the problem as a stimulus to integrate tools, solutions, methodologies, processes of teaching/learning, epistemological issues.

Information systems may be crucial to help teachers at different levels:

- To make easier and quicker tabulation and then the presentation of results back to the students (results given back after long time are useful only for marking. It cannot influence a learning process);
- To give output of different nature to teachers, classrooms and single students.

REFERENCES

Binet, A., and Simon, 1917, *A method of mesuring the devolpment of intelligence of young children*. Medical Book, Chicago.

Bloom, B. S., 1956,*Taxonomy of educational objectives*. McKay, New York.

Bruner, J., 1990, *Acts of meaning*. Harward University Press, Cambridge, MA.

Council of Europe, 1996, *Modern languages: Learning, teaching, Assessement. A Common European Framework of reference*. CC Lang (95) S Rev. IV, Strasbourg.

Minati, G., 1998, *Sistemica - etica, virtualità, didattica, economia*. Apogeo, Milano, pp. 62-63.

Osterlins, S. J., 1998, *Constructing Test Items*. Kluwer Academic, New York.

Peroni, F., 1997, La valutazione nella Scuola Secondaria Superiore. *OPPI Documenti* **74**: 10.

Vertecchi, B., 1984, *Manuale della valutazione*. Editori Riuniti, Roma.

von Foerster, H., 1991, Letologia. Una teoria dell'apprendimento e della conoscenza vis-a-vis con gli indeterminabili, gli indecidibili, inconoscibili. *Conoscenza come Educazione*, Franco Angeli, Milano.

APPENDIX 1: A SELECTION OF INTERNET WEB SITES USED DURING THE COURSE

(compiled by Anna Renna)

http://www.cede.it - Centro Europeo dell'Educazione. - National Institut for the valutation of the educational system. ADAS (Archivio Docimologico per l'Autovalutazione delle Scuole).

http://istruzione.it - Site of Ministero della Pubblica Istruzione.

http://www.pisa.oecd.org - The OECD's Programme for International Student Assessement (PISA) is a new, regular survey of 15-year-olds which assesses aspects of their preparedness for adult life. The first round of the survey is taking place in 32 countries during 2000, in collaboration between governments and the OECD. Between 4.500 and 10.000 students will be surveyed in each country.

http://pdp.it - Site of BDP (Biblioteca di Documentazione Pedagogica) with different links to various experiences.

http://members.xoom.it/laguidara - On this web page you can find valutation frames of the process teaching/learning, measuring tests grids.

http://www.provveditorato.bergamo.it/stresa - Progetto Stresa (effectiveness auto-valutation in the school) and different materials about tests.

http://www.smartlitesoftware.com - Information about WinAsks, educational software for the creation, management and analysis of multimedial tests and soundings. www.winasks.com.

http://digilander.iol.it/lucianopirri/index.html - Tests and random problems whit immediate answer, for at distance learning too.

http://www.apple.com/it/education/LezioneD/TestWeb/Test.html - To download, try and built tests.

http://www.psi.uniroma1.it/SCUOLA_VALUTAZIONE/siti_valutazione.html - Scuola di specializzazione in valutazione psicologica. Segnalazione di siti di riferimento.

http://www.comune.fe.it/apis/risorse/4_valutazione_processo_insegnamento_apprendimento.html - The valutation of the process teaching/learning, throught the lessons of Valentina Zappaterra (University of Ferrara).

http://www.polimi.it/mappa.shtml - Site of the Politecnico di Milano, where is possible see the ammission tests.

Systemic Approach and Information Technology

Reactive Navigation based on Self-Organised Visual Information

G. TASCINI, A. MONTESANTO, P. PULITI, and N. RABASCINI
Institute of Computer Science, University of Ancona
E-mail: tascini@inform.unian.it

1. INTRODUCTION

To build a visual servoing system to behaviour-based robotics navigation we need to understand how the whole visual features of the could be grouped in meaningful agglomerations of information also noting the action to associate to the particular visual input.

From a systemic point of view, according to the conception of Varela (1991), the sensor processes and the motors processes evolve together and they are seen like inseparable and mutually informative. This paradigm allows to reconsider the concept of representation too. Brooks (1991)shows the results of searches in situated robotica in which an intelligent system was decomposed in modules of parallel and independent activity (the subsumption architectures). In this case the intelligence of an autonomous agent is seen like a biological feature, the *enaction* (Varela, 1991).

The behaviour-based robotics considers the autonomous agents like embodied systems built to reach inside or external purposes through their actions, in continuous interactions with the environment. So the intelligence should be considered like the agent ability to interact with the environment (Maturana and Varela, 1987) more that the ability of internally model it. The concept of interactions carries to study the emergence of behaviors. The emergence is the bridge between the need of a complex and adaptive behavior both to the macro-level (the observer point of view) and the

Emergence in Complex, Cognitive, Social, and Biological Systems, Edited by
Minati and Pessa, Kluwer Academic/Plenum Publishers, New York 2002

modules of multiple actions based on the situation to the micro-level (that of the agent) (Chantemargue and Hirsbrunner, 1999).

The autonomous navigation is one of the most important tasks for a mobile robot. Typically, a robot is equipped with sensors and effectors, and in every moment it must control its own movement in order to avoid obstacles and to reach a target. In the case of autonomous agent, the main source of knowledge is contained in a set of sensory inputs that the agent perceives from the surrounding environment. Thus, the problem of collision-free navigation learning can be described as a function learning problem (Keabling, 1991). The agent has to learn a mapping from different sensor situations to actions it must perform in order to achieve its task.

Usually systems use sonar or range detector to build a map from the sensor data to the objects position to self-locate and to navigate correctly (Krose, 1998). Whit sonar is possible to recognise the material of an object simply processing the echo signals and this is a further information in object recognition (Ecemis and Gaudiano, 1999). Typically, however, sonar and range detectors are used to have a potential function of the objects' distance for reactive navigation based on the high-potential value motion (van Dam *et al.*, 1996). In some systems different kinds of sensor are used in a sensor fusion mode to obtain the best information possible. In fact, if the system has a collision detector too, it can learn the navigation problem using collisions to correct its actions in a similar way to Reinforcement learning (Oja and Heikkonen, 1993).

To build intelligent systems able to develop tasks like a biological brain, we have to understand what characterize the best representations of the information and how they could be create through the self-organization and the learning (Ritter, 2000).

A good organization of information could be like the topographical maps of the biological brain. A possibility is that such maps are fully grown in a process of self-organization where the global order in a brain area emerges from local interactions around the units activated during the learning. That allows a meaningful representation of the input space (Kohonen, 2000).

The information contained in the visual sensor is surely the most important and complete, so this system is based only on this sensor, to show that it is sufficient for an autonomous navigation. The visual sensor is trinocular. Three cameras, one oriented in the same direction of motion, equip the system, the others turned respectively of 10° on the left and on the right. One of the main goals of a robot vision system is to gather visual information in front of the vehicle and to provide a robust description of the scene. This description may include object location information or may give information on road boundaries for the road-following task. The information is often incomplete or corrupted, so in a lot of works the Fuzzy logic is used, to manipulate the uncertainty (Saffiotti, 1997). In this work the navigation strategy is based on the knowledge representation coming from a Self-Organizing Map.

The main idea is to perceive the depth and the location of object from the differences in the images of visual sensor, in a similar way to human stereo vision and ocular movements. In this way emerge the concept of "empty space", the space without obstacles.

2. SYSTEM DESCRIPTION

The system is essentially composed of 4 modules: the segmentation map, the obstacle avoidance function, the goal recognition function and finally the motion control. In figure 1 we can see the input data in the format effectively processed by each module.

The three images are (fig.1) first segmented, then they are the inputs of the obstacle avoidance function, simultaneously only the central image is filtered, segmented and sent to the goal recognition function. Outputs of the last two modules are combined together by the motion control to calculate the new position and the new direction of the mobile robot. In the rest of this section we describe every module with more details, in the same order than the information moves from the input to the output of the system.

2.1 First Image Processing

The first images processing is an unsupervised segmentation based on space-chromatic distribution of pixel inside squared tesserae of 10×10 pixel. This kind of processing allows to obtain an image description when are reduced the redundancy of homogenous region and excessive details of heterogeneous region. To obtain this result we exploit the spontaneous clustering capacity of Self-Organizing Maps. In the system is used a two dimension S.O.M. of 10×10 neurons. The size of images is 150×100 pixel, and the size of squared tesserae is 10×10 pixel, so each image is splitted in 150 tesserae.

The map learns a set of patterns formed by nine images (Fig. 2) that are randomly acquired in the virtual environment.The map uses the Kohonen learning algorithm with discrete neighbourhood function and not mexican hat, to reduce the processing time. Every tessera of pattern images is the input of the map in a top-down, left-right order, in the same order the pixels of each tessera are sent to the input lines of the map. A pixel is represented by three numbers, the Red, Green, Blue value of its colour, so the input section of the map is formed by 300 input lines.

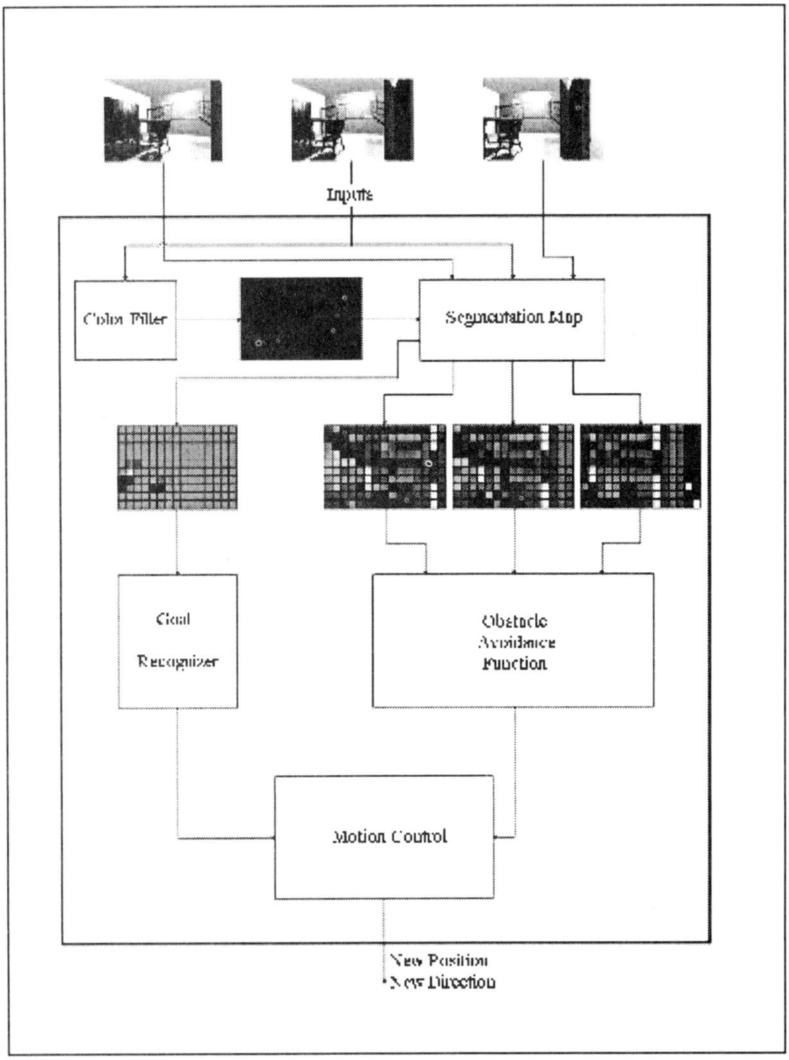

Figure 1. System's architecture.

The learning algorithm consists of two different phases, the first 1000 epochs for topological ordering, and the seconds 1000 epochs for fine tuning. During topological ordering the synaptic weight updating involves only the neurons inside the neighbourhood function around the winning neuron. The neighbourhood function size is initially set to 10 neurons to involve every neuron of the map, then it is linearly decreased every 100 epochs. The initial value of learning rate is $\eta_0 = 0.9$, then it is updated at every step in according to the following expression:

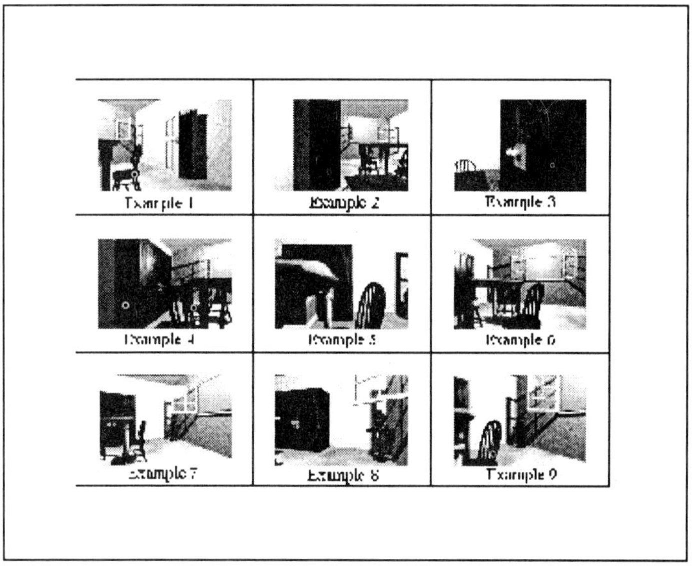

Figure 2. The learning set of the Kohonen SOM neural network.

$$\eta = \eta_0 \exp(-Epochs / \tau)$$

when $\tau = 200$.

The update value of weight vector is calculated as follows:

$$\begin{cases} w_j(n+1) = w_j(n) + \eta(n)[x(n) - w_j(n)] & \text{for} \quad j \in \Lambda_{i(x)}(n) \\ w_j(n+1) = 0 & \text{for} \quad j \notin \Lambda_{i(x)}(n) \end{cases}$$

$\Lambda_{i(x)}(n)$ is the neighborhood function around the winning neuron i at time n.

During the fine tuning the size of neighbourhood function is null, so only the winning neuron is updated, with a minimum learning rate value $\eta_{min} = 0.001$.

2.2 Obstacles Avoidance Function

The obstacles avoidance function is devised to perceive the depth of objects from the difference in the images acquired by the visual sensor. The obstacles avoidance function, shown in Figure 3, consists of two three-dimensional S.O.M. of 6×6×6 neurons, and a two-dimensional L.V.Q. of 24×24 neuron. The two Self-Organizing Maps are used to classify spontaneously 4 different possible situations: *Free Way, Left Free Way,*

Right Free Way, Blocked Way. The first-level spontaneous classification is then supervised by the Learning Vector Quantization, to get correct information for the next optimal motion strategy.

Each S.O.M. processes one of two values that identify every cluster in the segmented image (the winning neuron position of every cluster in the map). Using three-dimensional maps allows to obtain a very robust classification, that is necessary to solve this complex problem.

The L.V.Q. is organized like shown in Figure 3. It can classify both certain situations and uncertain situations, increasing the motion control strategy efficiency. The obstacle avoidance function learns 16 patterns for every situation, in all it learns 64 patterns. Every pattern consists of the three images acquired by the visual sensor, in Figure 4 are shown only the central images of all 64 patterns. The learning algorithm used for the Self-Organizing Maps is the same considered for the Segmentation Map. In this case some parameters have different values, topological ordering is 1200 epochs long, fine tuning is 2800 epochs long, neighbourhood function is initially set to 6 and then it is linearly decreased every 200 epochs.

In Figure 4 we can see that L.V.Q. consists of dark colour neurons and light colour neurons, firsts classify certain situations and seconds classify uncertain situations.

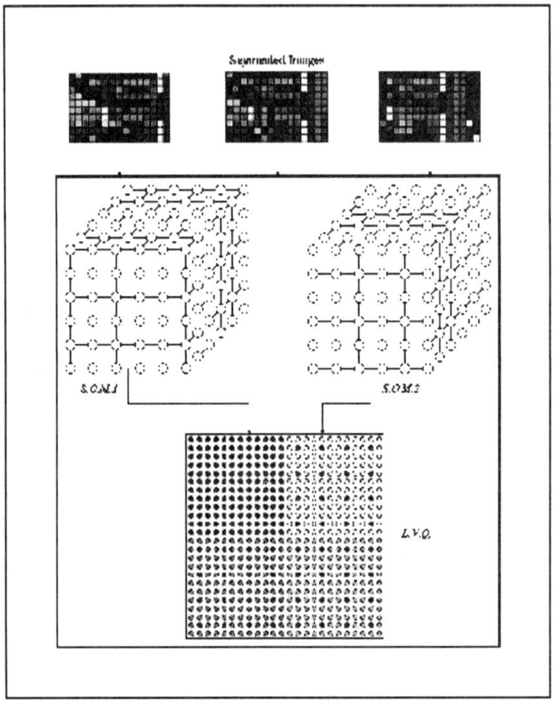

Figure 3. The obstacle avoidance function.

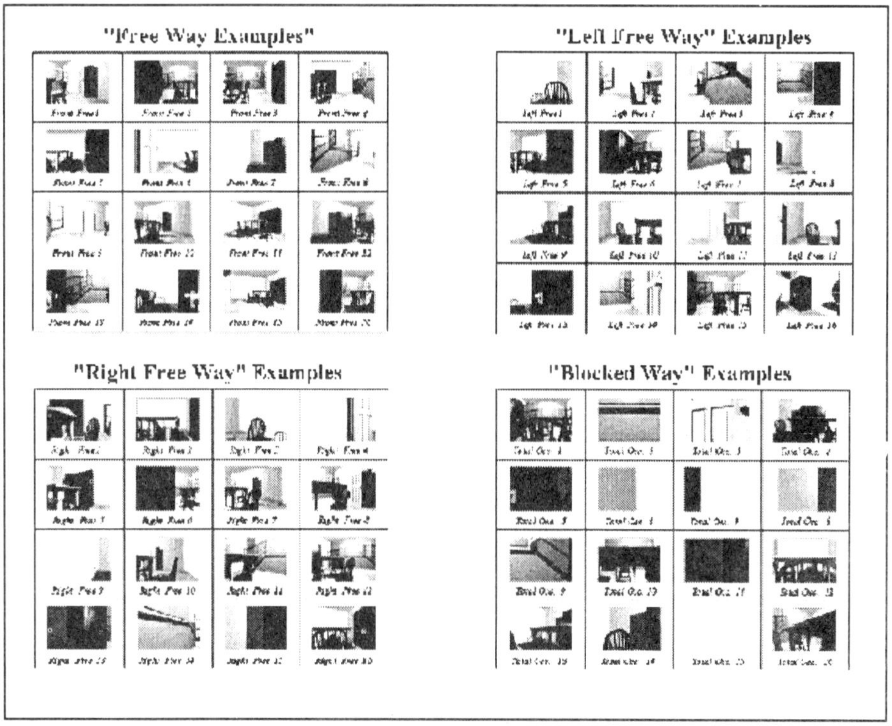

Figure 4. The central images of all 64 patterns of the trinocular sensor.

This result is obtained by a special learning practice: the length of learning is 50000 epochs, 2500 for ordering phase and 47500 for fine tuning phase. Furthermore the neighbourhood function initially set to 24 is linearly decreased every 100 epochs while the final step of ordering phase is 200 epochs long, so the direct neighbours of winning neurons are very similar to these. The certain or uncertain possible classification allows to the motion control to select long steps in the first case and more prudent steps in the second one.

2.3 Goal Recognize Function

Only the central image of visual sensor is processed to recognize the goal object, it is filtered, segmented by the same Segmentation Map previously seen, and then it is sent in input of a new map. The project scheme of this module is shown in Figure 5. The parameters of colour filter are experimentally setted on chromatic features of goal object, in this case a red vase.

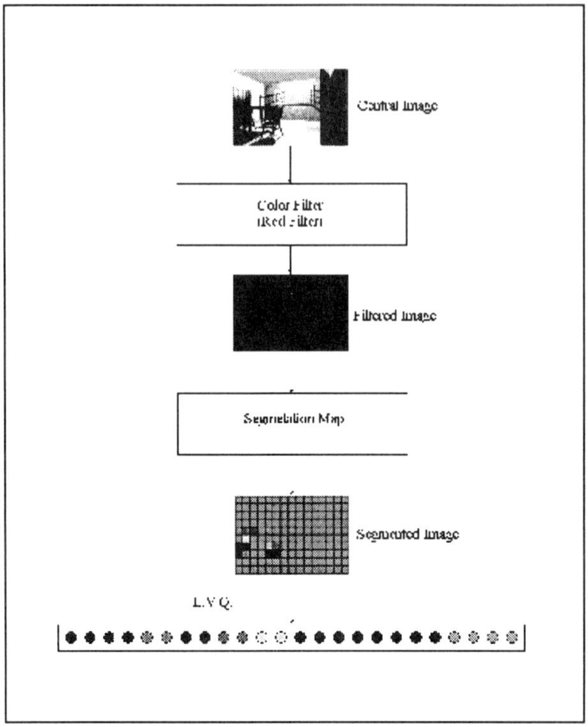

Figure 5. The scheme of the goal recognition.

The map is a one-dimensional L.V.Q. of 24 neurons, organized to classify 9 possible goal recognitions. These nine possibilities are: Central Far Goal, Left Far Goal, Right Far Goal, Central Medium Goal, Left Medium Goal, Right Medium Goal, Central Near Goal, Left Near Goal, Right Near Goal. The map learns 24 examples of goal in a similar way of an autoassociative memory, and during the execution phase it uses the *best-matching* algorithm and different selectivity criteria to establish if the input image is the goal.

The selectivity criteria are very strong for "*Far*" classes because goal object is small and incorrect recognitions are very probable, and then it becomes wider for other classes. The "*Central Near Goal*" class determines the end of a correct navigation. In Figure 6 are shown the 24 examples of goal.

2.4 The Motion Control

The motion controller is a set of rules that process two kinds of input data, output values of Obstacles Avoidance Function and output values of Goal Recognition Function, to establish the most efficient motion step.

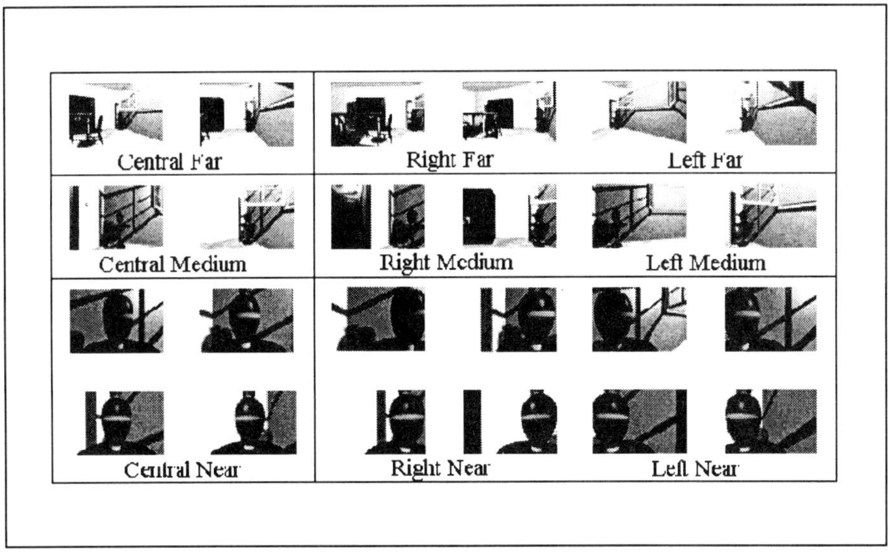

Figure 6. The 24 learning examples of goal.

Motion strategy presents also different priority levels between two previous kinds of data, so it is very complex, flexible and efficient too.

The motion controller presents also an internal memory device where are stored the critical positions acrossed by the robot during navigation. This allows to avoid that identical motion steps repeat themselves and that robot follows an infinite loop. In Figure 7 the motion strategy table is shown, divided in two parts, in the first goal is not recognized, in the second, on the contrary, goal is recognized.

The different priority levels between Obstacles Avoidance Function data and Goal Recognition Function data are resumed in the following.

a) If "*Central Far Goal*" class is recognized, the motion controller switches off the Obstacles Avoidance Function to the motion control until it does not classify a certain situation. While this condition persist robot goes forward of 100 mm. in the same motion direction.
b) If "*Central Medium Goal*" class is recognized, the motion controller switch off the Obstacles Avoidance Function to the motion control for good, and robot goes forward of 50 mm. in the same motion direction. This last strategy is used because in this situation surely there are not objects between robot and goal, so the quickest and shortest way to finish correctly the navigation is the straight line from robot to goal.
c) The "*Central Near Goal*" class determines the stop of robot and the end of a correct navigation.

Goal not recognized	
Robot goes forward of 100 mm. in the same direction.	Certain Free Way.
Robot goes forward of 50 mm. in the same direction.	Uncertain Free Way.
Robot turns left of 10° and goes forward of 50 mm.	Certain Left Free Way.
Robot goes forward of 25 mm. in the same direction.	Uncertain Left Free Way.
Robot turns right of 10° and goes forward of 50 mm.	Certain Right Free Way.
Robot goes forward of 25 mm. in the same direction.	Uncertain Right Free Way
Robot turns left/right of 30°, if it is a stored critical position the new angle is Alfa_New = -1.5Alfa_Old.	Certain Blocked Way.
Robot turns left/right of 15°, if it is a stored critical position the new angle is Alfa_New = -1.5Alfa_Old.	Uncertain Blocked Way.
Goal recognized	
Robot goes forward of 200 mm. in the same direction.	Central Far Goal
Robot turns left of 25°.	Left Far Goal.
Robot turns right of 25°.	Right Far Goal.
Robot goes forward of 100 mm. in the same direction.	Central Medium Goal.
Robot turns left of 25°	Left Medium Goal.
Robot turns right of 25°.	Right Medium Goal.
End of a correct navigation.	Central Near Goal.
Robot turns left of 25°.	Left Near Goal.
Robot turns right of 25°.	Right Near Goal.

Figure 7. The motion strategy table.

3. RESULTS

System has been tested by the simulation of different navigations into the virtual environment expressly designed for this purpose. In some navigations the environment was the same in which the system have acquired the images learnt by its maps, in some other navigations, on the contrary, the environment changed by moving some object. In Figure 8 are shown the results of different navigations.

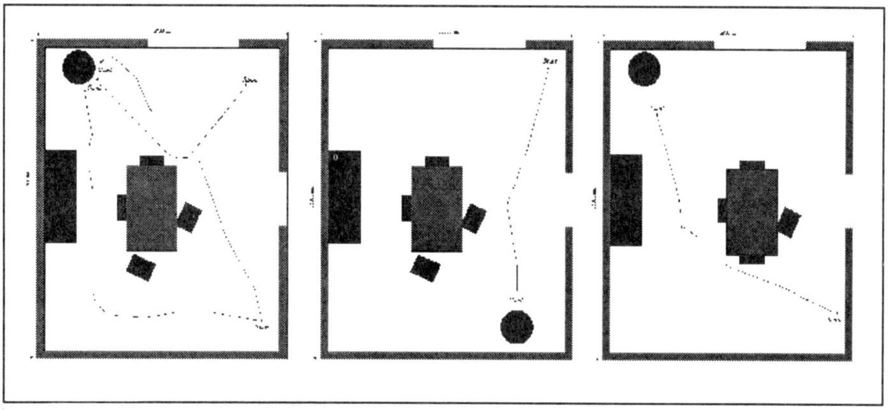

Figure 8. The results of three different navigation: the goal is the red dot.

4. CONCLUSIONS

The emergence of an autonomous navigation behaviour is based on the self-organising of visual information and on the interaction of the all 4 modules of the system. The results of simulation can be considered very appreciable, in fact the robot has correctly rich the goal either in standard conditions or in not standard condition (where standard conditions are the same environment in which the system has acquired the images learnt by its maps). The proportion of learnt information and total possible information is very little, equal to 0.000007446%, so results can be considered even better. In conclusion we can say that a vision-based autonomous reactive system could be really usefull both for goal recognition that for an efficient obstacles avoidance. The sub-symbolic approach for information analysis makes the system very smart by the mean of its generalization capacity, this is particularly emphasized by the very little percentage of off-line information learnt compared to system performances.

REFERENCES

Brooks, R. A., 1991, Intelligence Without Representation. *Artificial Intelligence Journal* **47**: 139-159.

Chantemargue, F., and Hirsbrunner, B., 1999, A collective Robotics application based on emergence and self-organization.

Dam, J. W. M., Van Kröse, B. J. A., and Groen, F. C. A., 1996, Neural Network Applications in Sensor Fusion for an Autonomous Mobile Robot. In *Reasoning with Uncertainty in Robotics* (L. Dorst, M. Lambalgen, and F. van Voorbraak, eds.), Springer, pp. 263-277.

Ecemis, M. I., and Gaudiano, P., 1999, Object Recognition with Ultrasonic Sensors. In *Proceedings of the 1999 IEEE International Symposium on Computational Intelligence in Robotics and Automation*, CIRA'99, Monterey, CA, pp. 250-255.

Kaebling, L. P., 1991, Foundations of learning in Autonomous Agents. In *Robotics and Autonomous System* **8**: 131-144.

Kohonen, T., 2000, Foreword. In *Faithful Representations and topographics maps. From distortion to information based self-organization* (M. M., van Hulle), John Wiley & Sons.

Krose, B., 1998, Environment learning and localization in 'sensor-space'. *Proc. of the 10th Netherlands/Belgium Conf. on Artificial Intelligence*, pp. 229-239.

Maturana, H., and Varela, F. J., 1987, *The tree of knowledge: the biological roots of human understanding.* New Science Library, Boston, MA.

Varela, F. J., Thompson, E., and Rosch, E., 1991, *The embodied mind: Cognitive science and Human experience.* MIT Press, Cambridge, MA.

Oja, E., and Heikkonen, J., 1993, Self-Organizing Maps for Visually Guided Collision-free Navigation. In *Proc. IJCNN-93*, Nagoya.

Ritter, H., 2000, Foreword. In *Faithful Representations and topographics maps. From distortion to information based self-organization* (M. M., van Hulle), John Wiley & Sons.

Saffiotti, A., 1997, Fuzzy Logic in Autonomous Robot Navigation. *Soft Computing* **1**(4): 180-197.

Web Usability: Meaning and Context. Over Data Structures

M. C. TOFONI*, and A. MONTESANTO#

*Corso Interfacoltà di Scienze della Comunicazione, Università degli studi di Macerata
E-mail: mctofoni@tiscalinet.it

#Istituto di informatica, Università degli studi di Ancona - E-mail: montesan@inform.unian.it

1. INTRODUCTION

Meaning in the text during the reading is given also from the impact the text itself has on the reader and from all those elements identifiable as components of the "vehiculum" of the message (Petöfi, 1994) or that are part of its text-design (Hesse-Luttich, 1999).

The methodology proposed in this paper is based on the observation of the user and on the study of the effects the web page has on him further to the observation of the authors during the phase of projecting/implementing the hypertext. The information carried by a hypertext realised according to this methodology should be always retrievable and easy to analyse so that it is always possible treating and organising the materials englobed in the web-site further to find in them different levels of reading of the information.

We hereby discuss in particular two structural levels of information in a web-site: (a) all what is visible within the navigable surface of the web-site, i.e. links, (b) the organisation of the same material in files and directories. The interaction of these two structures makes the general organisation of the hypermedia. In situations in which is important the control of the amount of the visualised information, there is the need of a systemic point of view to evaluate the importance of the nodes and the results of their connection.

Emergence in Complex, Cognitive, Social, and Biological Systems, Edited by Minati and Pessa, Kluwer Academic/Plenum Publishers, New York 2002

Implementing hypertext is becoming a widespread normal practice needing methodological indications especially if the implementation is the result of the collaboration of several subjects.

It will be helpful to establish some principles regarding:

1. User's necessity during the reading process:
   that is to say, to provide an enlarged definition of hypertext, for the reason that the hyperstructure of the information contained in a whole site is a sum of features that goes further the single page. The hyperstructure contains objects and their relations further to relations between pages. The hyperstructure of the author has to reflect the semantic structure of the target user for a high usability of the site. In so doing there is the need of a deep study (i.e. through scaling techniques) investigating users' semantic representation. Effects caused by the user on the page are not yet studied because of technical obstacles (it is not simple to "capture" the single event on the page especially when they are in "the Web". Programs thought for that function are not independent from hardware and software used). An important study is that concerning the time occurring between each click. The Knowledge of the time the user needs to decide on what link to click is not useless, especially if the investigation concerns the relation between the hyperstructure and time of clicking. The latter allows to study interpersonal semantic, link function, structures of exchange of information (link, multimedia metaphors), cognitive schemata and functions (perception, attention, memory ...).
2. Author's necessity during the implementation process, that is to say:
   - Providing some indication in the creation of the structure of an hypertext; both built up by a single person or a group;
   - Describing some methodologies for the developing of hypertext having HTML as source code.

2. ORGANIZING HYPERMEDIA

2.1 Projecting information in the page/s

While realising a hypertext an author has a particular schema of the topic coherent to the context. The context establishes the representation of information inside the hypertext. The same information can be organised differently to make different context and different interpretations.

The reader of the hypertext handles the information contained in the hypertext and follows several paths choosing different links thus creating a new text owning its own coherence related to the context in which he is moving (Landow, 1992; Bolter, 1991). Global meaning in the created text will be represented from the interaction between various subsystems. The text must maintain a strong internal coherence so that its aim will be not misunderstood and the user navigates easily inside the hypertext.

Hypertexts contain a structure made up by its links. Those are elements belonging to the concrete structure of the hypertext and to the cognitive schema of who writes and organises the information following his own thought. These are elements of the reader's cognitive structure while he is looking for any kind of information. The choice of the words to link up is extremely important because they are able to carry a knowledge that goes further the words themselves. This knowledge arises from the interaction between the single words inserted in a particular context.

The map given by the whole of the links within the single page in the hypertext can be organised so that to optimise specific aims. This implies that the visualisation of the map can be different for each aim or context. Because this visualisation is never totally clear or visible for the user there is the need to find a methodology allowing to build up primarily a structure of connections representing the target user's cognitive structure in that context. In so doing it is possible to realise hypertexts with highest degree of consistency for the functionality of the structure in the system. It is necessary to study the hyperstructure to help users to find appropriate paths in the information space (Brusilovsky and Eklund, 1999). In addiction to this, to find the users' cognitive schemata in a precise context there is the need of precise experimentation to give a statistic validation to the hyperstructure of the web-site. The latter allows the evaluation of the hypertext's usability and the optimising of the same hyperstructure.

It is important to examine the influence the hypertext has on the information it englobes and the implications it has on its author and on its users (Musch, 1995). Hypertexts thought for free navigation, infact, contain both the cognitive representation of the author and the user.

The cognitive schema becomes concrete in the screen layout that shows the hypertext spatial structure with the links as supporting points. They are basic components both in the hypertext and in the cognitive structure, besides being concrete perceptive elements and semantic relation between two documents.

Being the hyperstructure characterised by two fundamental variables (spatial distance and semantic distance) it is possible to think that they influence the so called "spatial distance". That is to say the time necessary to shift the attention from a link to another in the same page and the time needed to evaluate the link to choose.

It has been evidenced (Bonfiglio *et al.*, 2001) that different spatial and semantic hyperstructures cause different times on the user's decision about which link to click. This proves that the effect of an excessive approach or estrangement of the spatial distances between links corresponds to an increase of the time of choice of the user. The same happens to times of choice in relation to different semantic distances, especially in association to the optimal spatial distance. The context investigated in this work is that of a user who navigates in a hypertext made up of words, without pictures, who has the purpose to assemble how much information possible on the same argument. The methodology used in the first phase of the work is based on an analysis of multidimensional scaling to extract from a text the quantitative values of the semantic distance between words. In this way it is possible to individuate the words more suitable for a generic user and to represents links in a hypertext. Several hypertexts have been realised using these words with different values of the spatial and semantic distance, with the aim of investigating the effect of various hyperstructures on the user.

It is so possible to reach a union between comprehension and usability of the hypermedia instrument. Experimental evidences demonstrate that it is possible to explicit the user's cognitive structure through two variables (spatial and semantic distance between links). Besides it is also possible to optimise the same structure studying systemically the effects it has on the subject's comprehension of the text because also of its relation with the context.

In conclusion it is possible to affirm that in implementing hypertexts with high degrees of usability it is necessary to take into account the cognitive representation of the user. The same influences the physical structure of the hypertext through the choice and the organisation of the links. Furthermore it has been proved that the spatial disposition of the links and their semantic value strongly contribute to the user's comprehension of the hypertext. A consequence is that further to the comprehension there is also the usability of the hypermedia instrument.

2.2 Projecting files and directories' structure

The web-site, as a complex system, owns features as abundance and heterogeneity of information, interconnection of information, variety of experiences and professionals working on its organisation and/or facing its reading/navigation. In addiction to this, it provides different levels of reading of the information: besides the information in the pages it is possible to find an information given by the files and directories' structure

The general structure and the organisation of the material derive primarily from several graphical, technical topic choices. Secondly, while implementing, the author needs a valid instrument to orientate himself in the web of files he is creating. Thus for, after the choice of the materials, the next step is the map. The map is the only instrument able to manage the complexity of a hypertext. It indicates the topic organisation, possible paths, the degree of fragmentation and connection of the information (Landow, 1993; Petofi and Rossi ,1997; Rossi, 2000).

There is more than one map to describe a hypertext:

- It is possible to realise several maps of the same material;
- It is possible to zoom in and out the same material with different maps.

The first map to realise is the conceptual map of the chosen topic. It contains all the arguments to be developed and their connections. Nodes and links will make it. Each node can be zoomed in by more detailed maps. A node can be:

- An argument linked to other arguments;
- A lexia linked to other lexias;
- The lexia and the medial materials it contains.

Working on a hypertext structure is comparable to work on several maps, or more visualization, of the same map. The same is dynamic and expandable. It is possible to modify a map because the medium hypertext is itself something able to create links.

While the web of connections get larger the initial map change its shape and remains recognisable only in its main structures and in key links. It is then necessary to "foretell" the deformation to avoid loosing the first project.

The solution usually lays in a modular project considering each node as an expandable module. According to Lévi (1992) in a hypertext each node or link can be made of a net and so on, recursively. A structure like that is similar to a puzzle where each node is a card of the puzzle and puzzle itself. The graphical visualisation given by the tree of files and directories in a screen of a computer is usually limited to the portion of screen visualised. A systemic approach in the structuring of information makes possible, instead, to keep constant the quality and the quantity of information even when the user shift his point of observation of the information.

In the translation of the conceptual map in an electronic map each single media material (image, sound, video) will be file, each lexia an HTML file, each macro-argument a directory, file and directories will be nodes and the links the lines in the map.

The choice of an editor (i.e. Dreamweaver) enabling to see the files and directories' organisation will render the work easier. To work on a layout like the one described make the implementation easier because it is always possible to re-organise files and directories working only on the layout itself. To establish some principles according to realise the organisation of the material allows controlling their physical structure and the same organisation. A possible way is to create primarily all the folders that will contain the files. The existence of a folder is determined from the symmetry of the organisation and not from the number of files contained, it depends from exigencies of the project.

Inside the mentioned directories it is possible to create sub-folders with names indicating the materials contained.

This process involves both the project and the implementation. While adding materials and links to the hypertext it will be sufficient to add some nodes to map that will be reflected in the electronic mirror of file and folders in the screen.

Concerning files, as for directories, there is the need to follow some basic assumptions ruling their development and disposition inside the site.

The first care regards the name of a file. The name given to a file it is not only an element able to distinguish it from another file. To give a name to a file is like giving a title to a text. It must refer both to technical and rhetoric/textual exigencies.

The name, together with the position, will provide useful indications on the content of the node and on the information englobed.

In this way the structure becomes agile and plastic (while creating a link to understand which document we can anchor without opening it make things easier and faster and reduces possibility of error) and facilitate later interventions.

The structure in a hypertext depends on the communicative aim and the metaphor chosen. Common metaphors are the book, the film, the game, the encyclopaedia. For this aim a graphical project is very important. Once decided the graphic aspect of the principal nodes adding pages will be easy. It will be possible to work on the same file like a pattern and doing some simple variations. A useful operation can be that of creating a folder called "patterns". It will contain all the files which is possible to use as patterns.

The following phases will be material searching and preparation, text writing, deciding iconic and sound materials, organisation and catalogue of the materials.

Following the procedures described, from the directories' organisation to the site structure, the more difficult moment in the realisation of a hypertext will be the project and not the implementation.

2.3 Writing co-operative hypertext

In many articles concerning hypertext realisation the problem of how to organise a group working together, also being in different places, is developed as follows:

- List of the necessary instruments (chat line, web forum, E-mail, web conference, FTP ...);
- Catalogue of information and actions to follow;
- Description of each node to realise, description of the portion of screen.

Besides it is necessary to establish some principles that rule the co-operation, the division of the work, specific competence for each member in the group. The degree of freedom of each component of the group must be defined to allow the hypermedia comes to life.

Each subjects involved in the project of a hypertext have a luggage of different experience and competence to exploit.

In the developing of a hypertext these experiences lead to diverse ways of sharing the job which can be also integrated. The difference is not a limit but richness, advantage and resource. Finally it is possible to affirm that, in realising a professional product, the presence of different professional experiences becomes necessary.

3. CONCLUSION

The realisation of web-sites with the highest degree of usability, or in other words with the highest comprehension of the contextual information contained, goes through the visualisation of the information that allows schemata formation (Spence, 2000). The coherence in a text for several users passes through the minimisation of the distances between their cognitive representations on the argument.

The methodology suggested in this paper shows how, through a systemic approach, it is possible to investigate primarily the cognitive schema of a reader of a hypertext, taking into account the context. It shows also how it is possible to increase the comprehension expanding the visualisation of the information to the entire structure of files and directories maintaining the same coherent. The coherence to the context of the web-site as a global system is given from the interaction between several components where the division in files and directories is an over-structure of a meaningful information for the target user.

Generally the two methodologies are not strictly connected. It is possible infact to have a clear and well-organised information in the surface of the pages even though the structure underneath is a chaotic whole of files. The integration of the two modalities increase the usability of the web-site besides opening a concrete hypothesis of co-operative writing and "open-web", given from the possibility to add pages to the existent structure without distorting it.

The big problem is the necessity to not alter meaningfully these structures so that, as coherence allows user to move easily in the site, the coherent visualisation of the information makes the context explicitly comprehensible.

REFERENCES

Bolter, J. D., 1991, *Writing Space. The computer, hypertext and the history of writing.* Lawrence Erlbaum Associates Ltd, Hillsdale - Hove - London.

Bonfiglio, N., Montesanto, A., Penna, M. P., and Pessa, E., 2001, Rappresentazione cognitiva dell'utente negli ipertesti. *AIP2001*, Alghero.

Brusilovsky, P., and Eklund, J., A study of user model based link annotation in educational hypermedia. *Journal of Universal Computer Science*, **4** (4).

Chesebro, J. W., and Bertelsen, D. A., 1996, *Analyzing Media: Communication Technologies as Symbolic and Cognitive Systems.* The Guilford Press, New York.

Hesse-Luttich, E. W. B., 1999, Wissenshaftskommunikation und Textdesign. *Spachtheorie and germastiche Linguistik*, **9** (1): 3-17. University of Illinois Press, Chicago.

Landow, G. P., 1992, *Hypertext, the convergence of contemporary critical theory and technology.* The John Hopkins University Press, Baltimore.

Lévi, P., 1992, *Le tecnologie dell'intelligenza.* A-traverso, Bologna.

Musch, J., *A case study of potential system-user interactions with the IDEAS system.* Research Report, German Center for Artificial Intelligence, Kaiserslautern.

Petöfi, J. S., 1994, Testologia Semiotica e filosofia. In *Il testo filosofico. Analisi semiotica e ricognizione storiografica* (G. Marrone, ed.), Epos, Palermo.

Petöfi, J. S., and Rossi, P. G. L., (eds.), 1997, Sistemi segnici e loro uso nella comunicazione umana 4. Combinatoria ed ipertestualità nella ricerca e nella didattica. *Quaderni di ricerca e didattica* XVIII, Pubblicazione del Dipartimento di Filosofia e Scienze Umane, Università di Macerata.

Rossi, P. G., (eds.), 2000, *Dal testo alla rete. Multimedialità e didattica.* Tecnodid Editrice, Napoli.

Tofoni, M. C., 2000, Organizzare l'ipermedia. In *Dal testo alla rete. Multimedialità e didattica* (P. G. Rossi, ed.), Tecnodid Editrice, Napoli.

Membrane Systems for Computing

GIANCARLO MAURI, and CLAUDIO ZANDRON
DISCO - Universita di Milano-Bicocca
Via Bicocca degli Arcimboldi 8, 20126 Milano - Italy

1. INTRODUCTION

The *membrane systems* (also called *P Systems*) are a new class of *distributed* and *parallel* computing devices of a biochemical inspiration, recently introduced in (Paun, 1998).

Such systems are based upon the notion of *membrane structure*, which is a structure composed by several cell-membranes, hierarchically embedded in a main membrane called the *skin membrane*. A plane representation of a membrane structure can be given by means of a Venn diagram, without intersected sets and with a unique superset. The membranes delimit *regions* and we associate with each region a set of *objects*, described by some symbols over an alphabet, and a set of *evolution rules*. The objects evolve according to the evolution rules, which can modify the objects to obtain new objects and send them outside the membrane or to an inner membrane. The evolution rules are applied in a maximally parallel manner: at each step, all the objects which can evolve should evolve. Moreover, it is also possible to define priority relations among evolution rules.

The evolution rules can modify the membrane structure too, by *dissolving* the membranes. When a membrane is dissolved, all the objects in this membrane remain free in the membrane placed immediately outside, while the evolution rules of the dissolved membrane are lost. The skin membrane is never dissolved.

A computation device is obtained: we start from an initial configuration, with a certain number of objects in certain membranes, and we let the system

Emergence in Complex, Cognitive, Social, and Biological Systems, Edited by
Minati and Pessa, Kluwer Academic/Plenum Publishers, New York 2002

evolve. If a computation *halts*, that is no further evolution rule can be applied, the result of the computation is defined to be the number of the object in a specified membrane (or expelled through the skin membrane). If a computation never halts (i.e. one or more object can be rewritten forever), then it provides no output.

An up-to-date bibliography of the area and other useful resources can be found at the following web address[1].

2. MEMBRANE SYSTEMS: DEFINITIONS

In what follows, we refer to Rozenberg and Salomaa (1997) for formal languages elements.

A membrane structure is a construct consisting of several membranes placed in a unique membrane; this unique membrane is called a skin membrane. We identify a membrane structure with a string of correctly matching parentheses, placed in a unique pair of matching parentheses; each pair of matching parentheses corresponds to a membrane. A membrane identifies a region, delimited by it and the membrane immediately inside it. We can distinguish elementary membranes, that is membranes which do not contain other membranes inside and non-elementary membranes. If we place multisets of objects in the region from a specified finite set V, we get a super-cell. A super-cell system (or P-system) is a super-cell provided with evolution rules for its objects.

Formally, a *membrane system* (or *P system*) of degree n, $n > 1$, is a construct

$$\Pi = (V, \mu, M_1, \dots, M_{n,} (R_1,\rho_1), \dots, (R_n,\rho_n), i_0)$$

where:

- V is an alphabet which elements are called *objects*.
- μ is a membrane structure consisting of n membranes, with the membranes and the regions labelled in a one-to one manner with elements of a given set; In this section, we will use the labels $1, 2, \dots, n$.
- M_i, $1 \leq i \leq n$ are strings representing multisets over V associated with the regions $1, 2, \dots, n$ of μ.
- R_i, $1 \leq i \leq n$ are finite sets of evolution rules over V associated with the regions $1,2, \dots, n$ of μ; An evolution rule is a pair (u,v), usually written in the form $u \rightarrow v$, where $u \in V^+$ and $v=v'$ or $v=v'\delta$, where v' is a string over

[1] http://bioinformatics.bio.disco.unimib.it/psystems

$(V \times \{here, out\}) \cup (V \times \{in_j \mid 1 \le j \le n\})$,

and δ is a special symbols not in V. The length of the string u is called the *radius* of the rule $u \to v$.

- ρ_i, $1 \le i \le n$ are partial order relations over R_i, $1 \le i \le n$, specifying a *priority* relation among rules of R_i (the priority relation is non-circular; that is, the graph associated with it cannot contain cycles).
- $i_0 \in \{1, 2 \dots, n\} \cup \{\infty\}$. If i_0 is a number between 1 and n then it specifies the *output membrane* of Π; if $i_0 = \infty$ or if it is omitted, then the output is read in the outer region.

When presenting the evolution rules, the indication "here" is usually omitted. The multisets $M_1, M_2, \dots, M_n$ can be empty, and the same is true for any of the sets R_i and ρ_i, $1 \le i \le n$.

The use of priority relations can prevent the use of another rule as it follows: if a rule with a higher priority is used, then no rule of a lower priority can be used, even if the two rules do not compete for objects.

The $(n+1)$-tuple $(\mu, M_1, M_2, \dots, M_n)$ constitutes the *initial configuration* of the system. In general, any sequence $(\mu', M'_{i1}, M'_{i2}, \dots, M'_{ik})$ with $k \le n$ and μ' a membrane structure obtained by removing from μ all membranes different from $i_1, \dots, i_k$ (of course the skin membrane is not removed), with M_j strings over V, $1 \le i \le k$, and $\{i_1, \dots, i_k\} \subseteq \{1, 2, \dots n\}$, is called a *configuration* of Π.

It should be noted that the membranes preserve the initial labelling in all subsequent configurations, thus the correspondence between membranes, multisets of objects and sets of evolution rules is well specified by the subscript of these elements.

We can pass from a configuration to another one by using the evolution rules in parallel to all objects which can evolve, obeying the priority relations. For two configurations

$$C_1 = (\mu', M'_{i1}, \dots, M'_{ik}), \quad C_2 = (\mu'', M''_{j1}, \dots, M''_{jk})$$

of Π, we say that we have a *transition* $C_1 \Rightarrow C_2$ if we can pass from C_1 to C_2 by applying the evolution rules appearing in $R_{i1}, \dots, R_{ik}$ in the following manner:

Consider a rule of the form $u \to v$ in a set of rules R_{ih}. We look to the region of μ' associated with the membrane i_h If the multiset identified by u is included in the multiset $m(M'_{ih})$ (i.e. all the objects in u, with the multiplicities specified in u, appear in $m(M'_{ih})$), and if there is no another rule of an higher priority which can be applied at the same time, then we apply the rule $u \rightarrow v$. All objects to which the rule *can* be applied *must* be the subject of the rule application: all objects in u are "consumed", that is, the multiset identified by u is subtracted from the multiset $m(M'_{ih})$. In other words, we start to examine the rules in the decreasing order of their priority and assign objects to them. A rule can be used only if there are copies of the objects needed to apply such rule and which were not "consumed" by rules of a higher priority and, moreover, there is no rule of a higher priority, irrespective which objects it involves, which is applicable at the same step.

The result of using the rule is determined by v:

- If an object appears in $\boldsymbol{v}$ in a pair (a, here), then it will remain in the same region i_h.
- If an object appears in v in a pair (a, out), then a will exit the membrane ih and will become an element of the region immediately outside it (thus, it will be adjacent to the membrane i_h from which it was expelled). In this way, it is possible that an object leaves the super-cell itself: if it goes outside the skin of the system, then it never comes back.
- If an object appears in v in a pair (a, in_q), then a will be added to the multiset $m(M'_q)$, providing that a is adjacent to the membrane q. If (a,in_q) appears in v and the membrane q is not one of the membranes immediately inside the region i_h, then the application of the rule is not allowed.

If in v we have the symbol δ, then the membrane i_h is *dissolved* and, at the same time, the set of rules R_{ih} (with its associated priority relation) is removed. The multiset $m(M'_{ih})$ is added (in the sense of multisets union) to the multiset associated with the region which was immediately external to the membrane i_h. The skin membrane cannot be dissolved, because this means that the super-cell is lost and we do no longer have a correct configuration of the system.

The operations are done in a maximal parallel manner, for all possible applicable rules $u \rightarrow v$, for all occurrences of multisets u in the region associated with the rules and for all regions at the same time. No contradiction appears in the case of multiple membrane dissolving, or because of simultaneous appearance of symbols of the form (a,out) and δ. If at the same step we have (a,in_q) outside a membrane i and δ inside this membrane, then we assume that a is introduced in membrane i at the same time when it is dissolved, thus a will remain in the region placed outside membrane i; that is, from the point of view of a, the effect of (a,in_i) and δ is the same as the rule (a,here).

Hence, starting from the initial configuration C_0 of our system, we get a sequence of transitions. Such a sequence of transitions in a P system Π is called a *computation* with respect to Π. We say a computation $C_0 \Rightarrow C_1 \Rightarrow$

$\ldots \Rightarrow C_m$, $m \geq 0$ is *successful* if and only if there is no rule in C_m which can be applied to the objects present in C_m, i.e. the computation *halts*. On the contrary, the computation is *unsuccessful* if it can continue forever, that is, there exists a configuration C_{m+1} such that $C_m \Rightarrow C_{m+1}$. Note that it is not necessary to have $C_m \neq C_{m+1}$.

In this way, a P system Π can be seen as a computing device which nature is distinguished on the basis of the output membrane. If $i_0 \in \{0, \ldots, n\}$ then the system Π works in *internal mode*. Otherwise it works in *external mode*: we observe the system from outside and we collect the object expelled through the skin membrane. Using these objects, in the order they are ejected, we form a string. When several objects are expelled at the same time, any permutation of them is considered. Working in this way, we can associate a string or a set of strings with each computation, that is, a language is associated with the system. By $L(\Pi)$ we denote the language generated by Π in the way just described.

A P systems Π is said to be *cooperative* if there are rules Π with the radius at least two; in the opposite case, it is called *non-cooperative*. In particular, a system is said to be *catalytic* if there are objects $c_1, \ldots, c_n$ specified in advance, called *catalysts*, such that the rules of the system are either of the form $a \rightarrow v$, or of the form $c_i a \rightarrow c_i v$, where a is a non-catalyst object and v contains no catalyst. The catalyst are present in the evolution rules but they are not modified by the rule application. Moreover, there are no rules for the separate evolution of catalysts.

The biochemical interpretation of the previous description is the following. We have a cell, delimited by a skin (the cell membrane). Inside, we have cell-organs and free molecules, organized in a hierarchical way. The molecules and the organs float randomly in the "cytoplasmic liquid" of each membrane. Under specific conditions, the molecules evolve (alone or with the help of certain catalysts). This is done in parallel, synchronously for all molecules (a universal clock is assumed to exist). The new molecules can remain in the same region where they have appeared, or can pass through the membranes delimiting this space, selectively. Some reactions modify not only the molecules but the membranes, too; in particular certain chemicals are produced which dissolve the membrane. When a membrane is broken, the molecules previously placed inside it will remain free in the larger space newly created, but the evolution rules of the former membrane are lost. If the external membrane is broken, then the cell ceases to exist.

3. THE POWER OF MEMBRANE SYSTEMS

In this section, we will give a brief overview of known results about membrane systems, for both systems with internal and with external output. In particular results are given with respect to the computational power of such systems, considering the use of the various features (priority relations among the rules, catalysts and dissolving membrane action) and the number of membranes used. We refer to (Dassow and Paun, 1999; Paun, 1998), and (Paun, Rozenberg, and Salomaa, 2000) for details.

In the following, the set of Parikh images of languages in a family FL is denoted by $PsFL$.

3.1 Membrane Systems with Internal Output

We denote by $PsP_m(\alpha,\beta,\gamma)$ the family of the Parikh vectors $Ps(\Pi)$ generated by P systems with internal output of degree at most m, $m \geq 1$, where $\alpha \in \{Pri,nPri\}$ (with or without priority relations among evolution rules), $\beta \in \{Coo,nCoo\}$ (with or without cooperative rules) and $\gamma \in \{\delta, n\delta\}$ (with or without dissolving membrane action). The union of all families $PsP_m(\alpha,\beta,\gamma)$, $m \geq 1$, is denoted by $PsP_*(\alpha,\beta,\gamma)$.

Directly from the definitions, we have the following relations:

Lemma 3.1

1. $sP_m(\alpha,\beta,\gamma) \subseteq PsP_{m-1}(\alpha,\beta,\gamma)$, *for all* $m \geq 1$ *and for all possible* α,β,γ.
2. $PsP_m(\alpha',\beta',\gamma') \subseteq PsP_{m+1}(\alpha,\beta,\gamma)$, *for all* $m \geq 1$ *and for all possible* α,β,γ *such that* α',β',γ' *are equal to* α,β,γ *or to* $n\alpha$, $n\beta$, $n\gamma$, *respectively, when* n *is not already present.*

When the dissolving membrane action is not used, the hierarchy induced by the number of membranes collapses to level two:

Theorem 3.1 $PsP_*(\alpha,\beta,n\delta) \subseteq PsP_m(\alpha,\beta,n\delta)$, *for all* $m \geq 2$ *and for all possible* $\alpha \in \{Pri, nPri\}$ *and* $\beta \in \{Coo, nCoo\}$.

When we use priority relations and cooperative rules, we obtain powerful systems: two membranes suffice to obtain computational completeness.

Theorem 3.2 $PsRE = PsP_*(Pri, Coo,\gamma) = PsP_2(Pri,Coo,\gamma)$, *for* $\gamma \in \{\delta, n\delta\}$.

Other results concern systems which do not make use of priority relations. This is the case of the following:

Theorem 3.3 $PsMAT \subseteq PsP_2$ $(nPri, Coo, \gamma)$, for $\gamma \in \{\delta, n\delta\}$.

Theorem 3.4 $PsCF = PsP_m$ $(nPri, Coo, n\delta) = PsP_*$ $(nPri, nCoo, n\delta)$.

It is not known whether or not the hierarchies PsP_m $(Pri, nCoo, \delta)$, PsP_m $(nPri, Coo, \delta)$ and PsP_m $(nPri, nCoo, \delta)$ collapse; in (Dassow and Paun, 1999) it is conjectured that at least the hierarchy $PsP_m(nPri, nCoo, \delta)$ is an infinite one.

3.2 Membrane Systems with External Output

We denote by $LP_m(\alpha,\beta,\gamma)$ the family of languages $L(\Pi)$ generated by P systems with external output of degree at most m, $m \geq 1$, where $\alpha \in \{Pri, nPri\}$, $\beta \in \{Cat, nCat\}$ (with or without catalysts) and $\gamma \in \{\delta, n\delta\}$. The union of all families $LP_m(\alpha,\beta,\gamma)$, $m \geq 1$, is denoted by LP_* (α,β,γ). By $PsLP_m(\alpha,\beta,\gamma)$ and pLP_* (α,β,γ) we denote the Parikh sets and the permutation closure of such languages, respectively.

Directly from the definitions, we have the following relations:

Lemma 3.2

1. $LP_m(\alpha,\beta,\gamma) \subseteq LP_{m+1}(\alpha,\beta,\gamma)$, *for all* $m \geq 1$ *and for all possible* α,β,γ.
2. $LP_m(\alpha',\beta',\gamma') \subseteq LP_{m+1}(\alpha,\beta,\gamma)$, *for all* $m \geq 1$ *and for all possible* α,β,γ *such that* α', β', γ' *are equal to* α,β,γ *or to* $n\alpha$, $n\beta$, $n\gamma$, *respectively, when* n *is not already present.*

Lemma 3.3 $LP_m(\alpha,\beta,\gamma) \subseteq LP_{m+1}(\alpha,\beta,\gamma)$, *for all* $m \geq 1$ *and for all possible* α,β,γ.

We give now some results, from (Paun, Rozenberg and Salomaa, 2000), to compare the families $PsLP_m(\alpha,\beta,\gamma)$ with the families PsP_m (α,β,γ). Some of them concern the case when the output membrane of P systems is, in the initial configuration, an elementary membrane (thus, no object can reach this membrane from an internal region). We denote PsP'_m (α,β,γ) the family of languages generated by such systems.

Theorem 3.5

1. $PsP'_m(\alpha,\beta,\gamma) \subseteq PsLP_m$ (α,β,γ) *for all* $m \geq 1$ *and for all possible* α,β,γ
2. $PsLP_m(\alpha,\beta,\gamma) \subseteq PsP'_{m'}$ (α,β,γ) *for all* $m \geq 1$, $m' = m+1$ *and for all possible* α,β,γ. If $\gamma = n\delta$ *then* $m' = m$
3. $PsP_m(\alpha,\beta,\gamma) \subseteq PsLP_m$ (Pri, β, γ) *for all* $m \geq 1$ *and for all possible* α,β,γ.

The previous theorem says that, by collecting the output observing the system from outside, we get the same result we obtain in an internal output membrane; of course, this is true modulo the ordering. In fact, with a P system with external output we can get strings, with a precise ordering of their symbols, and not only Parikh vectors.

If we consider the Parikh sets associated with the families of languages generated by P systems which use both priority and catalysts, we have the following:

Theorem 3.6 $PsLP_2(Pri, Cat, n\delta) = PsLP_2(Pri, Cat, \delta) = PsLP_*(Pri, Cat, n\delta) = PsLP_*(Pri, Cat, \delta) = PsRE$.

Other results concern classical families of languages, such us families in the Chomsky area:

Theorem 3.7

1. $REG \subset LP_1(nPri, nCat, n\delta)$
2. $pCF \subseteq LP_1(nPri, nCat, n\delta)$
3. *All languages in* $LP_*(nPri, nCat, n\delta)$ *are semilinear.*

4. STRINGS IN MEMBRANE SYSTEMS

As we have seen, the basic variant of P systems uses objects of an atomic type. Each object is a unique entity, and the application of an evolution rule on it leads to the substitution of the object with a completely new one. Nonetheless, it is quite natural to consider the use of more complex molecules for codifying the information, which leads to more complex data structure. For instance, we can consider complex molecules such as DNA or proteins; the action of the chemical compound on such molecules leads to new molecules which differ, with respect to the starting ones, in some parts only. In this way, objects are still intended as a whole but we have an insight view on their structure.

Such a variant was introduced already in (Paun, 1998) as *Rewriting P system* (or *RP systems*), and then it was the subject of several papers (among others, Dassow and Paun, 1999; Paun, Rozenberg and Salomaa, 2000; Martin, Vide and Paun, 2000). In particular, this variant considers objects which can be described by finite strings over a given finite abet, instead of elements of a finite alphabet. The evolution of an object will correspond to a transformation of the string: the evolution rules are given as context-free rewriting rules.

4.1 Definitions

Formally, a Rewriting Membrane System of degree n, $n \geq 1$, is a construct

$$\Pi = (V, \mu, M_1, \dots, M_n, (R_1,\rho_1), \dots, (R_n,\rho_n), i_0)$$

where:

- V, μ and i_0 are defined as in an usual P system
- M_i, $1 \leq i \leq n$, are finite languages over V.
- R_i, $1 \leq i \leq n$, are finite sets of context free evolution rules. They are of the form $X \rightarrow v(tar)$, where X is a symbol of V and $v = v'$ or $v = v'\delta$. v' is a string over the alphabet V, δ is a special symbol not in V and $tar \in \{here, out, in_m\}$, $1 \leq i \leq n$ represents the target membrane, i.e. the membrane where the string will be sent after the application of the rule.
- ρ_i, $1 \leq i \leq n$, are partial order relations over R_i.

The membrane structure μ and the finite languages $M_1, \dots, M_n$ constitute the initial configuration of the system. We can pass from a configuration to another one by using the evolution rules in parallel on all strings which can be rewritten, obeying the priority relations. The strings generated are collected in a designated membrane, the output one, or we can consider the language constituted by the strings expelled through the skin membrane (this is the case when $i_0 = \infty$ or when i_0 is omitted).

Note that each string is processed by one rule only; the parallelism refers to processing simultaneously all available string by all applicable rules. If several rules can be applied to a string, then we take only one rule and only one possibility to apply it and consider the obtained string as the next state of the object described by the string. It is important to underline the fact that the evolution of strings is not independent, but interrelated in two ways:

1. If we have priorities, a rule r_1 applicable to a string x can forbid the use of another rule, r_2, for rewriting another string y which is present at the same time in the same membrane. Then, we can apply r_2 on y only if r_1 is not applicable to it and to the string x' obtained from x by using r_1.
2. Even without priority, if a string can be rewritten forever, in a membrane or in an itinerary through several membranes, and this cannot be avoided, then all strings are lost, because the computation never stops.

Moreover, a rule specifies the target membrane, as previously said. If a rule contains "*here*" as target membrane (or the target is not specified), it means that the string obtained after the rule is applied will remain in the same region where the rule is applied. If the specified target is "*out*" and the thickness of the membrane is 1, the string will be sent to the region placed immediately outside. Finally, if the specified target is "in_m", where m is a label of a membrane adjacent to the region where the rule is applied, then the string is sent to that membrane. Note that there is an important difference with respect to the P systems: a string is a unique object and it is communicated between two regions as a unique entity. Its symbols do not follow different itineraries, as in the case of multisets; of course, it is possible that several strings are present at the same time in the same region, and they can follow different itineraries.

If a rule contains the special symbol δ, then that membrane is dissolved and it is no longer recreated; the strings in the membrane become strings of the membrane placed immediately outside, while the rules of the dissolved membrane are removed.

We denote by $L(\Pi)$ the language generated by Π and by $RP(Pri, \delta)$ the family of languages generated by rewriting P systems using priority and dissolving membrane action. When one of the feature $\alpha \in \{ Pri, \delta \}$ is not used, we write $n\alpha$ instead of α. If only systems with at most n membranes are used, then we add the subscript n to RP.

We can consider systems in which a terminal alphabet (or output alphabet) $T \subseteq V$ is added. We obtain, in this case, Extended *Rewriting P systems* or *ERP* systems.

4.2 The Power of Rewriting Membrane Systems

In order to describe the results related to Rewriting membrane systems, we have to introduce the so-called *matrix grammars*. Very shortly, such a grammar is a construct $G = (N,T,S,M,F)$, where N, T are disjoint alphabets, $S \in N$, M is a finite set of sequences of the form $(A_1 \to x_1, \dots, A_n \to x_n)$, $n \geq 1$, of context-free rules over $N \cup T$ (with $A_i \in N$, $x_i \in (N \cup T)^*$, in all cases), and F is a set of occurrences of rules in M (N is the nonterminal alphabet, T is the terminal alphabet, S is the axiom, while the elements of M are called matrices).

For $w, z \in (N \cup T)^*$ we write $w \Rightarrow z$ if there is a matrix $(A_1 \to x_1, \dots, A_n \to x_n)$ in M and the strings $w_i \in (N \cup T)^*$, $1 \leq i \leq n+1$, such that $w = w'_i$, $z = w_{n+1}$, and, for all $1 \leq i \leq n$, either $w_i = w'_i A_i w''_i$, $w_{i+1} = w'_i x_i w''_i$, for some $w'_i, w''_i \in (N \cup T)^*$, or $w_i = w_i{+}1$, A_i does not appear in w_i, and the rule $A_i \to x_i$ appears in F. (The rules of a matrix are applied in order, possibly skipping the rules in F if they cannot be applied - one says that these rules are applied in the *appearance checking* mode.)

The language generated by G is defined by $L(G) = \{w \in T^* \mid S \Rightarrow^* w\}$. The family of languages of this form is denoted by MAT_{ac}. When $F = \Phi$

(hence we do not use the appearance checking feature), the generated family is denoted by *MAT*.

It is known that $CF \subset MAT \subset MAT_{ac} = RE$, all inclusion being proper. All one-letter languages in the family *MAT* are regular, see (Hauschild-Jantzen 1994).

Let us consider RP systems without priority relations and without dissolving membrane action. The family of languages generated by such systems with a single membrane is exactly the family of CF languages; using two membranes, we can generate non CF languages:

Theorem 4.1
1. $CF = RP_1(nPri, n\delta)$.
2. $CF \subset RP_2(nPri, n\delta)$.

If we add a membrane to systems with a single membrane, then the power of such systems is increased. Hence, a natural question is about the power of systems where we do not put restrictions on the number of membranes that can be used. Is it possible to obtain universality in this way, without adding any other feature? The answer is negative, even if we use extended systems. In fact, in (Ferretti-Mauri-Paun-Zandron 2001) it is shown that if we do not make use of priority relations nor of membrane dissolving action, then RP systems do not have enough power to generate all RE languages. The class of languages generated by such systems is exactly the class of languages generated by matrix grammars without appearance checking:

Theorem 4.2 $MAT = ERP_4(nPri, n\delta) = ERP_*(nPri, n\delta)$.

On the contrary, if we make use of priority relations to control the application of the rules, then we obtain the full computational power, with a very simple membrane structure.

Let us denote by *ORD* the family of languages generated by context-free ordered grammars. Obviously, we have $RP_1(nPri, n\delta) = ORD$. It is known that if we use priority relations among evolution rules, then two membranes are enough to get a system able to generate any RE language, thus a computationally complete system.

Theorem 4.3 $RE = RP_2(nPri, n\delta)$.

In view of these results, the dissolving membrane action is not necessary when we use priority relations: if we have a single membrane, then nothing can be dissolved (remember that the skin membrane cannot be dissolved); to obtain the full computational power, it suffices to use a membrane structure with two membranes. It is not known if the power of rewriting P systems can be increased using the dissolving membrane action without priority relations among the rules.

4.3 Regulated Rewriting in Membrane Systems

Of course, priority relations is not the only way to obtain universality results. The same results can be obtained by considering different possibilities to apply the rewriting rules which are classic in the formal language theory. In particular, in the case of imposing that each string is rewritten in the *leftmost* possible position, we get a characterization of recursively enumerable languages by systems with seven membranes.

We denote by $L_{left}(\Pi)$ the language generated by a system Π in this way and by $ERP_m(left)$, $m \geq 1$, we denote the family of all such languages, generated by systems with at most m membranes (if the degree of the systems is not bounded, then we replace the subscript m by *). We have the following:

Theorem 4.4 $RE = RP_7$ (left) $= ERP_6(left)$.

Another classical restriction which can be considered is the case when the rules are applied with respect to some *forbidding conditions*. This means that the rules are of the form $\langle X \rightarrow x;\ F \rangle$, where $F \subseteq V$, and the rule $X \rightarrow x$ can be applied only to the strings which do not contain any symbol from F (when $F = \emptyset$ this means that the rule is applied without any restriction, in the free mode). We denote by $L_{forb}(\Pi)$ the language generated by a P system Π using such rules, and by $ERP_m(forb)$, $m \geq 1$, we denote the family of all such languages, generated by systems with at most m membranes. Two membranes are shown to suffice in the case of using *forbidding conditions* associated with rules:

Theorem 4.5 $RE = RP_2(forb) = RP_*(forb) = ERP_2(forb) = ERP_*(forb)$.

If we consider the application of the rules with *permitting conditions*, then the rules are still of the form $\langle X \rightarrow x;\ F \rangle$, where $F \subseteq V$, but now the rule $X \rightarrow x$ can be applied only to the strings which contain all symbols in F (again, $F = \emptyset$ means no restriction, hence the free application of the rule). We denote by $L_{perm}(\Pi)$ the language generated by a P system Π using such rules and by ERP_m $(perm)$, $m \geq 1$, we denote the family of all such languages, generated by systems with at most m membranes. As previously said, if the degree of the systems is not bounded, then we replace the subscript m by *.

Surprisingly enough, the use of permitting conditions with ERP systems not only does not lead to a characterization of RE languages, but it simply does not increase the power of rewriting P systems: the class of languages generated by such systems coincides again with the class of languages generated by matrix grammars without appearance checking.

Theorem 4.6 $MAT = ERP_{*}(perm) = ERP_2(perm)$.

5. DIFFERENT KINDS OF COMMUNICATION: ELECTRICAL CHARGES AND GEMMATION OF MOBILE MEMBRANES

The communication of objects by labelled membranes is a simple way to represent the communication of molecules within a cell. In fact, in a real cell a lot of molecules (in particular ions) are moved according to different rules, for instance by means of electrical charges. A membrane is composed by a thin sheet of nonconducting material surrounded on both sides by electrically conducting material, hence it can store an electric charge. Using this electric potential power, it is possible to move molecules against the concentration gradient.

Another type of cellular-level communication is by means of vesicles, that is little parts of a membrane that transport molecules by budding from the source membrane and by fusing with the target membrane. This two kinds of communication were the subject of, among others, (Besozzi, Zandron, Mauri and Sabadini, 2001; Paun, 2000a; Petre and Petre, 1999; Zandron, Ferretti and Mauri, 2001). We briefly introduce, in this section, this two types of communication.

5.1 Communication by Means of Electrical Charges

This kind of communication of objects through membranes was proposed in (Paun 2000a): evolution rules do not specify the label of the membrane where the objects are sent; instead, "electrical charges" are used, associated to membranes and to objects: they can be marked with "positive" (+), "negative" (–) or “neutral”. An object marked with + (respectively –) will enter a membrane marked with – (respectively +), nondeterministically chosen from the set of membranes immediately inside to the membrane delimiting the region where the object is produced. The neutral objects are not introduced into an inner membrane.

In other words, the rules are of the form $X \rightarrow v(p)$ where X is a symbol of V and $v = v'$ or $v = v'\delta$. v' is a string over V, δ is a special symbol not in V and $p \in \{out, 0, +, -\}$. Note that the substring (p) is not inserted into the sentential form as subword. It is only used to communicate the objects through the membranes.

The use of this kind of communication does not decrease the computational power of P systems. In particular, in (Paun 2000a) it is shown that a variant of P systems which make use of Electrical Charges characterize the computable enumerable set of numbers.

Moreover, the feature of communication by means of electrical charges can give further advantages. For example, it allows us to control the communication of objects with more general communication rules. In fact, a rule does not have to specify the precise label of the inner membrane where an object have to be sent, but only a subset of membranes (positive or negative ones). By exploiting this, it is possible to obtain Membrane Systems with structures of a very simple form: for instance, in (Zandron, Ferretti and Mauri, 2001) the following normal form was proposed for rewriting P systems which make use of electrical charges:

Theorem 5.1 *Every Recursively Enumerable language can be generated by a RP system with electrical charges (with priority relations and without dissolving membranes action) in which each membrane contains exactly two rules and the skin membrane contains only elementary membranes.*

5.2 Communication by Means of Gemmation of Mobile Membranes

Cellular membranes are selectively permeable to many small substances as water and ions, but not to bigger ones as proteins. Such substances are communicated inside or outside the cell by means of vesicles, which are little parts of a membrane, encased on their cytosolic face by a specific protein that causes their budding from the membrane. When the vesicle fuses with its target membrane, the carried proteins are introduced inside it, where they can be modified by other chemical reactions. Many cellular compartments use this kind of communication, in particular this is the case of the Golgi apparatus, a stack of distinct elementary membranes (i.e. membranes without other membranes inside) where, in sequence, many proteins are stepwise modified and then sent to another Golgi-region. Specifically, only the substances that have completed their “maturation path” inside the current region can be communicated by a vesicle to the next destination (for example, proteins must reach their exact folding before entering a budding vesicle).

All these natural features were considered in P systems with Gemmation of Mobile Membranes, introduced in (Besozzi, Zandron, Mauri and Sabadini, 2001). These systems have a simple membrane structures (the skin membrane can only contain elementary membranes) and the operations on string-objects are of a biochemical inspiration: mutation, replication and splitting rules. A so-called *meta-priority* is defined, between the set of *classical evolution rules* and the set of *pre-dynamical rules*, which are the rules that give rise to the *gemmation* of mobile membranes (that is, the budding of vesicles in the cell). The meta-priority is needed to the aim of simulating the completion of the maturation path of an object. After a pre-dynamical rule has been used, the phases of gemmation and fusion of mobile membranes take place.

In (Besozzi, Mauri, Paun and Zandron, 2001) it is shown that such systems which make use only of pre-dynamical rules suffice to obtain universality. This is quite a surprise: there is no need of other kind of rewriting rules, nor in/out communication rules, and, more important there is no need of other features which are usually needed in other variants to increase the generative power, such as, for example, priority relations among rules. In fact, let us denote by $EGemP_m(Dyn)$ the family of languages generated by extended gemmating P systems of degree m, for $m \geq 1$, which make use of pre-dynamical rules only. We have the following:

Theorem 5.2 $EgemP_9(Dyn) = EgemP_*(Dyn) = RE$.

6. MEMBRANE SYSTEMS AND COMPUTATIONAL COMPLEXITY: MEMBRANE DIVISION

One important feature of membrane systems concerns the intrinsic parallelism of the model; at each step of computation, all the objects which can be used as a subject of a rewriting rule evolve in parallel. Moreover, it is quite easy to create exponentially many objects in linear time, by making use of rules of the form $a \to aa$. Thus, one natural question arises: do the basic variant posses enough parallelism to solve NP-complete problems in polynomial time? In (Zandron, Ferretti and Mauri, 2000) it is shown that this is not the case, unless $P = NP$:

Theorem 6.1 *Every deterministic P system can be simulated by a deterministic Turing machine with a polynomial slowdown.*

One possibility to increase the speed of computation of such systems, in order to gain enough power to efficiently solve NP-complete problems, is to trade time for space. In (Paun, 2001b) one considers the idea of creating new membranes by division of existing membranes. Starting from a membrane, it is possible to obtain two exact copies of that membrane, with the same set of objects and rules of the starting membrane (except for the object used to apply the rule). In particular, two different types of division rules were considered: division rules for elementary membranes and division rules for non-elementary membranes. In (Paun, 2001b) it is shown how to solve a well-known NP complete problem, the Satisfiability problem (SAT), in polynomial time with respect to the input length, using P systems with membrane division and symbol objects, while in (Krishna and Rama, 1999) it is shown how to solve the Hamiltonian Path Problem (HPP) in linear time, using a variant of such the previous system where a membrane can be divided in an arbitrary number of copies, and not only in two copies.

The solutions presented in these papers make use of division for both elementary membranes and non-elementary membranes, and, in this last case, the division only depends on the properties of the inner membranes. To avoid this, in (Paun, 2000) a solution for HPP was proposed where the division for non-elementary membranes is controlled by objects only, while in (Zandron, Ferretti and Mauri, 2000) it is shown that the previous problems can be solved in polynomial time by P systems which make use of division for elementary membranes only.

A different approach to obtain exponential space to solve NP-complete problems in polynomial time is to consider string objects which can be replicated. For details on this topic, we refer to (Castellanos, Paun-Rodriguez and Paton, 2000; Krishna and Rama, 2001; Manca, Martin, Vide and Paun, 2001; Martin, Vide, and Mitrana, 2000).

7. CONCLUSIONS

In this paper, we have introduced the basic ideas related to membrane computing and we presented some results of the area. A lot of research topics deserve to be further investigated. In the following, we will point out some of these topics we think are important for the future development of the area.

Of course, one fundamental question concerns the possibility of implementing membrane systems either in a bio-lab (*in vivo* or *in vitro*) or on an electronic computer, possibly using a dedicated hardware. Even if the bio-lab realization has not yet been approached, Tom Head of the Binghamton University (New York) did several DNA computing experiments trying to simulate a membrane system using the so-called *aqueous computing* (Head, 2001). Even if this is just a simulation using circular DNA molecules, different basic operations concerning membrane systems were considered, such as membrane division, creation of new membranes or separation of string-objects which satisfy a given property in a new membrane. For what concern the possibility of implementing P systems on an electronic computer using existing media, some attempts have already been carried out by different authors. Their work can be found in (Baranda, Castellanos, Molina, Arroyo and Mingo, 2000; Suzuki and Tanaka, 2000; Baranda, Castellanos, Arroyo and Gonzalo, 2001; Ciobanu and Paraschiv, 2001). An approach to implement a membrane system on a purposely designed architecture can be found in (Stefan, 2000).

The implementation of such a systems would benefit of a theoretical model which consider as many as possible features we have in a real cell. For instance, we think it would be interesting to consider operations like endocytosis and fagocytosis. Such operations correspond to the operation of moving a membrane, together with its contents, through another membrane as a whole, instead of moving, as in the basic variant, only the contents of the membranes. Systems which make use of this operation were considered in (Atanasiu, Martin and Vide, 2001), showing that such operation can be useful to avoid the division of membranes under the influence of inner membranes, and to implement recursive functions. We think that this operation deserves further investigations as well as other types of communication which take place at a cellular level, such as communication by means of ionic pumps, symport and antiport or budding of vesicles, that, up to today, have been only partially investigated (Besozzi, Zandron, Mauri and Sabadini, 2001; Martin, Vide, Paun and Rozenberg, 2001).

Another fruitful idea is that of consider systems obtained combining membrane systems with other computational models. Some works already explored the power of systems obtained combining Membrane systems with Splicing systems (see, for instance, Paun, Rozenberg and Salomaa, 2000; Paun and Yokomori; 1999). Other possibilities can be obtained combining Membrane and Lindenmayer systems, for example using the rules in parallel in a different way, by applying the rules to each symbol of the string that can be rewritten, as in Lindenmayer systems or considering Quantum Membrane systems, that is Membrane systems which use q-objects.

Other research topics and other problems related to the area can be found in (Paun, 2000b) and in (Paun, 2001a).

REFERENCES

Atanasiu, A., Martin-Vide, C., 2001, Recursive calculus with membranes. *Pre-Proceedings of Workshop on Membrane Computing*, Curtea de Arges, Romania.

Baranda, A. V., Castellanos, J., Arroyo, F., and Gonzalo, R., 2001, Towards an Electronic Implementation of Membrane Computing: A Formal Description of Nondeterministic Evolution in Transition P Systems. *Proc. 7th Intern. Meeting on DNA Based Computers* (N. Jonoska, N.C. Seeman, eds.), Tampa, Florida, USA, 273-282.

Baranda, A. V., Castellanos, J., Molina, R., Arroyo, F., and Mingo, L.F., 2000, Data structures for implementing transition P systems in silico. *Pre-proc. Workshop on Multiset Processing*, Curtea de Arges, Romania, and TR 140, CDMTCS, Univ. Auckland, 21--34.

Besozzi, D., Mauri, G., Paun, G., and Zandron, C., 2001, Gemmating P Systems: Collapsing Hierarchies. (submitted).

Besozzi, D., Zandron, C., Mauri, G., and Sabadini, N., 2001, P Systems with Gemmation of Mobile Membranes. In *Proc. of ICTCS 2001* (A. Restivo, S. Ronchi Della Rocca, L. Roversi, eds.), LNCS 2202, Springer-Verlag, Berlin, pp. 136-153.

Castellanos, J., Paun, G., and Rodriguez-Paton, A., 2000, P Systems with Worm-Objects. In *IEEE 7th. Intern. Conf. on String Processing and Information Retrieval, SPIRE 2000*, La Coruna, Spain, pp. 64-74

Ciobanu, G., and Paraschiv, D., 2001, Membrane software. A P system simulator. In *Pre-proc. Workshop on Membrane Computing*, Curtea de Arges, Romania, and Technical Report 17/01 of Research Group on Mathematical Linguistics, Rovira i Virgili University, Tarragona, Spain, pp. 45 50.

Dassow, J., and Paun, G., 1999, On the Power of Membrane Computing. *J. Univ. Computer Sci.* **5** (2): 33-49.

Dassow, J., and Paun, G., 1989, *Regulated Rewriting in Formal Language Theory*, Springer-Verlag, Berlin.

Ferretti, C., Mauri, G., Paun, G., and Zandron, C., 2001, On three Variants of Rewriting P Systems. In *Pre-Proceedings of Workshop on Membrane Computing*, Curtea de Arges, Romania, and Technical Report 17/01 of Research Group on Mathematical Linguistics, Rovira i Virgili University, Tarragona, Spain, pp. 63-76.

Hauschild, D., and Jantzen, M., 1994, Petri nets algorithms in the theory of matrix grammars. *Acta Informatica*, **31**: 719-728.

Head, T., 2001, Aqueous computations as membrane computations. *Romanian J. of Information Science and Technology* **4** (1-2).

Krishna, S.N., and Rama, R., 1999, A variant of P systems with active membranes: Solving NP-complete problems. *Romanian J. of Information Science and Technology* **2** (4).

Krishna, S.N., and Rama, R., 2001, P Systems with Replicated Rewriting. *J. Automata, Languages and Combinatorics* (to appear).

Manca, V., Martin-Vide, C., and Paun, G., 2001, On the Power of P Systems with Replicated Rewriting. *J. Automata, Languages, and Combinatorics* (to appear).

Martin-Vide, C., and Mitrana, V., 2000, P Systems with Valuations. In *Unconventional Models of Computation* (I. Antoniou, C.S. Calude and M.J. Dinneen, eds.), Springer-Verlag, London, pp. 154-166.

Martin-Vide, C., and Paun, G., 2000, String Objects in P Systems. In *Proc. of Algebraic Systems, Formal Languages and Computations Workshop*, Kyoto, RIMS Kokyuroku, Kyoto Univ., pp. 161-169.

Martin-Vide, C., Paun, G., and Rozenberg, G., 2001, Membrane Systems with Carriers. *Theoretical Computer Sci.* (to appear).

Paun, A., 2000, On P Systems with Membrane Division. In *Unconventional Models of Computation* (I. Antoniou, C.S. Calude and M.J. Dinneen, eds.), Springer-Verlag, London, pp. 187-201.

Paun, A., and Paun, M., 2001, On Membrane Computing Based on Splicing. In *Where Mathematics, Computer Science, Linguistics, and Biology Meet* (C. Martin-Vide and V. Mitrana, eds.), Kluwer, Dordrecht, pp. 409-422.

Paun, G., 1998, Computing with membranes. *J. of Computer and System Sciences*, **61** (1): 108-143, and *TUCS Research Report No 208*, (http://www.tucs.fi).

Paun, G., 2000a, Computing with membranes - a variant: P systems with polarized membranes. *Intern. J. of Foundations of Computer Science* **11** (1): 167-182.

Paun, G., 2000b, Computing with membrane (P-systems): Twenty-six research topics. *CDMTCS TR 119*, Univ. of Auckland.

Paun, G., 2001a, Further Research Topics About P Systems. In *Pre-Proc. of Workshop on Membrane Computing*, Curtea de Arges, Romania, Technical Report 17/01 of Research Group on Mathematical Linguistics, Rovira i Virgili University, Tarragona, Spain, pp. 243-250.

Paun, G., 2001b, P systems with active membranes: attacking NP complete problems. *J. Automata, Languages and Combinatorics* **6** (1): 75-90.

Paun, G., Rozenberg, G., and Salomaa, A., 2000, Membrane computing with external output. *Fundamenta Informaticae* **41** (3): 259-266.

Paun, G., and Yokomori, T., 1999, Membrane computing based on splicing. In *Proc. of Fifth Intern. Meeting on DNA Based Computers* (E. Winfree and D. Gifford, eds.), June 1999, MIT, pp. 213-227.

Petre, I., 1999, A normal form for P systems. *Bulletin of EATCS* **67**: 165-172.

Petre, I., and Petre, L., 1999, Mobile Ambients and P Systems. In *Workshop on Formal Languages*, FCT99, Iasi, 1999, *J. Universal Computer Sci.* 5 (9): 588-598.

Rozenberg, G., and Salomaa, A., (eds.), 1997, *Handbook of Formal Languages*. Springer-Verlag, Heidelberg.

Stefan, G., 2000, Membrane computation with cellular automata on a dedicated hardware. In *Workshop on Multiset Processing*, Curtea de Arges, Romania.

Suzuki, Y., and Tanaka, H., 2000, On a LISP implementation of a class of P systems. *Romanian J. of Information Science and Technology* **3** (2): 173-186.

Zandron, C., Ferretti, C., and Mauri, G., 2000, Solving NP-Complete Problems Using P Systems with Active Membranes. In *Unconventional Models of Computation* (I. Antoniou, C.S. Calude and M. J. Dinneen, eds.), Springer-Verlag, London, pp. 289-301.

Zandron, C., Ferretti, C., and Mauri, G., 2001, Two Normal Forms for Rewriting P Systems. In *Proc. Third. Intern. Conf. on Universal Machines and Computations* (M. Margenstern and Y. Rogozhin, eds.), Chisinau, Moldova, 2001, Springer-Verlag, pp. 153-164.

Zandron, C., Ferretti, C., and Mauri, G., 2001, Using Membrane Features in P Systems. *Rom. Journ. of Information Science and Technology* **4** (1-2): 241-257.

General Systems

Transforming Leadership Through Coalitions: Building the Ethics of Sustainable Development in Globalization

LARRY A. MAGLIOCCA[*], and GIANFRANCO MINATI[#]
[*]*The Ohio State University*
[#]*AIRS – Associazione Italiana per la Ricerca sui Sistemi*

1. THE PROMISES AND RISKS OF GLOBALISATION

Globalisation, for better or worse, seems to have become reality in the last two decades without fanfare or referendum. The meaning of globalisation also seems to be ambiguous and contradictory. It has the promise of new markets for developed countries at the same time it is blamed for the loss of jobs. It is hailed as the source of new capital for developing nation while it is excoriated for increasing an overwhelming debt for the people of these countries to repay. It is either praised or damned by various political groups or movements without reference to one side or another of its contradictions. Due to recent public demonstrations and violence in Seattle, Washington and Genoa, Italy, it has become an international symbol that serves as a catalyst and conduit for latent frustration and anger.

What is globalisation? According to the International Monetary Fund (2000/2001), economic "globalisation" refers to the increasing integration of economies, particularly through trade and financial flows, and also the movement of people (labour) and knowledge (technology) across international borders. There are also broader cultural, political and

Emergence in Complex, Cognitive, Social, and Biological Systems, Edited by
Minati and Pessa, Kluwer Academic/Plenum Publishers, New York 2002

environmental dimensions of globalisation that follow this increasing integration.

Weisbrot (1999) summaries many of the contradictory viewpoints about globalisation among economists, business leaders, government leaders and the general public. In the development of globalisation, many of the journalists and other writers usually take a cautious but positive viewpoint to globalisation. An example of a recent book by Thomas Friedman (1999), a New York Times journalist, is a prominent example of the type of journalistic coverage currently provided in the United States; a brief synopsis of the key points is provided below. But public polls have shown the majority of the public in the U.S. have serious reservations about globalisation. For instance, the real wages of American workers have dropped over the last two decades. In addition, through agencies like the World Bank and IMF, the American public perceives an increasing responsibility and concern about involvement in other countries economic and fiscal decision-making. The most notable example of IMF intervention occurred in the recent fiscal crises in Asia. The IMF intervention, according to Weisbrot, increased the instability and poverty levels of several Asian countries through their inappropriate handling of fiscal policies. Related concerns such as child labour, destabilizing developing countries through sudden capital flight, cultural hegemony by the powerful countries on the developing nations, and continuing environmental degradation that seems destined to slide downward. Weisbrot summaries the heart of the problem from many critics: the dominant countries continuously change the rules of the globalisation game so that power and income are redistributed to the elite; indeed, governance and influence by the elected bodies of the general public have diminishing influence while unelected bodies such as IMF, World Bank, and transnational corporations now make the critical decisions behind closed doors.

Thomas Friedman's book, The Lexus and the Olive Tree, identifies globalisation as the opportunity for world economic and political development, led by the United States, in the aftermath of the end of the Cold War. Primarily this book is a series of anecdotes that illustrates the author's point of view. Symbolically, the collapse of the Berlin Wall of the Cold War is replaced by the global symbol of potential economic and political integration, the Internet. In essence, the integration of information, capital, and technology are creating a global village. According to Friedman, "...the two biggest threats to today's global financial system—crises triggered by 'bad lenders' and 'bad borrowers'." The question of sustainable development is never directly raised as an issue of limits to growth, but rather on the role of the United States as a stable power structure. He advocates the provision of a social safety net for displaced workers as a major solution to the potential problems of globalisation. If his work represents a consensual view of globalisation of the United States, there is a "blind spot" to sustainable development issues.

With the recent negative international attention to globalisation, important issues such as sustainable development are forced on to the agenda of international leaders of business and government. The most vocal stakeholder groups will be those people who believe globalisation is having or will have a negative impact on their future. As international leaders respond to the discordant voices concerning globalisation, will they adopt a "fix-it" strategy or a change strategy? The fix-it strategy is a form of problem solving that considers the issues in piecemeal fashion to repair the obvious flaws. This type of "incrementalism" or "muddling-through" approach attacks the symptoms of systemic situations, not the deep drivers of change. The change strategy, however, is a wholistic and participative process of working on the critical, deep drivers of systemic change. The change strategy approach requires a unique type of political leadership, identified as transforming leadership, which will be advocated here. However, an international coalition of leaders would require an emergent ethic of sustainable development based on values and aspirations of a critical mass of stakeholders. This critical mass of stakeholders (identified as the "cultural creatives" by Ray and Anderson (2000)) would need to advocate a change strategy approach to globalisation that seriously considers the future stakeholders of the planet. Although business and government leaders declare professed ethics of sustainable development, the ethics-in-practice embedded in nationalistic self-interests, reflects the values and interests of international business, permits only a negotiated, satisficing approach to globalisation.

Perhaps the frustration may be due to the lack of a forum for public debate or dialogue. It appears to the public that globalisation, as an international movement by governments and large, multi-national corporations, never appeared to be a choice on their part. Alienated from participation, globalisation arouses a cynical attitude about its purposes and long term outcomes for a significant and growing portion of the international community. The question becomes: As the international community has started down the path of globalisation, what are the emergent ethics of international development and how do the international stakeholders affected by these developments participate in its development? How has political leadership (using James MacGregor Burns' (1978) concepts of transactional and transformational leadership) affected the emergent ethics of globalization? How will future stakeholders of the planet be considered through the emergent ethics of sustainable development in an environment of finite resources?

The purpose of this article is to identify how the traditional role of political leadership (in the form of transactional leadership) contributes to the disjointed effects of globalization and to propose the viability of transforming leadership as a source of change strategies of sustainable international development. James MacGregor Burns (1978) in his award-winning book, Leadership, identified a type of leadership whereby national leaders and followers elevated each other to new levels of moral consciousness, shared purposes and commitment to action. However, his leadership studies focused primarily on transforming leaders and their followers within a specific nationalistic context, e.g. Gandhi in India and Lenin in the Soviet Union. Could the same type of transforming leadership that is successful in a nationalistic setting assist the international communities efforts come to terms with globalization? Transforming leadership evolves from the needs, values, and aspirations of leaders and followers in situations of pluralistic views and conflictual choices. Transactional leadership, on the other hand, flows ubiquitously from a millennium of barter and negotiations that establish decisions by the powerful with reference to the immediate competitive business environment and the gain of competitive advantage at whatever costs.

The stakes are high in globalization for the entire international community, but particularly for those stakeholders of the planet who are yet to be born. If globalization proceeds in its current "free markets" path of exploitive economic development and the destabilizing effects of environmental degradation and cultural hegemony, the future stakeholders will bear the costs. The costs may well be impossible to meet or reverse in environmental degradation and a worldwide culture of infinite materialism. Transactional leadership is the default path into this future; it is the favored (and many times only) style of leadership of business and government. For instance, the recommended solution of the IMF (2000/2001) to its own concerns of the destabilizing influence of globalization is to refer to the default path of open market economies as the solution to the most difficult problems in the long run. For instance, it is acknowledged that gains are not evenly gained between or within certain countries. These very countries are the ones who will be unable to provide a safety net for displaced workers because of limited resources. The professed ethics of the IMF are, in reality, a default to the open markets philosophy of dealing with complex issues whether it fits or not:

… workers in declining older industries may not be able to make an easy transition to new industries. What is the appropriate policy response? Should governments try to protect particular groups, like low-paid workers or old industries, by restricting trade or capital flows? Such an approach might help some in the short-term, but ultimately it is at the expense of the living standards of the population at large. Rather, governments should pursue policies that encourage integration into the global economy while putting in place measures to help those adversely affected by the changes. The economy as a whole will prosper more from policies that embrace globalization by promoting an open economy, and, at the same time, squarely address the need to ensure the benefits are widely shared.

Transforming leadership would become a significant shaping force of the future of globalization if the emergent ethics of globalization would embed an explicit concern for future stakeholders. However, from our studies of ethics as an emergent property of the loosely coupled system known as globalization, the prospects are not encouraging for such ethics to emerge without the determined efforts of an international transforming leadership coalition.

2. PROFESSED, PRACTICED AND EMERGENT ETHICS OF GLOBALIZATION

Normative ethics is intended as rational inquiry that describes how people in a particular social system ought to behave in relation to others and their environment. Normative ethics derived from the expressed values of its stakeholders are a powerful context for change processes when activated by a collective purpose. However, one must recognize that all social systems have an ethic of interactive relationships whether "professed" or "in practice". In times of change, ethics in practice may be temporarily in flux as emergent ethics. At this nexus of time and opportunity, emergent ethics may serve as the powerful foundation for systems change work under the proper type of leadership. The practiced and the emergent ethics evolve from the micro and local social rules of the components of the social system (e.g., business organizations). The practiced and the emergent one may be considerably different from the professed ethics, as will be identified in international efforts to deal with sustainable development of globalization. This method of identifying practiced and emergent ethics is not intended to attribute deception or dissembling by the major social agents of the system (although there may be instances of this type of deception consciously practiced).

It is important to note here that in nexus of opportunities for change one must be sure that stakeholders make the distinctions between practiced ethics and professed ethics. Without the presence of international leadership, the professed ethics of various international gatherings or coalitions of stakeholders have at least two types of effects. In the best scenario, professed ethics can be the genuine expression of stakeholders of an idealized future that expresses a strong belief in certain values. In the worst scenario, professed ethics can be used as a posturing device that exasperates and inflames the underlying conflicts in values without any intention of opening up a dialogue about those values. In the globalization process, transnational corporations may perceive it to be in their self-interests to espouse a professed ethic of sustainable development or, at least, consideration or sensitivity to the environmental and human problems of their efforts. They may even make token efforts to demonstrate these professed ethics for public consumption. However, the ethics-in-practice do not substantially change from one of exploiting the short-term opportunities of low environmental and human costs to the bottom line of their balance sheets. One might assume with some degree of confidence that the professed ethics of many transnational businesses to be unconsciously deceptive if one perceives the structural restraints that many business enterprises operate under in implementing their actual ethics-in-practice (Minati, In Press).

Gini (1998) identifies "structural restraints" that shape business ethics. Ethics-in-practice develop through the habits and culture of various contexts of business operation that are directly influenced by organizational leaders. One restraint is called "organizational totalitarianism" by Schwartz (1990). He identifies the dark side of ethics-in-practice as an approach to conduct to be accepted without reflection by its practitioners as the rules of the game. In *Narcissistic Process and Corporate Decay*, Schwartz identifies the rules of business as the "law of the jungle" that requires the micro social rules of getting ahead by all of the means necessary. This "organizational totalitarianism" requires conformity of its practitioners and preservation of this approach as the status quo. Success through competition is the sole justification of the ends. A related concept is proposed by Jackall (1988) through an "institutional logic" that reduces business ethics to unreflective loyalty to the organizational leaders to create "patrimonial bureaucracies". It becomes "...the creation of a private moral universe...that is totalitarian (self-sustained), solipsistic (self-defined), and narcissistic (self-centered)." (Gini, 1998, p. 369). Gini acknowledges the significance of this structural constraints to a moral business ethic. He concludes that without moral leadership, ethics in business and organizational life will not emerge or be sustainable. Leadership in the business culture creates the barriers and holds the keys to resolution of an emergent ethic of sustainable development.

Thus, the social system of activities we call globalization inherit the micro, local rules of thought and behavior that engender a continuum of competitive ideas. Under the influence of Hayek's (1944) seminal work, *The Road to Serfdom*, economics and political philosophy of free market systems and democracy have become intertwined. The competitive idea of free market regulation assumes a superior quality of self-governance through competition that is used as justification of economic policies and practices counter to the broader interests of stakeholders who are profoundly affected. The ethic of free market regulation as the primary determinant of international development has profound implications particularly for future stakeholders (yet to be born) who have no voice in the decisions being made. Hayek's proposed the benefits of Nineteenth Century liberalism—unfettered support for the conditions that enable individual initiative, rather than bureaucratic central control over economic planning—to Twentieth Century economic planning, clearly anticipating globalization, which excludes a central thesis of this article:

Planning on an international scale, even more than is true on a national scale, cannot be anything but a naked rule of force, an imposition by a small group on all the rest of that sort of standard and employment which the planners think suitable for the rest ... To undertake the direction of the economic life of people with widely divergent ideals and values is to assume responsibilities which commit one to the use of force; it is to assume a position where the best intentions cannot prevent one from being forced to act in a way which to some of those affected must appear highly immoral. (pp. 244-245)

Hayek points out the exceedingly difficult task of an international planning effort to regulate transnational concerns with widely varying "local" values. Hayek makes a powerful case for the protection of individual initiative through "free market regulations" rather than central economic planning; however, he also identified the central role of "values" and the concept of moral standards. *Is it possible to advocate for the values and ethics of sustainable development without assuming the counterproductive aspects of central economic and political control?*

Without consensus by the international community around the emergent ethics of globalization to support sustainable development, the only available exogenuous methods of control relay upon national and international law. A means for regulating important practices (if not usually the underlying values and ethics) is through regulations of national and international law which should manage interactions among the business corporations engaged in globalization. Interactions among countries should be disciplined by international law. International conferences are organized to establish rules to discipline relationships and interactions on some topics among countries. The general concept is to establish standards generally accepted by all countries, achieving some type of global consensus. International conferences have been organized on subjects related to development in an effort to set standards. These include such notable efforts as:

- The United Nations Declaration on the Human Environment at Stockholm (1972),

- Declaration on Environment and Development at Rio de Janeiro (1992),
- United Nations International Conference on Population and Development at Cairo (1994),
- Third Conference of the Parties at Kyoto, Japan (1997),
- Conference of the World Trade Organization at Seattle (1999), and
- Conference of the World Trade Organization at Genoa, Italy (2001).

In an interview with Cornish (2001) Maurice Strong discusses his new book *Where on Earth Are We Going*. There are intimate, behind-the-scenes accounts by the person in the United Nations most responsible for the convening the Stockholm Conference (1972) and the Rio Declaration (1992). The difficulties in the Stockholm Conference were set in the context of cold war politics. This tended to obscure the latent conflict of practiced ethics by the individual countries involved. However, continuing to this day, the distrust and antagonism among the industrialized and developing countries as to the meaning and responsibilities for sustainable development have been obvious and undiminished. In essence, how would an equitable solution be gained given the vast developmental differences between countries and their needs and resources? Free of the overt antagonisms of the cold war, the Rio Conference in 1992 had an unparalleled opportunity with the heads of 116 governments in attendance. The Rio Conference or "Earth Summit" was very productive in the professed ethics of the participants. It produced two historical conventions, one on climate change and the other on biodiversity. In addition, *Agenda 21* set out a comprehensive framework for a new global partnership for sustainable development. Strong concedes in a recent interview (Cornish, 2001) that the Rio agreements require the national governments to act though they continue to resist in spite of their "professed ethics" in the documents. Strong's comments present his estimation of the consequences of this inability to act:

We are literally the architects now of our own future ... The next 30 years will be decisive. If we have not made the shift toward a sustainable world by the end of the second or third decade of this century, then the costs and risks of doing so will be so great that it will be unlikely to be made at all ... I am somewhat pessimistic when I look at the evidence ... But operationally I have to be optimistic. (p. 30)

The attempt to establish rules for interacting national growths and developments fail mainly because the expected scenarios do not appear to be acceptable for economic reasons. The ethics-in-practice of the participants works against achieving general consensus whenever the competition of the free market is threatened. The normative ethic of free market regulation as the ultimate justification for decision-making in globalization processes continues to suppress consideration of the emergent ethic of sustainable development.

3. THE VALUE BASE FOR THE EMERGENT ETHICS OF SUSTAINABLE DEVELOPMENT

One of the early internationally visible efforts to raise concerns about the future of humankind was the Club of Rome. The Club of Rome's Executive Committee choose during the 1970s and 1980s to develop system forecasting and simulation of world problems to investigate the concerns about unrestrained population growth, industrialization, desertification, and resource depletion. As the Club's President, Aurelio Peccei (1981) reports, the modeling and simulation of these interacting and interdependent problems was necessary to combat the prevailing notion of unrestrained growth and unlimited resources. The ethics-in-practice of the business culture, believing in exploitive and unlimited resources, would be the basis of a worldwide catastrophe by 2025. Incidentally, this timeline closely resembles the three decades timeline predicted by Maurice Strong. Through a series of published reports by the Club of Rome members, two major concepts were established that assisted in the value base and ethics of sustainable development (Peccei, 1977). One was the introduction of the "problematique" concept that proposed many of the world's problems were interactive and interdependent; attempts to resolve them required a wholistic, not piecemeal, approach. The "problematique" concept makes the international leadership effort even more challenging since it prohibits the human tendency to isolate and "tinker" with single elements of the larger globalization process. Secondly, the concept of limits to growth that are imposed by the realization of the finite biosphere of Earth. Peccei, in speaking for the Club of Rome members, concludes that international political leadership is essential for the required change and all humanity must be prepared to learn our way out of the anticipated crises. These pronouncements about the limits of growth, although surrounded by a great deal of controversy, did have an impact of questioning unlimited and unregulated growth and development on a finite planet. It also is instructive of the type of international leadership effort that would be required to match the difficult situation.

Sustainable development was broadly introduced for the world to consider by the Brundtland Report, *Our Common Future*, in 1987, and elaborated in the context of the United Nations. This significantly built on a trend initiated in the 1970s and 1980s that focused on the fragility of the planet. In the Brundtland Report, sustainable development was introduced in reference to the configuration of bonds and resources passed to future generations. In this new concept of sustainable development the kind of input prepared for the systems to be changed by future generations is taken into consideration. It becomes apparent that future systems may not be changed on assumption of the constant availability of growth processes and resources. Although science and technology are inexhaustible sources and generators of resources in terms of future change of globalization systems, but the interdependent nature of most environmental problems such as pollution, wastes, consumption of non-endless resources, become irreversible modifications on the Earth and on life on the planet.

Ray and Anderson (2000) claim a new international group of approximately 50 million people has adopted the concepts of sustainable development, but is unable to express these new values as practical actions to be taken by the international community. This group, called the "cultural creatives", has incorporated into their individual, corporate, and association memberships a deep desire to preserve the planet for future generations. Most of them conceive of the changes in a constructive manner of altering our old values of "through away consumerism" to one of a "greening consciousness". They are not necessarily anti-business. Over many years of surveys, interview data, demographic studies and other types of data, Ray and Anderson believe the new emergent value base is now achieving a critical mass to support practical actions:

Our studies show that huge majorities of Americans want ecologically sensible 'green' products and services, but most aren't willing to pay a premium to get products that should not have been destructive in the first place. Similarly, they want business to stop inflicting environmental damage, and they believe that a better use of technology could do the job without destroying the economy... They want actual leadership and proactive initiatives from all levels of government that aren't sellouts to big business... But nobody's delivering... no institutions have been set up to translate public sentiment favoring ecological sustainability into practical action... Green activists, accustomed solely to raising awareness, have not constructed these institutions. (p. 338)

When the ethic-in-practice of free market regulation as the self-managing mechanism of globalization, the status quo of uncontrolled development would seem to be the outcome. The national political leadership of each country defaults to the ethics-in-practice of its business leaders. The ability to act requires a political leadership capable of dealing with real change in an inherently conflictual, and most importantly international, situation. Is there the will and wherewithal of international political leadership to build an emergent ethic of sustainable development? How could this be accomplished?

4. POLITICAL LEADERSHIP IN AN ERA OF RAPID, GLOBAL CHANGE

Leadership is about real change, according to Burns (1978) and Rost (1990), whose works have substantially permitted the researchers in leadership studies to distinguish leadership from management. Also, as Burns (1978) points out, real change involves a context of conflicting values and perspectives; it is the very essence of why leadership is necessary. Management will suffice in times of relative stability. Burns' two types of leadership, transactional and transforming, both share this initial context. However, how and on what basis leaders and followers interact in this context of conflicting values and perspectives makes the significant difference between the two approaches to leadership. Transactional leadership is about trade-offs and negotiations that does not involve significant engagement of leaders and followers around their conflicts or differences in perspectives. In the transactional scenario of change, individual participants or organizations negotiate their "best deal" in an environment of balanced power and self-interests. Transforming leadership is about engaging leaders and followers in a dialogue about their differences so that group learning may occur. In the best lead situations, group learning elevates all the participants, both leaders and followers, to a higher level of moral consciousness. A shared sense of what is good for all becomes the basis for a shared or collective purpose that may direct their activities in finding a lasting solution to their jointly held issues.

The current global stability, far from the equilibrium, of the system "world" looks to be not sustainable. That's because the stability of the system is based on such an unbalanced configuration of interacting subsystems. To keep stability between unbalanced subsystems requires energy and resources greater than the available and produced ones by the system. This approach may be applied to different kinds of systems like micro social systems, companies or people with disabilities. *In economics it applies when the different systems interact only looking to maximize their particular interest without any global ethics caring about the global effects in space and time. In the circumstances of a scenario where everyone looks only to their own self-interest, transforming leadership may be the only answer to develop an emergent ethic of sustainable development.*

In the "business as usual" mindset of political leadership and international business interests, the transactional leadership approach to globalization strives to negotiate mutually agreeable solutions to the many disjointed effects of globalization. These are at best satisficing attempts to deal with complex international problems; that is, in the complexities of negotiating solutions to "messy" problems, the solution that appears in the short term to alleviate the problem would be implemented rather searching for a more long term, optimal solution. Importantly, they are based on the implicit or "fuzzy" values, ethics-in-practice, and power of each individual countries political and business approaches to development. The needs and aspirations of the stakeholders' voices are omitted. The vision and collective purpose (if one exists) of globalization may be based on the narrow self-interest of each participant in this new community. If the voices of current stakeholders are not heard in the international decision-making arena, the voices of future stakeholders - yet to be born - will never be heard. These future stakeholders have a vested interest in sustainable development in a global environment of finite resources.

The concept of transforming leadership emerged for Burns in his study of specific leaders and followers in the context of a particular country and culture. For instance, he explored the psychological and sociological development of specific transforming leaders, such as Gandhi and Lenin, in the cultural environment of their followers. The leaders emerge as they found a way to engage followers in expressing their needs, values, and aspirations in constructive actions of fulfilment. In doing so, leaders and followers were both transformed. As Burns indicates, leaders and followers elevated each other to higher levels of moral consciousness of the common good. An emergent ethic would develop that would guide the attitudes and behaviors of those involved in such movements.

Transforming leadership is the type of leadership necessary to promote a constructive engagement of all concerned toward a more sustainable future. However, transforming leadership occurred in specific national settings within the common culture of the leaders and followers. If the concept and practice of transforming leadership is a potential approach to globalization, rather than the current transactional leadership approach which is exclusive to the culture of powerful interests, what form might it take and how would it be implemented on an international basis?

5. THE TRANSFORMING LEADERSHIP CHALLENGE

Transactional leadership appears to be the guiding principle of globalization. One negotiated settlement through transactional leadership has been the divestiture of many important decision-making functions to "technocrats". As Weisbrot (1999) points out, the actual operatives of globalization have been placed in the hands of non-elected bodies such as the International Monetary Fund (IMF), and the World Bank, and other agencies that regulate business arrangements under various treaties.

Globalization has become bureaucratized with an ethics-in-practice that reflects the transnational business culture. Ordinary citizens across the international community feel incapable of communicating or acting upon their concerns since there is not central or controlling body.

The challenges to developing transforming leadership across international boundaries are formidable. Although we can't do anything about how globalization evolved, the future is an open possibility for human learning and leadership. There are, however, very significant challenges to the leadership approach espoused here. First, there is the challenge of developing transforming leadership among widely different groups or associations with differing cultures and aspirations. The emergent ethics of sustainable development would require a moral and holistic view of the common good from many currently incompatible viewpoints. How much does transforming leadership depend upon a similar cultural background among leaders and constituents? Burns (1978) did not address this issue of the potential complications of differing cultures and identities that would be part of an international leadership movement. Second, there are the "true believers" on both sides of the issues of globalization who want to prevail over their adversaries. They might be open to "debate" but never to dialogue. This type of adversarial approach to the issues of globalization does not permit solutions that design a new future for everyone. Finally, the complexity of globalization will require a considerable length of time to shape around the ethics of sustainable development. Is there the will and commitment to persevere?

A beginning response to these complex issues may be fashioned by examining the critical systemic blocks to action to the process of designing new solutions. Manzini (1995) has identified human factors that impede the design process of working together. It begins with the human condition of self-referentiality and the "culture of doing" inherent in Western culture. As a species we find it extremely difficult to think of issues like globalization beyond a self-referential sense of our own purposes and intentions. This may block or eliminate our ability to learn about different perspectives of globalization that will enable us to progress to a higher level of understanding that includes many others beside ourselves. Many times it takes a crisis to force this recognition that we need to look at the situation from an alternative viewpoint. In the circumstances of globalization, the needs of future generations will be difficult to take seriously enough to act in the present. In addition to this thinking tendency is the powerful combination of the Western "culture of doing". This culture (perhaps synonymous with the business culture) believes there are clear objective solutions (usually tackled one at a time) that are generalizable to the entire system and technology can be the instrument to obtain this change. This culture of doing ignores the complexity (especially the human issues) of globalization and chooses to ignore the limitations to growth.

These initial conditions to the challenges of globalization are the very strengths of the transforming leadership approach. Transforming leaders understand the necessity of correctly discerning their constituents needs, values, and aspirations. They are also astute in identifying the obvious or hidden conflicts of needs and values and building a consensus toward action. In transactional leadership, conflicts are negotiated into compromise situations that may serve self-interests but lack a focus on the general welfare. Out of these conflictual situations transforming leaders are given the opportunity to shape emergent ethics that will support consensual decision-making. It fundamentally begins with listening and speaking through the process of dialogue. Dialogue allows the exploration of different meanings and understandings of the same event from multiple viewpoints. It is a leadership of engagement rather than negotiating or brokering piecemeal. Most importantly, transforming leadership is about mobilizing people around issues important to them. Transactional leadership processes in globalization will continue to emphasize technical solutions that leave people and their values out of the loop.

Transforming leadership has been a remarkable process for leaders and their constituents or collaborators on a national basis. The national basis of transforming leadership assumes a shared culture and identity among leaders and constituents. One open question of the viability of this approach may be asked based on globalization to date: Have enough people on an international basis share a similar value base and ethics of sustainable development? According to Ray and Anderson (2000) there may be somewhere around 50 million individuals who profess this belief. We are also beginning to see certain leaders in the business world professing the cost efficiency of developing sustainable business practices. Product designers and marketers are also shifting from a "throw away" approach to consumerism to a sensitivity of renewable and aesthetically pleasing products (Buchanan and Margolin, 1990). There is a "greening" of consumerism and growing public awareness of environmental limits, for instance such as ozone depletion, radical climate shifts and desertification, and depletion of "fishing" stocks to name a few. It is our belief that there are enough people across the globe who will be receptive to the potential of this dialogue.

Given the issues about the specific cultural context that unites leaders and constituents, how would transforming leadership develop in such an environment? We believe that transforming leadership is already taking on a different form in the international arena of globalization. It is occurring through the development of coalitions, rather than individual transforming leaders, among historically different groups in different countries. As different groups wish to be heard and seek the political strength to be heard, they are banding together. If these leadership coalitions seek more common platforms to represent and express their concerns, such as sustainable development, the focus and strength of their dialogue will increase. As John Kotter (1996) has indicated in developing leadership coalitions, you need to involve all in the dialogue who can also stop or resist your efforts. That is, the natural tendency to want to speak to those of like viewpoints must not be selection criteria for participants; the dialogue that will mobilize a leadership coalition must include many different people with diversity of opinions. The only requirement is the coalition members must be willing to listen to each other and reason together.

Magliocca and Christakis (2001) have demonstrated three principles for establishing transforming leadership across multiple organizations that can achieve the above requirements. The process of creating transforming leadership occurs through the process of a disciplined approach to dialogue that encompasses these requirements:

- *The collective purpose principle.* [Building the] context of collective purpose invites commitment by stakeholders ...
- *The intent of real transformation principle.* The content of design activity must focus on intended real transformation, generated and interrelated by the stakeholders, that successfully integrates the needs and values of stakeholders within a consensual design solution ...
- *The powershift principle.* The process of languaging through focused and open dialogue is the means to establish equitable power relations among the stakeholders based on knowledge, reason, and persuasion through the emergence of a consensual linguistic domain. (p. 266)

Building transforming leadership coalitions are possible across the international community involved in globalization. This could be a "grassroot movement" that approaches globalization as a engagement of many stakeholders focused on sustainable development. This article has met its purposes if an awareness is achieved about an alternative way to challenge the ahistorical, ethics-in-practice of the business and political cultures currently directing globalization. Although major international meetings focused on globalization processes are currently being met with large numbers of demonstrators, it will not be long before the events of such meetings will become less visible. The insistence should be on a constructive, long term dialogue about the human needs, values and aspirations of globalization as part of the operational equation. Our concern is that the small part of the globalization operational structure may be "underground" to avoid demonstrators' confrontations.

We believe there is a constructive approach to developing an emergent ethic of sustainable development within the process of globalization. It requires a human approach to the issues by engaging their best thinking and commitment to sustainable development through learning. As differing needs, values, and aspirations are shared and integrated into a new moral and spiritual consciousness, coalitions of leadership groups in many different nations can begin to educate others to an awareness of an impossible future for the next generations.

REFERENCES

Buchanan, R., and Margolin, V., (Eds.), 1990, *Discovering Design: Explorations in Design Studies*. University of Chicago Press, Chicago.

Burns, J. M., 1978, *Leadership*. Harper & Row, New York.

Cornish, E., 2001, The global struggle to save the environment. *The Futurist*, **35** (2): 26-30.

Friedman, T. L., 1999, *The Lexus and the Olive Tree*. Farrar, Straus, Giroux, New York.

Gini, A., 1998, Moral leadership and business ethics. In *Leading Organizations: Perspectives for a New Era* (G. R. Hickman, ed.), Sage Publications, Thousand Oaks, CA.

Hayek, F. A., 1944, *The Road to Serfdom (Fiftieth Anniversary Edition)*. University of Chicago Press, Chicago.

International Monetary Fund (IMF), 2000/2001, *Globalization: Threat or Opportunity?* IMF, Washington, D.C. (www.imf.org/external/np/exr/ib/2000/041200.htm#VIII).

Jackall, R., 1988, *Moral Mazes*. Oxford University Press, New York.

Kotter, J. P., 1996, *Leading Change*. Harvard Business School Press, Boston.

Magliocca, L. A., and Christakis, A. N., 2001, Creating transforming leadership for organizational change: The Cogniscope system approach. *Systems Research and Behavioral Science* **18**: 259-277.

Manzini, E., 1995, Prometheus of the everyday: The ecology of the artificial and the designer's responsibility. In *Discovering Design: Explorations in Design Studies* (R. Buchanan and V. Margolin, eds.), University of Chicago Press, Chicago.

Minati, G., (In press). *Ethics as Emergent Property of the Behavior of Living Systems*. UNESCO.

Peccei, A., 1977, *The Human Quality*. Pergamon Press, Oxford.

Peccei, A., 1981, *One Hundred Pages for the Future: Reflections of the President of the Club of Rome*. Pergamon Press, New York.

Ray, P. H., and Anderson, S. R., 2000, *The Cultural Creatives: How 50 Million People Are Changing the World*. Harmony Books, New York.

Rost, J. C., 1991, *Leadership for the Twenty-First Century*. Praeger, New York.

Schwartz, H. S., 1990, *Narcissistic Project and Corporate Decay*. New York University Press, New York.

Weisbrot, M., 1999, *Globalization: A Primer*. Center for Economic and Policy Research, Washington, D.C., (www.ceprinet/GlobalPrimer.htm).

Meaning Extraction from the Analysis of Video-Registrations of Human Movements

M. P. PENNA[°,#,*], D. ZANDONÀ[*], A. MONTESANTO[^], and
N. S. BONFIGLIO[#]
[°]*Facoltà di Scienze della Formazione, Università di Cagliari - E-mail: npenna@hotmail.com*
[#]*Facoltà di Lettere e Filosofia, Università di Pavia*
[*]*LUMSA, Corso di Laurea in Scienze della Comunicazione*
E-mail: davide.zandona@infinito.it
[^]*Istituto di Informatica, Facoltà di Ingegneria, Università di Ancona*

1. INTRODUCTION

The human skill of giving a meaning to perceived movements attracted the interest of many researchers aiming to reproduce such an ability through artificial systems. Their general goal was to design artificial devices able to monitor an environment and to recognize actions or behaviours occurring within it. Some proposals and implementations were based only on a typical engineering approach and tried to discover more and more powerful techniques for action recognition, engaging in the research of the best possible methods to detect suitable image features. Often underlying these techniques there is the requirement of existence of individual properties that can be extracted from each frame of an image sequence. These approaches accomplish the motion understanding by recognizing a sequence of static configurations.

Other researches, then, take into consideration the operation of biological visual systems, particularly the human one, aiming to understand the neurophysiological and cognitive mechanisms underlying the biological movement recognition (Logothetis, Pauls and Poggio, 1995; Bulthoff and Edelman, 1992; Gyulai, 2000). One hypothesis that brings forth this research

Emergence in Complex, Cognitive, Social, and Biological Systems, Edited by
Minati and Pessa, Kluwer Academic/Plenum Publishers, New York 2002

trend is the assumption that a mental three-dimensional reconstruction of the scene, based on data coming from the perceived images, precedes action recognition (Goncalves, DiBernardo, Ursella and Perona, 1995). Nevertheless some experiments that use blurred images, that deprive the observer of the necessary two-dimensional references for a mental three-dimensional reconstruction of the stimulus, show that movement is recognised within blurred video-sequences but not within single blurred frames. These data suggested the hypothesis that biological visual systems recognise global spatio-temporal configurations (Bobick and Davis, 1996). Originated from these findings and these hypotheses, many studies have been done to search for movement prototypes to be stored in the memory of artificial systems for a further matching with input data (Giese and Poggio, 2001; Emering, Boulic, Balcisay and Tholmann, 1997; Dittrich 1993).

The aim of the present work was to study how much the attribution of meaning to video registrations of human movements depends on factors such as dynamical features, shapes of the framed objects and spatial relationships occurring between them. To this regard, the experimental psychology (Bernstein, 1967; Johannson, 1980; Runeson, 1994; Dittrich, 1993) evidenced a considerable specialization of biological visual systems in recognizing movements, and underlined the high relevance of this circumstance in explaining how human subjects build a semantics of the environment.

In studying the attribution of meaning to video registrations we took into consideration a system (fig.1) made up of three subsystems: the image, the observer, and the action semantics. With reference to it we formulated the following three hypotheses:

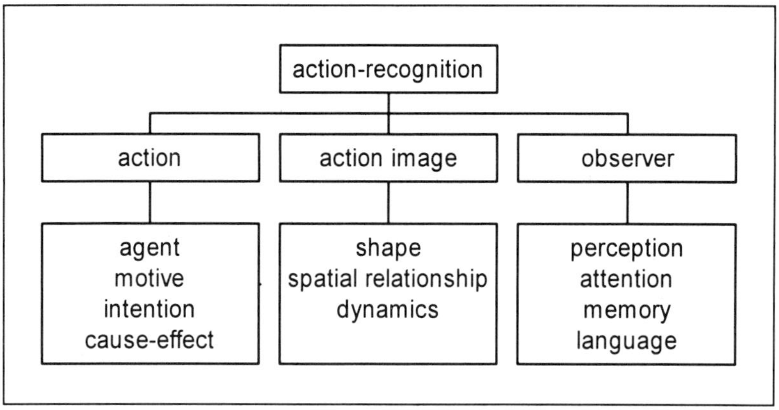

Figure 1. Action-recognition as a system.

- *There is a visual semantics of dynamics.* Human beings attribute a meaning to spatio-temporal patterns and regard as a fundamental factor for action recognition the way in which a movement is produced in time.
- *Human beings use general mental schemata to recognize dynamics.* Movement is recognized as a kinematical structure by means of movement schemata. Each movement scheme used to recognize an action represents a class of movements and different actions can belong to the same class.
- *A very small body of visual information can be enough to recognize an action in an univocal way.*

2. METHODS AND EXPERIMENTAL STRUCTURE

2.1 Experimental subjects

80 university students (mean age = 23) with normal or corrected to normal vision.

2.2 Stimuli and materials

Stimuli were movies and slides, each depicting an action, projected without colour and audio on a 18 inches television screen located 80 cm away from the subject. Movies were recorded from real scenes shot at a fixed distance by a fixed camera on Super VHS stands and acquired on a Power Pc G4 Apple. Slides were meaningful frames extracted from the movies. A frame was considered meaningful if it got the same interpretations inspired by the clear movie.

Slides and movies were quantized on four levels: 0, 1, 2, 3 (fig. 2, 3, 4, 5) by means of a mosaic technique (a square knit grid with variable side unit of the mosaic function supplied by Adobe Premiere 5.1 software) which masked the shapes of the framed objects and the spatial relationships occurring between them.

The actions to be filmed were chosen on the basis of a resemblance between their dynamics in such a way as to assemble them into 3 classes(class typologies 1, 2, 3; see table 1). Each class contained 4 action typologies, similar in dynamics but associated to different meanings if shown at the level 0 of cryptizing. The distinction between the action typologies proposed in the table 1 is purely structural, that is there are no qualitative differences between them.

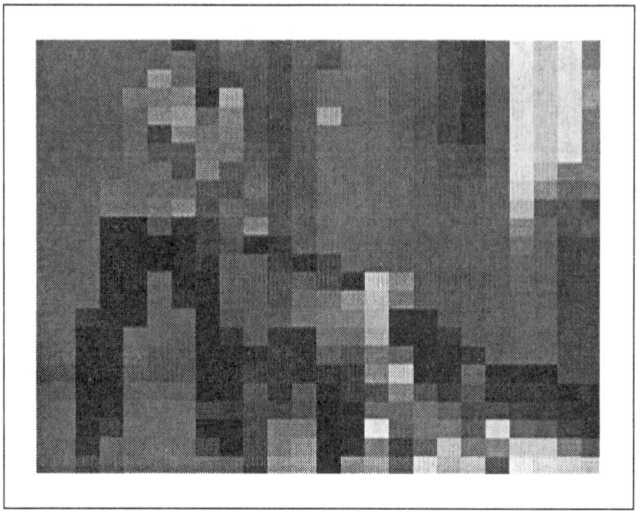

Figure 2. Beating: slides at compression level 3.

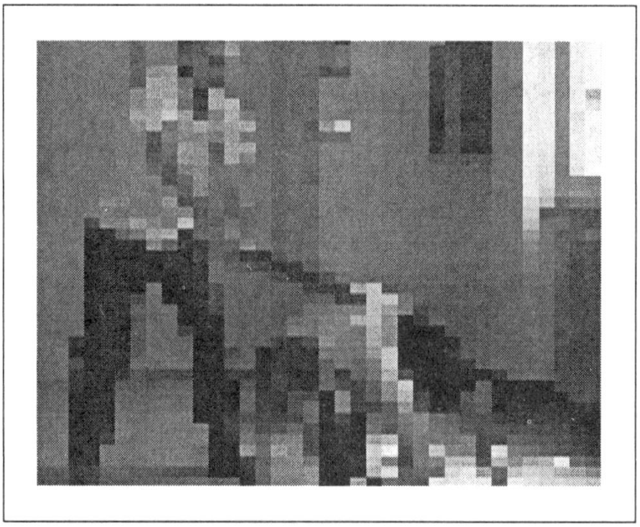

Figure 3. Beating: slides at compression level 2.

Table 1. Presented actions divided into three class typologies and 4 action typologies

	CLASS TYPOLOGY 1	CLASS TYPOLOGY 2	CLASS TYPOLOGY 3
Action typology 1	Woodcutter	Group dance	Rescue
Action typology 2	Tennis player	March	Porter
Action typology 3	Hoeing	Team match	Kidnapping
Action typology 4	Beating	Brawl	Interview

Figure 4. Beating: slides at compression level 1.

Figure 5. Beating: slides at compression level 0.

Each of these 12 actions was projected under 8 different conditions: 4 cryptizing levels × 2 categories of materials (film/slide), so that the total number of stimuli was 96. The duration of each movie was about 900 msec, whereas each slide projection lasted 600 msec.

2.3 Experimental procedure

Each experimental session was made up of 12 movies and 12 slides, randomly selected in order to avoid the influence of a projection on the following ones, plus two trial stimuli. Each stimulus was preceded by a white number appearing in the middle of the screen and was followed by an interval of 20 sec black screen. Subjects were asked to participate individually in an experiment on visual perception during which they had to interpret some scenes projected on a screen. They were told to push a button every time they think to have understood what was happening in a scene. After they pushed the button they had to verbalize their interpretation. Subjects could give several interpretations, within the time intercurrent between the beginning of a scene and the projection of the next white number.

3. RESULTS

To verify the hypotheses we proposed before we resorted to a statistical analysis of subject answers and of the associated reaction times. Such an analysis was based on: 1) a nested analysis of variance 2) a one way analysis of variance. The subject sample was subdivided into four groups (between factor) that were distinguishable only by the representative nature of the given stimulus; except for that, each group was characterized by identical experimental conditions (repeated measure). This kind of distribution allowed us to avoid unwanted experimental effects caused by the repeated presentation of the same cryptizing or of the same dynamics or of the same action.

3.1 Time reaction analysis

Reaction time analysis was based on a 3 (class typology) × 4 (action typology) × 2 (stimulus typology) factorial experimental design.

The CLASS TYPOLOGY factor refers to the three typologies above-mentioned (see tab.1), the ACTION TYPOLOGY factor was divided in four levels (see tab.1), the STIMULUS TYPOLOGY had two levels (slide or movie).

We found significant differences for CLASS TYPOLOGY ($F = 6,6$; $df = 2$; $p \leq 0,01$), ACTION TYPOLOGY ($F = 27,2$; $df = 3$; $p \leq 0,001$), STYMULUS TYPOLOGY ($F = 3698,2$; $df = 1$; $p \leq 0,001$) as shown in the following graphs.

The considerable differences of reaction times between the three dynamics typologies evidenced that there were pattern classes better recognizable than others, even if we can't, at the moment, evaluate the answer precision (graph 1).

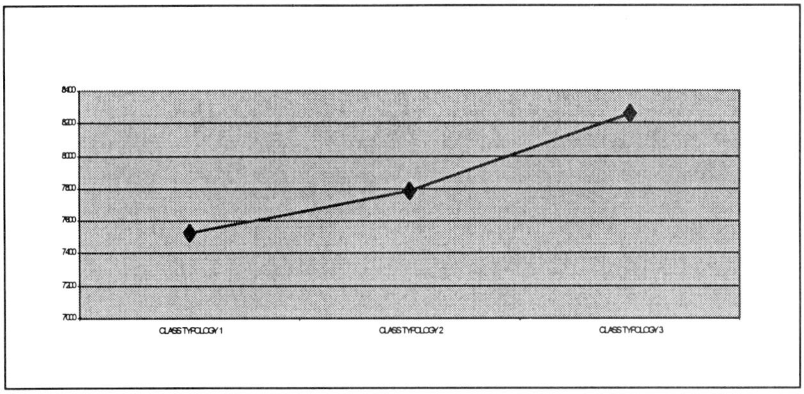

Figure 6. Graph 1. Average reaction times, referred to the three different class typologies.

As shown in the following graph 2, there were differences between the two action typologies 2 and 4 and the other two 1 and 3; but, as there was no qualitative difference between the four action typologies, this difference can depends only on the kind of representation associated to the stimulus.

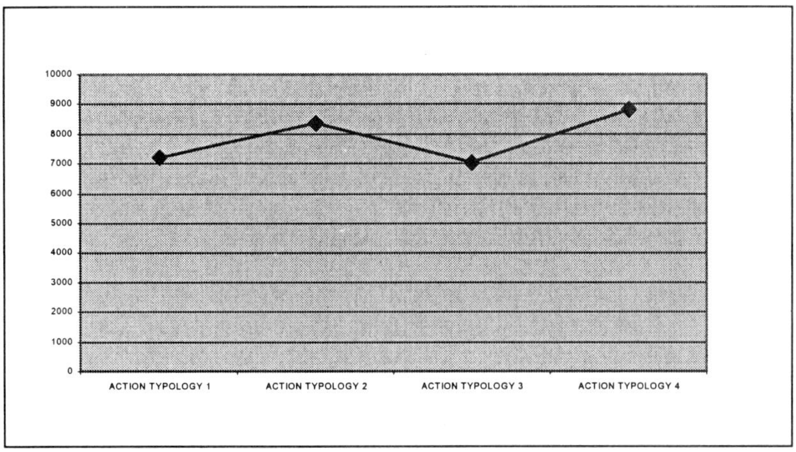

Figure 7. Graph 2. Average reaction times related to the four action typologies.

Graph 3 clearly shows how reaction times to movie stimuli are considerably lower than to slide stimuli.

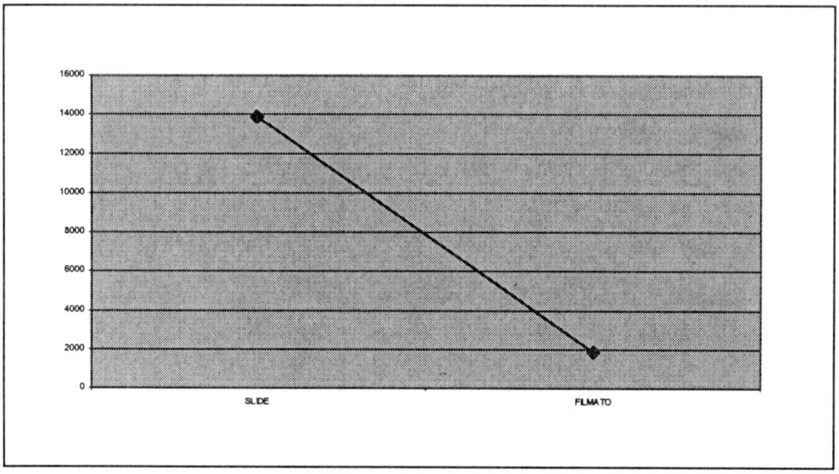

Figure 8. Graph 3. Average reaction times for each stimulus typology.

The following tables show the existence of significant interactions between the considered factors. Within table 2 we have the quantitative description of the phenomenon just presented in graphs 2 and 3, that is that the reaction times are higher for classes 3 and 4 and raise in interaction with action typologies 2 and 4.

Table 2. Average reaction times resulting from the interaction CLASS TYPOLOGY × ACTION TYPOLOGY (F = 11,08; df = 6; p ≤ 0,001).

	CLASS TYPOLOGY 1	CLASS TYPOLOGY 2	CLASS TYPOLOGY 3
Action typology 1	5701,66875	8120,51875	7779,96875
Action typology 2	8022,7625	7620,63125	9461,88125
Action typology 3	7401,15	7203,8625	6490,00625
Action typology 4	8960,13125	8201,78125	9312,31875

The same trend, but related to presentation typologies, is shown in table 3. We can observe, in fact, the lower reaction times associated to the movies, and the difference between the classes 2 and 3.

Table 3. Average reaction times resulting from the interaction CLASS TYPOLOGY × STIMULUS TYPOLOGY (F = 6,9; df = 2; p ≤ 0,01)

	SLIDE	FILM
Class Typology 1	13212,2375	1830,61875
Class Typology 2	13700,5375	1872,859375
Class Typology 3	14716,83438	1805,253125

The same phenomenon characterizes the interaction with action typology (table 4).

Table 4. Average reaction times resulting from the interaction ACTION TYPOLOGY × STIMULUS TYPOLOGY (F = 14,3; df = 3; p ≤ 0,001).

	SLIDE	FILM
Action Typology 1	12671,3375	1730,1
Action Typology 2	1730,1	1873,458333
Action Typology 3	12430,64583	1632,7
Action Typology 4	15540,77083	2108,716667

Table 5, finally, confirms preceding trends and shows that slide stimulus typology has higher reaction times, particularly for actions 2 e 3 and classes 3 and, much more, 4.

Table 5. Average reaction times resulting from the interaction CLASS TYPOLOGY × ACTION TYPOLOGY × STYMULUS TYPOLOGY (F = 6,3; df = 6; p ≤ 0,001)

		SLIDE	FILM
Class Typology 1	Action Typology 1	10014,6375	1388,7
	Action Typology 2	14273,075	1772,45
	Action Typology 3	13221,5125	1580,7875
	Action Typology 4	15339,725	2580,5375
Class Typology 2	Action Typology 1	14236,4875	2004,55
	Action Typology 2	13480,9	1760,3625
	Action Typology 3	12519,1375	1888,5875
	Action Typology 4	14565,625	1837,9375
Class Typology 3	Action Typology 1	13762,8875	1797,05
	Action Typology 2	16836,2	2087,5625
	Action Typology 3	11551,2875	1428,725
	Action Typology 4	16716,9625	1907,675

3.2 Analysis of Subject Answers

This analysis was based on a 3 (CLASS TYPOLOGY) × 4 (ACTION TYPOLOGY) × 2 (STIMULUS TYPOLOGY) × 2 (ANSWER TYPOLOGY) factorial experimental design.

The above-mentioned factors had the same properties as the factors of the preceding analysis, with the addition of the ANSWER TYPOLOGY factor, made up of two levels that referred to the given answer typology; these levels coded the presence of a correct answer, or of a wrong answer related to the action typology associated to the stimulus (e.g. when beating was interpreted as hoeing).

Significant differences were found for the CLASS TYPOLOGY (F = 6,9; df = 2; p ≤ 0,01), the ACTION TYPOLOGY (F = 4,22; df = 3; p ≤ 0,01), the STIMULUS TYPOLOGY (F = 2013; df = 1 p ≤ 0,001) and the ANSWER TYPOLOGY (F = 1541,2; df = 1; p ≤ 0,001). The differences between averages are shown in the following graphs.

The answer typology analysis followed the trend previously shown in graph 1, i.e. that it was more difficult to recognize correctly the class typology 3, probably owing to its very nature.

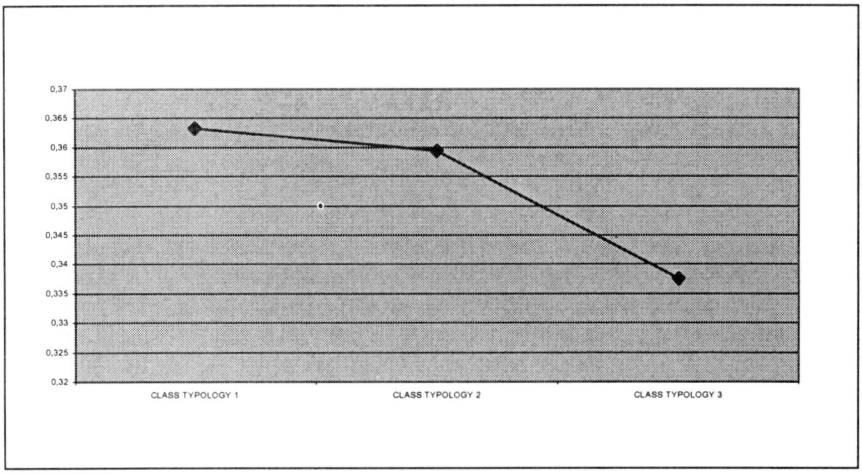

Figure 9. Graph 4. Average number of correct answers related to the class typology.

The Graph 5 shows a trend similar to the one evidenced by the graph 2. Action typologies 2 and 3 seem to have been more difficult to recognize correctly and, even in this case, the reason is to be find in the stimulus nature.

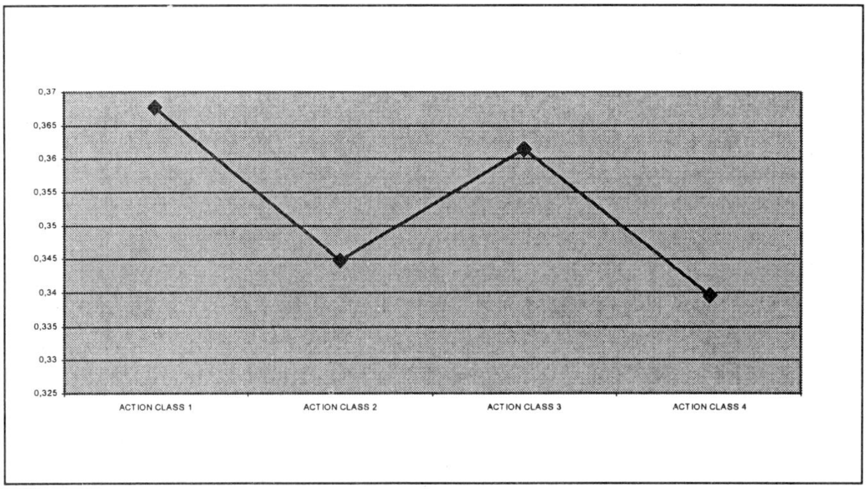

Figure 10. Graph 5. Average number of correct answers for each action class.

Besides, answers given to movie typology stimuli were more precise than answers given to slide typology (graph 6).

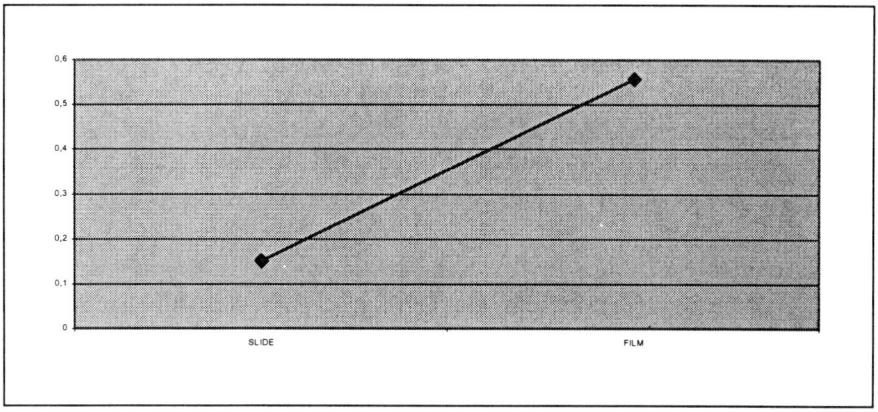

Figure 11. Graph 6. Average number of correct answers for film and slide typology.

Finally, the greater number of correct answers seems to have been referred to dynamics, and not to the action effectively depicted in the stimulus (graph 7).

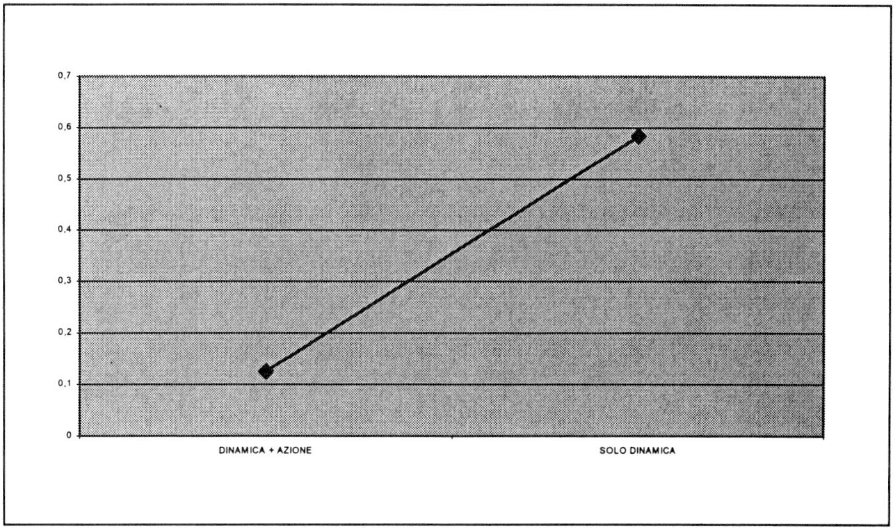

Figure 12. Graph 7. Average number of correct answers relative to action and to dynamics.

The significant interactions between the factors are shown in the following tables.

Table 6 shows that the quantity of correct answers referred to class typologies 1 and 2 was greater than the same quantity referred to class 3; this trend was marked with the action typologies 2 and 4.

Table 6. Average number of correct answers resulting from the interaction CLASS TYPOLOGY × ACTION TYPOLOGY ($F = 4,3$; $df = 6$; $p \leq 0,001$)

	CLASS TYPOLOGY 1	CLASS TYPOLOGY 2	CLASS TYPOLOGY 3
Action typology 1	0,384375	0,3625	0,35625
Action typology 2	0,35625	0,371875	0,30625
Action typology 3	0,365625	0,346875	0,371875
Action typology 4	0,346875	0,35625	0,315625

The slide typology seems to have produced interpretation mistakes both in relation with the class and the actions (tables 7, 8 and 9).

Table 7. Average number of correct answers resulting from the interaction CLASS TYPOLOGY × STIMULUS TYPOLOGY ($F = 7,5$; $df = 2$; $p \leq 0,001$)

	SLIDE	FILM
Class Typology 1	0,176563	0,55
Class Typology 2	0,151563	0,567188
Class Typology 3	0,121875	0,553125

Table 8. Average number of correct answers resulting from the interaction ACTION TYPOLOGY × STIMULUS TYPOLOGY ($F = 20,3$; $df = 3$; $p \leq 0,001$).

	SLIDE	FILM
Action Typology 1	0,179167	0,55625
Action Typology 2	0,10625	0,583333
Action Typology 3	0,197917	0,525
Action Typology 4	0,116667	0,5625

Table 9. Average number of correct answers resulting from the interaction CLASS TYPOLOGY × ACTION TYPOLOGY × STYMULUS TYPOLOGY ($F = 7,5$; $df = 6$; $p \leq 0,001$)

		SLIDE	FILM
Class Typology 1	Action Typology 1	0,24375	0,525
	Action Typology 2	0,15625	0,55625
	Action Typology 3	0,18125	0,55
	Action Typology 4	0,125	0,56875
Class Typology 2	Action Typology 1	0,13125	0,59375
	Action Typology 2	0,1375	0,60625
	Action Typology 3	0,2	0,49375
	Action Typology 4	0,1375	0,575
Class Typology 3	Action Typology 1	0,1625	0,55
	Action Typology 2	0,025	0,5875
	Action Typology 3	0,2125	0,53125
	Action Typology 4	0,0875	0,54375

Dynamics, finally, seems to have been more easily interpretable than action (tables 10 and 11).

Table 10. Average number of correct answers resulting from the interaction ACTION TYPOLOGY × ANSWER TYPOLOGY (F = 4 6,5; df = 3; p ≤ 0,001)

	Dynamics + Action	Only Dynamics
Action Typology 1	0,114583	0,620833
Action Typology 2	0,177083	0,5125
Action Typology 3	0,058333	0,664583
Action Typology 4	0,147917	0,53125

Table 11. Average number of correct answers resulting from the interaction CLASS TYPOLOGY × ACTION TYPOLOGY × ANSWER TYPOLOGY (F = 14,4; df = 6; p ≤ 0,001)

		Dynamics + Action	Only Dynamics
Class Typology 1	Action Typology 1	0,04375	0,725
	Action Typology 2	0,1875	0,525
	Action Typology 3	0,1125	0,61875
Class Typology 2	Action Typology 1	0,1375	0,55625
	Action Typology 2	0,19375	0,53125
	Action Typology 3	0,1625	0,58125
Class Typology 3	Action Typology 1	0,00625	0,6875
	Action Typology 2	0,2	0,5125
	Action Typology 3	0,10625	0,60625
Class Typology 4	Action Typology 1	0,18125	0,43125
	Action Typology 2	0,05625	0,6875
	Action Typology 3	0,10625	0,525

To verify the third hypothesis we performed a one-way analysis of variance taking into consideration the four levels of cryptizing as independent variable. Significant differences have been found in relation to the reaction time (slide F = 305,5; df = 3,316; p ≤ 0,001, movie F = 9,37; df = 3,316; p ≤ 0,001), to answers related to the situations in which the available information is relative only to dynamics (slide F = 4,98; df = 3,316; p ≤ 0,01, movie F = 118; df = 3,316; p ≤ 0,001), and to answers related to the actions (slide F = 302,1; df = 3,316; p ≤ 0,001, movie F = 70,2; df = 3,316; p ≤ 0,001). These differences are clearly shown in the three graphs below (graphs 8, 9 e 10) that report the four different cryptizing typologies on the abscissa and the average values both of the average reaction times (in seconds) and the average number of correct answers for the actions and for dynamics typologies on the ordinate.

Graph 8 shows the reaction times calculated on the subjects answers. It appears that the reaction times to the movie stimuli were lower than to slide stimuli, and that the difficulty related to the increase of the cryptizing level was produced from the slide stimuli and not from movie stimuli.

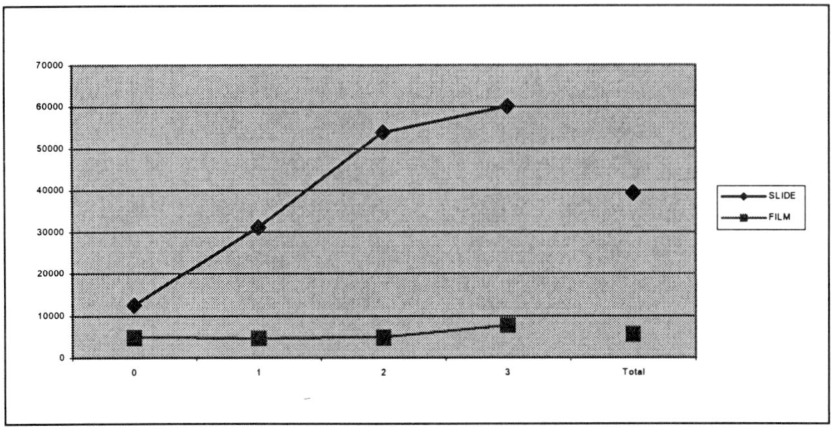

Figure 13. Graph 8 Average reaction times (in seconds).

By looking at the graph 9, containing the average number of correct answers given to the actions, it can be seen, besides the differences between movies and slides, how the number of correct answers was decreasing with the increasing of the cryptizing level.

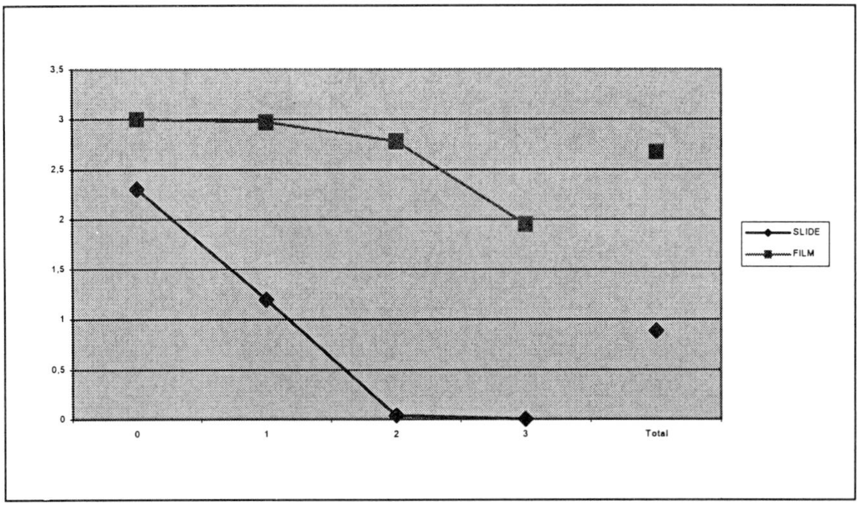

Figure 14. Graph 9. Average number of correct answers to the actions.

Graph 10 shows how, relatively to the correct dynamics, there were not particular differences between movies and slides and between cryptizing levels, except for the last cryptizing level (level 3).

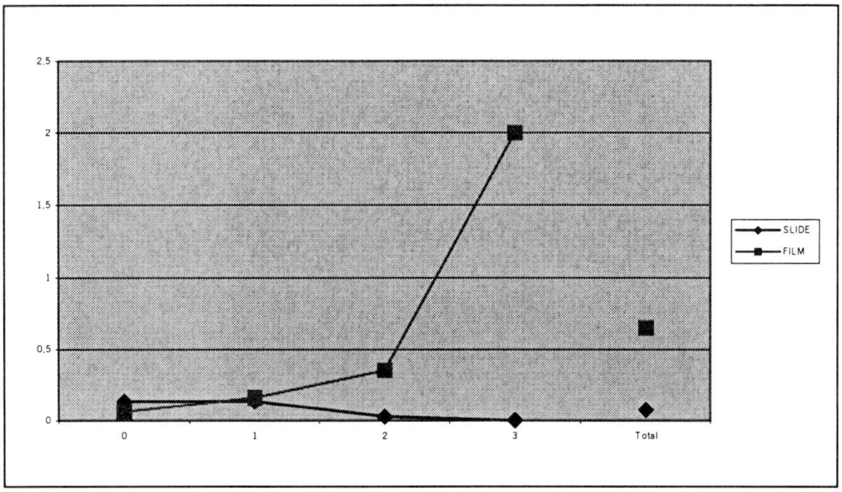

Figure 15. Graph 10. Average number of correct answers to only dynamical information.

By comparing the graphs 9 and 10, it is possible to see that the number of answers that correctly interpret the actions was greater than those that correctly interpret the dynamics typologies. This datum seems to contradict what has been previously reported relatively to the differences between dynamics and action (graph 7). This contradiction is justified by the fact that within the data from which we started to analyse the effect of cryptizing differences, and from which the above-mentioned graphs have been originated, the answers to actions have been separated from the answers to dynamics. This means that a correct answer given to an action was not considered as correct for dynamics too and vice versa, in spite of what could be intuitively imagined, i.e. that if a subject understands the action depicted in a stimulus he understands the dynamics too.

This separation, however, was introduced for a precise reason, that is to understand at what cryptizing level the subjects were beginning to confuse an action starting only from information relative to dynamics. In our data such a confusion seems to occur at cryptizing level 3, considered that the others levels don't show differences.

4. CONCLUSIONS

The fact that we consider the action-recognition as a product of the interaction of the sub-parts of a system (as it is shown in figure1) allowed us to account for the complexity of the phenomenon that was analysed. In fact, after a movement has been defined by means of an analysis of its physical parameters, it should be understood how and when this represented movement - for example through a movie - becomes an action recognised by an observer. The question is to understand what differences exist between movement and action when they became perceptive objects. In fact, before being recognised, action is a movement, that is a reorganization of visual space caused by a shape that in such a space prevails with a given strength. Both action and movement are spatio-temporal variations, so that it becomes necessary to study the movement in accordance with the observer cognitive representation, for discovering how movement is interpreted like an action.

Through a comparison of the action-recognition performances of subjects dealing with impoverished stimuli we found that actions are not recognised if they are depicted in compressed slides (levels 2 and 3). This happens because in this situation the image loses static diagnostic elements which are essential requisites for the interpretation of the scene, such as the shape of framed objects and spatial references needed for a three-dimensional reconstruction. In the case of compressed moving images, instead, data evidenced high performances of action recognition which can be explained by asserting the existence of a visual semantics of dynamics. Dynamics is in fact the only one factor which distinguishes movies from slides, because slides were meaningful frames extracted from movies themselves.

Nevertheless, the question is to explain how the subjects can make a semantics of a dynamics, and one hypothesis is that subjects recognise a spatio-temporal configuration. We suppose that observers utilize movement schemata, that is mental representations of movement that are compared with inflow data. Such an assumption puts the question if schemata are specific or general representations of movement. In other words: does each scheme represent precisely only one movement, so that there is one scheme for each action, or there are general schemata, in which one scheme roughly represents a certain number of similar movements? We showed to subjects some actions that show similar dynamics but are associated to different meanings, as dancing and fighting, and we found that at certain compression levels subjects can't distinguish between actions depicting similar dynamics and mistake, for example, a game for a violence and a violence for a game. Such recognition mistakes recur frequently (see graph 10) if movies are compressed at the highest compression level (levels 3), that is when subjects have to base their interpretation of the scene only on dynamics, and are explainable with the assumption that movement schemata are general. So, the movement scheme used in the recognition of an action can represent a class of dynamics made up of similar movements and thus different actions can belong to the same class. Nevertheless, the fact that actions are correctly recognised at low compression level (level 2) shows that a minimal information is sufficient to recognize an action in an univocal way and this latter could be present at high compression levels. The fact that, in correspondence to the same compression level, objects and actions are not recognized in the same way and with the same precision, suggests the hypothesis that there is another powerful semantics, based on some specific spatio-temporal relations that work as reference points for an analysis of the dynamics. Besides, such an hypothesis allows us to explain the differences found for the stimulus typologies (slide and movie). At last, the differences in recognition precision and velocity found between dynamics and action, demonstrate how both these mechanisms are necessary for the correct and efficient recognition of the action. So, if by means of the only dynamical analysis of a scene a lot of different actions are contemporaneously plausible interpretations of the scene itself, it is thank to the information associated with some spatio-temporal relations that the performed action is recognised unequivocally between all those that have been activated by the movement scheme. Thus, for not falling into recognition mistakes when scenes are ambiguous, the visual system must have both the types of information and has to be able to integrate them.

REFERENCES

Bernstein, N., 1967, *The coordination and regulation of movements*. Pergamon Press, London.

Blake, A., and Isard, M., 2000, *Active Contours*. Springer, Berlin.

Bobick, A.,and Ivanov, Y., 1998, *Action recognition using probabilistic parsing*. mit. Edu /pub/ techreports/ TR449.ps.Z.

Bobick, A F., and Davis, J. W. 1996, *An appearance-based representation of action*. Technical Report 369, M.I.T.

Brand, M., 2000, *SIGGRAPH Proceedings 2000* (in press).

Buelthoff, H. H., and Edelman, S., 1992, *Proc. Nat. Acad. Sci.* **89**: 60-64.

Dittrich, W. H., 1993, Action categories and the perception of biological motion. *Perception* **22** (1): 15-22.

Emering, L., Boulic, R., Balcisay, S., and Tholmann, D., 1997, Multi-level modelling and recognition of human actions involving full body. *First ACM Conf. On Autonomous Agents*, Los Angeles. pp. 504-510, ligwww.epfl.ch/emering.html.

Essa, I. A, and Pentland, A. 1997, *IEEE trans. PAMI* **10**: 757-763.

Giese, M. A. and Poggio, T., 2001, *Learning based representation of complex movements patterns*. www.ai.mit.edu.

Goncalves, L., Di Bernardo, E., Ursella, E., and Perona P., 1995, *Monocular tracking of the human arm in 3D*. ICCV.

Gyulai, E., 2000, Attribution of meaning from movement of simple objects. *Perceptual and Motor skills* (2): 27-35.

Logothetis, N. K., Pauls, J., and Poggio, T., 1995, *Curr. Biol.* **5**: 552-563.

Johansson, G., Hofsten, C., and Jansson, G., 1980, Event Perception. *Annual Review of Psychology* **31**: 27-63.

Runeson, S., 1994, Perception of biological movement: the KDS princple and the implementation of a Distal versus Proximal approach. In *Perceiving events and object* (G. Jansson and W. Epstein, eds.), Erlbaum, Hillsdale, New Jersey, pp. 383-405.

Multidisciplinary Techniques for Classifying Styles in Descriptive Texts: a Preliminary Study

M. NARDON, E. NOT, F. PIANESI, and M. ZANCANARO
IRST, Istituto per la Ricerca Scientifica e Tecnologica, I-38050 Povo TN, ITALY.
E-mail: {nardon, not, pianesi, zancana}@itc.it

1. INTRODUCTION

The style of a text seems to be defined by a combination of various elements that range from lexical choices to discourse organisation. In this respect, writing style is a systemic variable that needs a multi-disciplinary approach for this characterisation.

In this paper we present a preliminary study that elaborates a method for analysing texts by isolating the topic organization and the *stylistic indicators* that characterise a certain writing style. The ultimate aim of the our research is to identify a multi-disciplinary paradigm for categorizing texts.

In the literature there are very few systematic works investigating the linguistic differences of texts written for different audiences (Milosavljevic, 1997). In this study we were interested in investigating if it is plausible to measure this "change in tone" and in identifying classes of writing styles for a more objective classification of texts.

In the genre of narrative texts, it is common practise to distinguish at least four different categories: *children's, general, technical and telegraphic.* With our research, we applied different methodologies with which to collect empirical evidence for this classification. In the first phase of our study, 16 stylistic indicators were selected by observing the regularities that seemed to characterize various sample texts treating the same subject. These indicators range from lexical choices to rhetorical strategies. We then verified the actual reliability of the chosen stylistic indicators by measuring the statistical

Emergence in Complex, Cognitive, Social, and Biological Systems, Edited by Minati and Pessa, Kluwer Academic/Plenum Publishers, New York 2002

occurrence of each of them in the examined texts. Eventually, the indicators have been grouped using a standard clustering algorithm. These clusters appear to approximate the considered standard categories, suggesting that "writing styles" indeed exist.

In a second phase, we checked the hypothesis that different writing styles entail different ways of presenting topics. Our methodological approach is inspired to text analysis techniques *à la* van Dijk (van Dijk, 1972, 1985), which provide a method to isolate at different levels of abstraction the topics expressed by different portions of text and allow us to check whether there is a variation in the selection of the topics when a text is written for different audiences.

2. SELECTION AND CREATION OF THE TEXT CORPUS

A corpus of 90 texts in the domain of animal description has been collected from the library of the *Museo Tridentino di Scienze Naturali* (Natural Science Museum), which contains books and encyclopaedias for teenagers as well as general, technical, and scientific encyclopaedias.

Each text in the collection has been classified in one of the four "standard" categories according to its actual classification in the library. In particular, our corpus consisted of: 24 texts belonging to the children's category, 24 texts to the general category, 24 texts to the technical category and 18 texts to telegraphic one.

3. EVIDENCE FOR WRITING STYLES IN NARRATIVE TEXTS

The hypothesis of this study holds that a writer typically uses different writing styles according to the audience. For instance, a writer writing a book for children will tend, even if unknowingly or unconscientiously, to use a certain vocabulary or text structure geared toward inexperienced readers.

By observing the regularities that characterize the texts in this domain, we selected the following indicators:

- *Use of jargon, anthropomorphism*: presence of anthropomorphic description of the animal described such as -"the brown bear it's really a *cuddly* ..."-.
- *Use of technical terms*: presence of technical terms in the text.
- *Explanation of technical terms*: presence of an explicit explanation for the technical terms.

- *Writing in 1st singular person*: the text it is written in 1st person singular.
- *Writing in 3rd singular person*: the text it is written in 3rd person singular.
- *Use of figures and pictures*: presence of figures or pictures to enhance the text.
- *Use of deictic references to figures and pictures*: presence of phrases such as -"as you can see in the pictures ..."-.
- *Personification of the animal*: the animal described in the text is a particular one, such as -"the grizzly bear *Yoghy* lives in the zoo since..."-.
- *Non personification of the animal*: the text is describing the general characteristics of a kind of animal.
- *Presence of introduction/conclusion*: presence of an introduction and a conclusion at the beginning and at the end of the text.
- *Explicit interaction with reader*: presence of phrases such as -"do you know that...?"- in the examined text.
- *Non-interaction with reader*: absence of phrases such as the one above.
- *Synthetic writing style*: concepts are expressed only once;
- *Redundant writing style*: concepts are expressed more than once
- *Articulated writing style*: the description is developed like a story;
- *Bulleted writing style*: in the description concepts are simply listed.

A protocol was defined to instruct an annotator to judge, for each chosen indicator, whether it applies or not to a given text.

Table 1 shows the relative frequency, expressed in percentage, of each stylistic indicator annotated in each text category.

Table 1.

STYLISTIC INDICATORS	Children's category	General category	Technical category	Telegraphic category
Use of jargon, anthropomorphism	91.67	8.33	0	0
Technical terms	0	75	100	100
Explanation of technical terms	0	75	8.33	0
Writing in 1st person singular	41.67	0	0	0
Writing in 3rd person singular	58.33	100	100	100
Presence of figures and pictures	91.67	91.67	62.5	16.67
Presence of deictic references to figures and pictures	83.33	91.67	0	0
Personification of the animal	62.5	0	0	0
Non-personification of the animal	29.67	100	100	100
Presence of introduction/conclusions	100	100	20.83	0
Explicit interaction with reader	83.33	70.83	0	0
Implicit interaction with reader	16.67	29.17	100	100
Synthetic writing style	0	0	83.33	100
Redundant writing style	100	100	16.67	0
Articulated writing style	100	100	100	22.22
Bulleted writing style	0	0	0	77.78

The Kruskal-Wallis chi-square test has been applied to compare the distribution of the frequency of each stylistic indicator for each text category. All the differences were significant (alpha = 0.01). Table 2 shows the results.

Table 2.

STYLISTIC INDICATORS	Chi-square value	df	p-value
Use of jargon, anthropomorphism	70.4583	3	0
Technical terms	68.975	3	0
Explanation of technical terms	52.7643	3	0
Writing in 1st person singular	30.5938	3	0
Writing in 3rd person singular	45.0855	3	0
Presence of figures and pictures	34.5926	3	0
Presence of deictics to figures or pictures	62.3	3	0
Personification of the animal	48.95	3	0
Non-personification of the animal	48.95	3	0
Presence of introduction/conclusions	72.8316	3	0
Explicit interaction with the reader	55.1314	3	0
Implicit interaction with the reader	55.1314	3	0
Synthetic writing style	75.4879	3	0
Redundant writing style	75.4879	3	0
Articulated writing style	65.5789	3	0
Bulleted writing style	65.5789	3	0

By performing a *Partitioning Around Medoids Cluster Analysis* on the frequencies of the stylistic indicators in the whole corpus (considering the presence/absence = 1/0 as a *symmetric binary variable*), we obtained 4 clusters within which the variables (stylistic indicators) gather together, as shown in Table 3.

Comparing these clusters obtained with the frequencies of each stylistic indicator in each text category, we can notice that the two classifications are compatible. This provides evidence to the fact that there exist text categories characterized by different "writing styles".

However in applying the *clustering vector*, we noticed that not all the stylistic indicators we have chosen are discriminant for a certain writing style, but only some of them. In this sense it resulted that some of the texts initially collected as belonging to the *children's category*, cluster together with the texts collected as belonging to the *general category*, and this happens because these texts lacked some stylistic indicators, such as *Personification of the animal*, which statistically define the belonging to a certain cluster.We may conclude that it is impossible to classify texts by isolating single variables. It is necessary to consider the text as a result of multiple variables co-occurring. In this sense, we think that writing style it is an emergency of the combination and interrelation of many different elements.

Table 3.

	Cluster 1	Cluster 2	Cluster 3	Cluster 4
Use of jargon, anthropomorphism	1	0	0	0
Technical terms	0	1	1	1
Explanation of technical terms	0	1	0	0
Writing in 1st person singular	1	0	0	0
Writing in 3rd person singular	0	1	1	1
Presence of figures and pictures	1	1	1	0
Presence of deictics	1	1	0	0
Personification of the animal	1	0	0	0
Non-personification of the animal	0	1	1	1
Presence of introduction/conclusions	1	1	0	0
Explicit interaction with the reader	0	1	0	0
Implicit interaction with the reader	1	0	1	1
Synthetic writing style	0	0	1	1
Redundant writing style	1	1	0	0
Articulated writing style	1	1	1	0
Bulleted writing style	0	0	0	1

4. ANALYSIS OF THE MACROSTRUCTURE OF THE TEXT CORPUS

We now focus our attention on the way in which the topics are chosen and organized in the texts. Using the *propositionalisation* technique introduced by van Dijk (1972, 1985), we can skip some details of the grammatical or logical analysis of the single phrases[1]. According to this technique, a *macroproposition* is a text portion introducing a single topic. For each of the 90 texts of our corpus, macropropositions have been identified and associated to a topic identifier. This kind of analysis allowed us to identify the various topics treated in each text and to calculate the frequencies of each topic in the corpus. The schema in Figure 1 shows an example of macropropositions identified in texts belonging to different text category. The text highlighted in grey shows a macroproposition corresponding to the topic *Physical characteristics* in the 4 different text category. (The examples in Figure 1 are written in Italian consistently to the original language of our text corpus. The English translation it is reported in Appendix A.)

[1] A first attempt to use this technique to estimate the complexity of a text can be found in (Nardon, 1999).

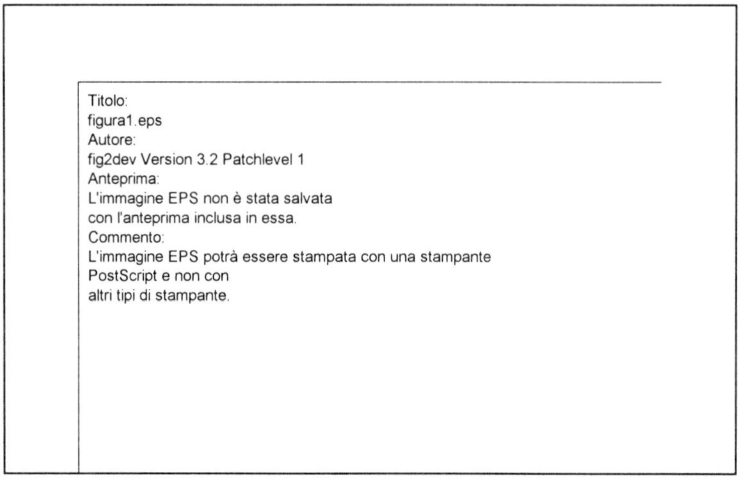

Figure 1.

Table 4.

TOPIC	Frequency in global corpus	Frequency in Children's category	Frequency in General category	Frequency in Technical category	Frequency in Telegraphic category
Habitat	81.11	79.17	79.17	79.17	83.33
Physical characteristics	93.33	91.67	95.83	95.83	100
Feeding habits	93.33	91.67	87.5	87.5	100
Diffusion	74.44	70.83	79.17	79.17	83.33
Technical data	78.89	41.67	87.5	87.5	100
Behaviour	86.67	83.33	83.33	83.33	88.88
Reproduction	82.22	45.83	91.67	91.67	100
Biological classification	77.78	45.83	87.5	87.5	100
Danger of extinction	48.89	45.83	54.17	54.17	44.44
Captivity	24.44	29.17	29.17	29.17	11.11

Table 4 shows the topics identified in our corpus with their relative frequencies, expressed in percentage, within the overall corpus and within each text category.

These topics have been handled as dependent variables for each individual text. We enumerated the relative frequencies of each topic in each text. The rating scale was 0 (absent) or 1 (present).

We found significant differences when analysing each single topic's frequency. We found that the group "children's category" has a lower frequency for the topic *technical data, reproduction* and *biological classification*, in Table 5 are reported the Kruskal–Wallis chi-square results (alpha = 0.01).

Table 5.

Comparison Between Text Categories	Chi-square value	df	p-value
Habitat	0.251	3	0.969
Physical characteristics	2.9137	3	0.4051
Feeding habits	2.9137	3	0.4051
Diffusion	1.0135	3	0.798
Technical data	27.8908	3	0*
Behaviour	1.0135	3	0.789
Reproduction	29.8045	3	0*
Biological classification	20.8196	3	0.0001*
Danger of extinction	0.5057	3	0.9176
Captivity	2.2904	3	0.5144

*significant results

One possible interpretation of these results is that writers consider these three topics overly difficult or demanding for young readers, and, therefore, indeed writers take the type of expected audience into account when selecting the content of a text.

5. CONCLUSION AND FUTURE INDICATIONS

Using our experimental protocol combined with the text analysis *à la* van Dijk, we found that it is possible to classify texts written for different audiences both by observing the presence/absence of the stylistic indicators in each text and by measuring the variation in the topic selection.

Future work could investigate the real effectiveness or salience of the stylistic indicators we have chosen by testing their psycholinguistic and computational reliability, possibly extending the preliminary analysis to text corpora belonging to other semantic domains.

It is important to keep in mind that this is a preliminary study, with the ultimate goal of proposing a new and multidisciplinary methodology to classify various written styles. By quantifying qualitative elements of a text it becomes plausible to use the obtained results in a computational setting. In particular the aim of this research is to extend and improve the *MacroNode formalism* (Not, Zancanaro, 2000), which is a particular approach to Adaptive Hypermedia generation.

REFERENCES

van Dijk, T. A., 1972, *Some aspects of text grammars.* The Hague, Mouton.
van Dijk, T. A., 1985, *Handbook of Discourse Analysis.* Academic Press, New York.

Milosavljevic, M., 1997, Augmenting the User's Knowledge via Comparison. In *Proc. of the 6th International Conference on User Modelling, UM 1997*, Sardinia, Italy.

Not, E., and Zancanaro, M., 2000, The MacroNode Approach: Mediating Between Adaptive and Dynamic Hypermedia. In *Proc. International Conference, AH 2000*, Trento, Italy, 28-30 Aug., pp. 167-178.

Nardon, M., 1999, Tesi di Laurea in Psicologia *Dalla comprensione di testi audio alla progettazione di una interfaccia uomo-macchina*. Facoltà di Psicologia, Università di Roma "La Sapienza".

APPENDIX A

Telegraphic category:

(...)

Description: very big and stocky, dark-brown hair. Young animals have a light brown collar. Length from head to tail average from 1,70-2,50 m; tail 6-14 cm; height on four limbs 90-100 cm; weight 105-265 kg.

Call: deep mumblings.

Habitat: (...)

Technical category:

(...)

At least 2 breeds are in danger of extinction.

The brown bear's fur ranges from light-brown to dark-brown, or even black, in colour. All brown bears are big, extremely strong and are the largest of carnivorous.

They live either alone or in small family groups and are active (...)

General category:

(...)

The American black bear is a big carnivorous that when standing on its back limbs can reach up to 2 meters in height. It has thick and ruffled fur with colours that range from black to brow, yellow and even white.

The babies in a same brood (...)

Children's category:

(...)

Hi, I'm Tom and I'm a very big and well-built animal (I can be up to 200 Kg in weight!)

Did you know that when we go into hibernation we don't eat (...)

Text in italic shows the macropropositions for the topic Physical characteristics.

The Religious Aspects of the Israeli-Palestinian Conflict in the West Bank: Systemic Implications

SABRINA BRAHMS
368 Museum Drive, Los Angeles, CA, 90065, USA - E-mail: sabejams@pacbell.net;
Consulting Faculty, Phillips Graduate Institute, Los Angeles, CA, USA
Doctoral Student, Saybrook Graduate School, San Francisco, CA, USA

1. INTRODUCTION

In this paper, I will outline some of the systemic issues associated with the Israeli-Palestinian conflict, specifically as they are expressed in the region of the occupied territories known as the West Bank. The Israeli-Palestinian conflict is a stellar example of multiple causality; therefore it is impossible to address all its aspects in one paper. I will consequently limit this analysis to the role of religious ideology. I am particularly interested in the fact that both Jews and Muslims revere Abraham as their father. They share a connection to this historical and religious figure, but this shared connection does not unite them.

The Tomb of Abraham is the tactile symbol of this shared homage as it is the burial place of Abraham and is considered a holy site by both Jews and Muslims. The tomb is located in the town of Hebron, a significant city in the West Bank and often an arena for violent outbreaks between Palestinians and Israelis. The Tomb of Abraham and the city of Hebron will provide the context to which the systems concepts will be applied.

Emergence in Complex, Cognitive, Social, and Biological Systems, Edited by
Minati and Pessa, Kluwer Academic/Plenum Publishers, New York 2002

2. BRIEF HISTORICAL OVERVIEW

Israel was declared a state in 1948 by United Nations resolution. The land that comprised the state did not include the area specific to this analysis, the West Bank, which refers to the lands that lie on the western bank of the Jordan River. From 1948-1967, the West Bank belonged to Jordan. However, in the Six-Day War of 1967, Israel captured these lands, thus extending the country's eastern border to the Jordan River. Since then, the Israeli government has hotly debated what to do with the land itself and the people who live there. The West Bank as well as the Gaza Strip has not been annexed into Israel, and thus, its residents do not have the formal rights of citizenship. The Palestinians who live there are currently living under military occupation.

The city of Hebron is named in honor of Abraham. This Hebrew name is derived from the Biblical passage that describes Abraham as a "kind friend". The Arabic name for the city is Al Khalil Al Rahman, which translates to "the friend of God", also referring to Abraham. Hebron lies approximately 20 miles south of Jerusalem. It is one of the four holy cities of Islam as are Mecca, Medina, and Jerusalem. It is also one of the four sacred cities to Jewish people, along with Jerusalem, Tsfat, and Tiberias, and was King David's capital city before he conquered Jerusalem.

As was mentioned earlier, it is in Hebron where the Tomb of Abraham is located. According to the Bible, Abraham purchased the land, and he and his wife Sarah are buried there. Abraham, also according to the Bible, is the father of monotheism as well as the ancient Israelites through his son Isaac. According to the Qu'ran, Abraham is of great significance to Islamic people as a holy man and the father of the Arab people through his son Ishmael.

In 1267 the Mamluk sultan, Baybars, ruled that Jews were not permitted to enter the mosque, which had previously been built over the burial spot. Exactly 700 years later, in 1967, because of the Israeli occupation of the city, Jews once again could gain access to the site. To this day, Muslims use the mosque during certain designated times and Jews during others.

3. CO-EVOLUTION OF RELIGIOUS ZEALOTRY

The beginning of the modern conflict over the West Bank commenced with the 1967 Israeli occupation. It continued when merely three weeks after the end of the Six-Day War, the Israeli government began to establish Jewish settlements in this territory. By 1992, there were close to 100,000 Israeli settlers in 180 communities in the West Bank (O'Sullivan, 1994). Palestinians were outraged by the encroachments of these settlements and over the years, the settler's presence has been a great source of tension in the West Bank. Sprinzak (1998) explains that these settlers maintain a particular religious ideology that fuels their zeal for settlement. Many believe that the success in the Six-Day war was a sign from God of the beginning of the messianic era. In particular, settlers in the Gush Emunim movement espouse the idea that the kingdom of Israel in the land of Palestine as defined by the Bible, will lead to the redemption of the world. It must be pointed out that this ideological concept of the sovereignty of redemption is a privilege exclusive to Jews.

Sprinzak (1998) explains that the world's perceptions of the State of Israel as a democratic island in the sea of fundamental Arab extremism has been justly eroded by the recent surge of Jewish religious extremism. According to statistics (Smooha, 1992), not only Arabs but Israelis have become more religious in the past thirty years as well, and both Islamic and Jewish fundamentalists have gained powerful political control. Although Gush Emunim is not an official political movement, it wields power with the right-winged political party that constitutes nearly fifty percent of the parliament.

Abu-Amr (1993) reviews the history of the Hamas movement. An offshoot of the Muslim Brotherhood, Hamas emerged in 1998 during the intifada as a more radical and confrontational approach than its parent organization to the Israeli oppression of Palestinians in the West Bank and Gaza. Like the Jewish settler movement, it originated in the occupied territories. One of Hamas' express purposes is to rival the PLO, which appeals to the nationalistic desires of Palestinians to have their own homeland. Instead, this religious Islamic group emphasizes the holiness, not only of the West Bank, but of the entire land of Palestine. It is stated in the Hamas Charter (1993) that no Muslim has a right to give up any part of Palestine because it is an Islamic trust, defined as a waaf, and must fight to defend it. Litvak (1998) explains that these charter statements represent a change in previous interpretations of religious texts, reinterpretations that are used to support this new, more radical ideology. According to Litvak, Palestine as a waaf, (religious land trust) is a recent invention.

This is a point of similarity with Jewish religious groups, who believe that as part of the messianic redemption, all the land of Israel must be returned to Jewish rule. Neither Jewish nor Islamic fundamentalist groups will accept the idea of dividing Israel into two separate states. The only acceptable idea to the Muslims is one state under Islamic religious rule, and the only acceptable solution for the Jewish settlers is a Jewish state where religious rule takes the place of the current secular system. Therefore we see that the system has co-evolved into one of greater intolerance, an intolerance which is supported by religious ideology.

4. CIRCULAR CAUSALITY OF RELIGIOUSLY SANCTIONED VIOLENCE

Hamas distinguishes itself from the Muslim Brotherhood in what it considers acceptable means of reclaiming lost Muslim land. The concept of jihad, which means struggle, a foundational precept of the Islamic religion, is generally interpreted by members of the Muslim Brotherhood as spiritual struggle, but for Hamas members, jihad necessarily means armed struggle. For Hamas, violence is a requisite means of redemption. Abu-Amr (1993) explains that this represents a shift toward increased violence in the Islamic resistance movement.

Karpin and Friedman (1998) point out a similar shift in the interpretation of Biblical texts that are used to substantiate the increased violence on the part of Jewish settlers. In this case, not only are Biblical texts used to support acts of violence against Arabs, but perhaps more noteworthy is the fact that such interpretations are also being used to condone violence against Jews who participate in the peace process. The most extreme example is found in the case of the assassination of Yithak Rabin. He was deemed a rodef (an obscure reference to a "dangerous pursuer" in Jewish legal texts) by some ultra-orthodox rabbis and it was this legal interpretation that was used by Rabin's assassin to justify his death based on his involvement in the Oslo negotiations with Arafat (Karpin and Friedman, 1998).

Shahak (1994) addresses the terrorism committed at the Tomb of Abraham in 1994. On February 25 of that year, a religious Jewish settler, an American medical doctor who immigrated to Israel to live in the West Bank, walked into the mosque that houses the tomb during the time of Islamic prayer and fired 111 bullets. He killed 29 people and wounded nearly 100 (Reich, 1996). Within minutes, he was beaten and killed by Islamic survivors. Shahak points out that Baruch Goldstein's act was not an isolated incident, but was connected to the larger ideological framework described above. The religious settler community in general claimed Goldstein as a hero and honored him as a martyr at his funeral.

In order for the current violence in Israel to cease, the systemic level of religious ideology as described above must be addressed. Because both Jewish and Muslim fundamentalist groups promote the use of violence as a method of serving God, they are realistically dangerous and contribute to the cycle of escalation. The secular nationalists, who until the outbreak of the current intifada were negotiating the Israeli-Palestinian peace process, must decide how to manage these fringe movements if they want to reach a successful peace agreement.

5. SYSTEMIC EQUILIBRIUM

The current situation between Palestinians and Israelis is obviously at far-from-equilibrium conditions. But in contrast to many other cases of far-from equilibrium states, where the level of system functioning in not sustainable for a long period of time, in this case, there is an interesting factor that seems to contribute to the longevity of such far-from-equilibrium conditions. Usually a system that is functioning in a far-from-equilibrium state is on the verge of great change or at least the possibility for systemic transformation is present, but with religious Muslims and Jews there is perhaps a stabilizer within each system that maintains the far-from-equilibrium state between the two groups. The stabilizing factor is that each religious group as its own system, operates under very close-to-equilibrium conditions. For instance, the Jewish and Islamic religious systems are highly stabilized by the religious laws governing daily life. In Israeli and Palestinian secular society, by contrast, the religious aspect of life have been eroded by modernity. In both Muslim and Jewish religious communities, however, religious law is strengthened external threats and provides security and structure in uncertain times. The less stable the relationship between Palestinians and Israelis, the more stable becomes each religious group's religious aspect of life.

6. NEGATIVE INTERDEPENDENCE

Israelis and Palestinians are interdependent based on their mutual interests and commitment to one small stretch of land. Because they live in such close proximity to each other, their lives necessarily co-evolve. Kelman (1999) suggests that, for the most part, this interdependence is negative. Although Kelman's analysis is not specific to the religious aspects of the conflict, his analysis addressed the problems that are manifest on the West Bank, and we shall see that his nationalist descriptions can be easily applied to the religious arena. His major point is that negative interdependence is expressed in terms of national identity. Both Israelis and Palestinians view their national identity to be mutually exclusive of the other. This also can be said in the case with religious identity. If one nation or religion is to succeed, the other must necessarily fail. Kelman also points out that each group actually creates its identity in relation to the conflict and therefore builds its identity based on the hatred it perceives from the other group.

Kelman, as an academic who is very active in the peace-building community, suggests that Palestinians and Israelis have the opportunity to transform negative interdependence into positive interdependence through a process of acknowledging the other's national identity. He advocates that this work be conducted in the forum of peace workshops. While I support the form of the workshops as well as their goal, I would like to suggest a necessary prerequisite to this mutual acknowledgement that Kelman has not addressed. My suggestions will follow in a discussion that will first review some aspects of systemic change as defined by Jackson (1988, 1995).

7. SYSTEMS METHODS FOR CREATING SYSTEMIC CHANGE

Jackson (1988) suggests three models that describe the nature of the relationships between system participants. The first he refers to as a unitary system in which the participants generally agree on the objectives of the system in which they function. The second system type he calls pluralist. In this situation the participants do not agree on the general system objectives but are both willing and able to work toward a consensual agreement that would benefit all system members. The third system type, Jackson refers to as coercive. In this model consensus is only achieved through the use of power and domination by some members of the system over the others.

The current situation between Israelis and Palestinians, particularly those who live in the occupied territories is clearly that of a coercive system. Israel is in a position of military occupation over the Palestinian people who live in the West Bank and Gaza. Border closures, house demolitions, and land confiscations are but some of the methods used by the Israeli army to oppress Palestinian residents. This assessment does not remove responsibility from either side for the use of violence against other human beings, but instead, it points to an important aspect of the conflict that has, as yet, not been mentioned in this paper. It is not enough to see the recursive and reciprocal nature of the relationship between Jews and Muslims in the West Bank and it is not enough to understand the tensions that are manifest at the Tomb of Abraham in Hebron only from the systemic concepts described earlier. As Jackson (1995) suggests, the power balances in any system must be addressed.

Although Kelman (1999) will surely acknowledge that the current situation between Palestinians and Israelis is unbalanced in terms of power, he does not address the issue directly as it pertains to the context of peace workshops. It is not enough to bring together Israelis and Palestinians so that they can transform their negative interdependence into positive co-existence. The power imbalance must be acknowledged and changed before such workshops can ultimately be affective. It is unreasonable to believe that Palestinians would arrive at a workshop, and build together with their

Israeli counterparts, a sense of positive co-existence, and then return home where they are unable to go to work because a closed border prevents them from doing so.

8. CONCLUSIONS

Jackson (1988) admits that not much work has been done to develop systems practices to work with coercive systems. He suggests as a preliminary measure that systemic change in such systems begins by focusing on what "ought to be" instead on "what is". It is my hope, that as a systems practitioner and a beginning practitioner in the field of peace-building in the Israeli-Palestinian conflict that I may further develop ways to work with the coercive system as I have outlined it above. I hope to do this by keeping in mind Kelman's idea of transforming negative interdependence into positive interdependence. In addition, I will develop my skills all the while considering the destructive impact of the power imbalance between Israelis and Palestinians as well as the profoundly antagonistic impact that the Jewish and Muslim religious communities have on any advances in the peace process.

REFERENCES

Abu-Amr, Z., 1993, Hamas: A historical and political background. *Journal of Palestine Studies* **22** (4): 5-19.

Charter of the Islamic resistance movement (Hamas) of Palestine, 1993, *Journal of Palestine Studies* **23** (4): 122-134.

Karpin, M., Friedman, I., 1998, *Murder in the name of God: the plot to kill Yitchak Rabin.* Metropolitan Books, New York.

Kelman, H. C., 1999, The interdependence of Israeli and Palestinian national identities: The role of the other in existential conflicts. *Journal of Social Issues* **56** (3).

Jackson, M.C., 1988, Systems methods for organizational analysis and design. *Systems Research* **5** (3): 201-210.

Jackson, M.C., 1995, Beyond the fads: Systems thinking for managers. *Systems Research* **12** (1): 25-42.

Litvak, M., 1998, The Islamization of the Palestinian-Israeli conflict: The case of Hamas. *Middle Eastern Studies* **34** (1): 148-163.

O'Sullivan, G., 1994, Settler terror. *Humanist* **54** (3): 45-47.

Reich, B. (Ed.), 1996, *An historical encyclopedia of the Arab-Israeli conflict.* Greenwood Press ,Westport, CT.

Shahak, I., 1994, The background and consequences of the massacre in Hebron. *Middle Eastern Policy* **3** (2): 63-76.

Smooha, S., 1992, *Arabs and Jews in Israel, Vol. 2: Change and continuity in mutual intolerance.* Westview Press, Boulder, CO.

Sprinzak, E., 1998, Extremism and violence in Israel: The crisis of messianic politics. *Annals of the American Academy of Political and Social Science* 555: 114-126.

Disciplinarity in the Pursuit of Knowledge

ARNE COLLEN
Saybrook Graduate School, 450 Pacific, San Francisco, CA 94133 USA

1. INTRODUCTION

From the perspective of systemics, Minati and Collen (1997) have described disciplinarity in terms of phases or forms of human activity to seek, develop, and produce knowledge. Disciplinarity manifests in four forms: singular, multiple, inter-relational, and boundary-breaking pursuits. Although there is the notion that knowledge resultant of each form is reflective of and delimited by its characteristic form, the presentation falls short in its depiction of the scheme as a whole to be a higher order conceptual system.

The purpose here is to reconsider and extend their presentation. By doing so, disciplinarity becomes a conceptual chameleon-like system, a way of systemic thinking, and a dynamic process by which human beings engage reflectively and communicatively with each other to come to know and share more informatively their worlds.

The premise taken in developing the conceptual system is to conceive disciplinarity as a conscious, creative, decision-oriented, dynamic, meaningful, and productive activity of the inquirer. The strategy behind this development is, at its core, an elementary methodology, which follows related publications (Collen, 1994; Collen *et al.*, 1996; Linstone, 1995). The defining constructs comprise the elements. These elements provide the base and platform of departure from which the relations among them are examined. This exercise leads one to explicate the conceptual system and then to consider some consequences and implications in regard to its application.

Emergence in Complex, Cognitive, Social, and Biological Systems, Edited by
Minati and Pessa, Kluwer Academic/Plenum Publishers, New York 2002

The organization is sequenced in four parts. First, disciplinarity requires some definition. Second, there is a synopsis of forms found in Minati and Collen (1997), with the addition of a fifth form not included in this source. Third, the five forms are brought together to be compared, contrasted, and integrated into a more wholistic conceptual system. Finally, select implications, having methodological, philosophical, pragmatic, psychological, and theoretical import, are discussed.

2. WHAT IS DISCIPLINARITY?

Dictionaries provide the narrower definitions of the construct. In contrast, broadened meanings are found in the meta sciences, under such phrases as the cognitive arts, cognitive sciences, the sciences of the artificial, sciences of complexity, design sciences, and systems sciences. After following some denotative transitions from discipline to disciplinary to disciplinarity, there is a more contemporary treatment among the forms of the construct.

To paraphrase *The Compact Oxford English Dictionary* (1991, p. 442), where *discipline* points to specific instruction, practice, and exercise of that which one is learning and has learned, such as a lesson or teaching, it can also mean a particular course or branch of knowledge, instruction, education, art, and science. Typically, it suggests a known direction for a particular course of action, such that the inquirer (learner, disciple) acquires the appropriate, proper, and correct course of action, namely the training effect (trained condition) of experience, having carried out the lesson or teaching.

Even more relevant to this paper is the denotative emphasis given to discipline defined as "a system or method for the maintenance of order" or "a system of rules" for the conduct of inquiry, practice, and training. However, this intention is not to carry a tone of inflexibility to the detriment of advancing the knowledge and skill of the discipline. Contrary to popular interpretation of the construct, discipline should not lead to rigidity. In fact, the opposite is the intended result. Disciplined inquiry, for example, should enable a researcher to use and communicate to other researchers a sufficiency of specific rules and procedures to further the inquiry without stultifying it. Discipline is taken to be a necessary and advantageous condition for scientific and methodological advancement, which is quite the contrary to those who would imply discipline is a straight-jacket-like or miltaristic-like activity. Suffice it to state that the practice of discipline importantly positions the inquirer to repeat the work of other researchers, demonstrate repeatedly the stability of findings, determine delimitations (the boundedness of generalization), and notice and subsequently explore anomalous and serendipitous events during the conduct of inquiry.

There is a last and equally popular definition that defines discipline as "the correction, chastisement, or punishment inflicted by way of training." As with the notion of rigidity, it leads to another misunderstanding often imparted to discipline in the sense being developed here. In other words, this lay definition is also not the preoccupation of disciplinarity in regard to research, science, and the pursuit of knowledge.

It follows that we can expect *disciplinary* to be "of or pertaining to the character of discipline." The relevant definition would heavily emphasize the learning and mental training involved in any focused investigatory pursuit.

The care taken to situate the denotations of discipline can be directly applied to *disciplinarity*. Simply stated, it refers to being in the state or condition of a discipline, or manifesting the characteristics of a discipline.

3. FORMS OF DISCIPLINARITY

The five forms to be discussed here are: mono, multi, inter, trans, and meta. This five-fold scheme is not meant to exhaust the possibilities. The subject matter domain includes terms like cross and pluri, to be found for example in the *International Encyclopedia of Systems and Cybernetics*, and Internet sources as the "Encyclopedia of World Problems and Human Potential," and the "International Center for Transdisciplinary Research."

Any scheme is for conceptual convenience to comprehend relations, systemicity, and complexity. We impose our conceptual divisions upon our observation and study of phenomena. Our taxonomies, classes, and schemata organize and group that which we study, but at the same time, they place boundaries that separate, albeit artificially, many of the linkages we revisit to integrate what we come to know, after acquiring more knowledge about what we observe. These boundaries come to be known by the focus, that is for illustrative purposes, the biological, cultural, economic, physical, political, psychological, social, and spiritual. Again, the previous terms are not meant to be complete but merely representative of the variety of perspectives possible.

3.1 Mono

As the inquirer comes to know one focus (perspective, domain, position), there is great temptation to specialize in that focus, gain confidence and comfort in this knowing, and resist foraging into neighboring fields when it may complicate and jeopardize that which one has invested much time and resources in acquiring. The inquirer becomes established and affiliated with the focus, which typically solidifies into mono-disciplinarity. The resultant expertise of course is to be valued as long as it remains uncompromised by ideology and methodolatry.

Mono-disciplinarity works with fragmented knowledge attained through effective pursuit strategies associated with its subject domain. Science has repeatedly proven itself to operate quite effectively by the process of separating, dividing, and specializing.

The observables, pedagogy, research methods and strategies, and theory of a discipline tend to be relatively homogeneous in relation to other disciplines. Descriptions of phenomena and organizations of them (taxonomies) tend to be widely shared among those affiliated with the subject matter domain of the discipline.

In the pursuit of knowledge of a relatively sparsely explored domain, this initial phase makes a great deal of sense, for without the establishment of the various disciplines, there is little reason to imagine the inquirer to have what is necessary to move toward more involved forms of disciplinarity.

To picture the prevalence of mono-disciplinarity is to comprehend the internal activities of each discipline, operating in parallel with others, yet not communicating and sharing knowledge actively with others. These solo pursuits are metaphorically like a room full of persons at their workstations, keeping to themselves, and when the need to communicate with another person does arise, it happens only with one whose affiliation is in common between inquirers.

Today the problems and phenomena are less likely with satisfaction to be addressed and studied, respectively, through one discipline alone. This realization compels an advance in our thinking beyond mono-disciplinarity.

3.2 Multi

One means to transcend the restraints of mono-disciplinarity is to gather together a set of experts representing their disciplines to examine a particular phenomenon or problem. The unilateral look and contribution of each representative conveys the process known as multi-disciplinarity. The result is a knowledge product considered more a collection of positions, views, and expositions of the nature of a roundtable and cafeteria-like display of what can be known about the focus. This is the general manifestation of multi-disciplinarity, and nothing more. At some stage, this bibliographic and encyclopedic form of disciplinarity has an important place as a depository and resource of knowledge as understood by those representatives at that time in human history. Where multi-disciplinary collaboration tends to be of some value in providing a broadened and deepened range of the contributions brought to bear by the participating parties, the products tend to be limited in moving the process toward more integrative work that considers the complexification and systemicity of phenomena and problems.

In this form of disciplinarity we can imagine for example, symposia, conferences, and consultations convening to solicit the views of the experts. Anthologies, proceedings, and edited books of readings make available to others the range of views on the focus of study and discussion. As to the metaphor of the room full of active workstations, sharing information through contributions to and use of institutional archives, databases, and web sites provides multi-disciplinarity in the human organization.

However, for the same reasons as mono-disciplinarity, multi-disciplinarity is severely limited. The participating disciplines in and of themselves need not achieve any systemic and wholistic form devoted to their focus.

Here the terms multi- and pluri are applied interchangeably to mean the simultaneous use of a set of relatively autonomous disciplines to address a focus, interest, problem, and issue, without requisites imposed for the linkage, synthesis, and integration of knowledge. For example, pluri-disciplinarity may mean the employment of a plurality of research methods, some of which may be stock-and-trade of a discipline outside the researcher's affiliated discipline. Finer distinctions among these two and other forms of disciplinarity may be found elsewhere, namely, Francios (1997) and the International Center for Transdisciplinary Research on the Internet.

However, the view taken here is that there is no mandate to restrict and define the forms solely according to centricity (requirement of a centralizing discipline), classification (systematic description and taxonomic aggregation of phenomena), formalization (extensive concretization of rules and procedures), method (infusion of methodology from one discipline into another), or theory (presence of a falsifiable body of propositions, axioms, and laws).

3.3 Inter

When the representatives of disciplines truly communicate with mutual respect and learning, there can be attained a form of disciplinarity well beyond separate and parallel contacts. Participants come to view their interests, issues, problems, questions, answers, and solutions with common concern and the seeds of collaboration germinate. Colonizing and condescending attitudes of the practitioners in a discipline give way to the recognition there is value in multiple perspectives and a need for a diversity of expertise to advance a step in a more integrative and synthetic fashion the knowledge of all disciplines. The delineation of common interest and the contribution from difference become accepted and important characteristics of disciplinarity.

We can imagine a Venn type diagram of overlapping circles, in which each circle represents a discipline. The circles overlap such that there is a common core, yet each circle also overlaps with its neighboring disciplines. This picture generates a rich set of possibilities for collaboration. An outstanding example of inter-disciplinarity is the specialty focus known as psychoneuroimmunology. More informed considerations of many human maladies may benefit selectively from what is known in biology, chemistry, cytology, epidemiology, endocrinology, hematology, neurology, and psychology. And this pursuit is meant of course to be inclusive of relevant inter-disciplines like biochemistry, neurochemistry, and neuropsychology.

Inter-disciplinarity begins with a particular tie or bridge that connects two disciplines. There are many examples where knowledge domains become interrelated, such as astrophysics, biochemistry, ecopsychology, psychohistory, and sociobiology. Two disciplines may join through infusion of one into the study of the other, or from symbiotic benefits of synthesizing aspects of their subject matter. For example in psychohistory, psychology brings a particular viewpoint by way of theoretical organizing frameworks and the emphasis on personhood to view historical accounts. As to a particular history, when infused with cartography, economics, political science, psychology, and sociology, it reaps many mutual benefits that enables the researcher to weave a richer collective tapestry of any human event, movement, and trend than could be conveyed solely through one of the disciplines.

To follow further the workstation metaphor, inter-disciplinarity would mean a higher level of interaction among workstations and additional layers of organization memory devoted to collaboration and synthesis of what the various contributors bring to the work flow. Electronic trails give evidence to the productivity, progress, and eventual information and knowledge products produced in the process.

3.4 Trans

As the considerations of the interrelations become more evident and intricate, it necessitates the realization that the complexification of the focus as well as involved relations among the disciplines compels us toward trans-disciplinarity. At this point, cybernetic and systemic constructs become paramount, such as positive and negative feedback loops, the co-evolving nature of any conceptualization of system with its context (environment), and the embeddedness of any conceptualization of system in regard to its subsystems and interrelations with adjacent, sister, and superordinate systems.

Like inter-disciplinarity, trans-disciplinarity fosters a common language, usable knowledge among disciplines, and shared methodologies. But trans-disciplinarity gears the endeavor more to match the systemicity and complexity of the focus. In distinction from the previous forms described, trans-disciplinarity takes the pursuit beyond any consideration and limiting factor, problem, and theory within and between the disciplines. Other forms of disciplinarity cannot achieve these aims very well.

It is instructive at this juncture to note the use of cross-disciplinarity as a construct. It is sometimes used to distinguish the basic form (mono-disciplinarity) and the more complex form (trans-disciplinarity) from the intermediate forms (multi- or pluri- and inter-disciplinarity). The prefix "cross" implies a breaking through, transgression, and perhaps the bridging of disciplinary boundaries, but it does not inform us about what happens and what is accomplished when the inquirer goes beyond the confines of particular disciplines. This prefix is not used in the present scheme here, because of the ambiguity of meanings among the intermediate forms. Although the phenomenon of crossing disciplinary boundaries is an important and often courageous professional undertaking worthy of study in its own right, cross-disciplinarity may refer to any of the intermediate forms, hence it is of limited usefulness in exploring the progression of forms as a conceptual system.

It is out of trans-disciplinarity that an integrated body of knowledge springs through the integrated work of collaborators among the participating disciplines. It is difficult to imagine such a body of knowledge from earlier forms of disciplinarity because of the absence of participation from key disciplines. Each core discipline has key parts of the puzzle to contribute and until the pieces are present, the new whole has little chance of being achieved.

From the integrated character of this new body of knowledge, reaped from the interrelations of the contributing disciplines, we can comprehend its systemicity. Systemicity is emergent from trans-disciplinarity. Systemicity is an emergent property from the complexification of the disciplines when integration occurs.

However, despite the wholistic realizations of trans-disciplinarity, the individual disciplines remain somewhat autonomous. This autonomy is evidenced through all forms of disciplinarity covered to this point in the pursuit of knowledge.

3.5 Meta

The last form of disciplinarity to be discussed is meta-disciplinarity. The boundaries distinguishing disciplines dissolve to a degree in inter- and trans-disciplinarity, but they become more than meaningless, in fact non existent, at the meta level. Meta-disciplinarity shows the interwoven qualities of the blended disciplines working collaboratively and collectively, making the pursuit of knowledge more than tangibly trans-disciplinary. In short, the extent of integration suggests a superordinate and emergent mono-disciplinarity of greater complexity.

Bateson (1979) was widely recognized for his reach to encourage and model meta-disciplinarity. More recent acknowledgement of Bateson's influence is given by Folk (1995), who points out that "a metapattern is a pattern so wide-flung that it appears throughout the spectrum of reality ... a pattern of patterns ..." (pp. viii-ix). The meta-disciplinarian seeks to synthesize, integrate, and convey the wholism of a body of knowledge, whose saliency may be characterized and thematized in terms of its meta-patterns.

By way of another illustration, at some point in the development of the subject domain, many will argue that psychoneuroimmunology will cease to be inter- and trans-disciplinary in character, and be truly meta-disciplinary, at which point its parent disciplines will become indistinguishable and meaningless to those who affiliate with this emergent meta-discipline.

As a horizon, one may take a similar view of those disciplines contributing to the study of consciousness. Such a cluster of disciplines appears to be undergoing the progression described in this paper. In other words, there are collectivities from among the sciences and various fields of study, which one might consider to be progressing toward meta-disciplinarity, such as the earth sciences, human sciences, and life sciences. These collectives serve to illustrate further some likely candidates for meta-disciplinarity.

4. SUMMARY AND INTEGRATION

The five forms of disciplinarity covered in the previous section may be summarized by means of a table and two figures. Each summary sets the stage for the subsequent discussion.

The table compares the types of disciplinarity in two important respects. From the table we can see a progression toward more systemicity and increasing complexity.

Table 1. Select properties distinguishing types of disciplinarity

Disciplinarity	Relation	Systemicity	Complexity
Mono	Within	-	0
Multi	Among	-	1
Inter	Between	+	2
Trans	Beyond	+	3
Meta	Above	+	4

The first figure is meant to convey that the forms of disciplinarity work as a conceptual system. We can move among the forms to study any given focus (interest, phenomenon, problem). The trend in the pursuit of knowledge today is a progression of a discipline from mono- to multi- and/or inter- to trans-disciplinarity. Meta-disciplinarity represents a culminating state for the process to begin all over again as knowledge continues to be revisited, revised, and reformulated. Meta-disciplinarity is the four forms taken as whole.

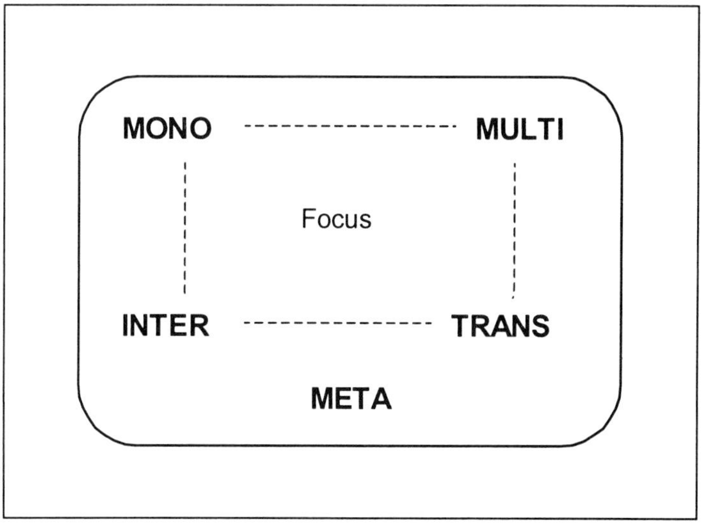

Figure 1. Meta-disciplinarity as a conceptual system.

The relations among the forms are relevant to working with them as a whole system. It is important to emphasis that as tempting as it may seem, the scheme does not unfold as a simple linear phase progression that enables an accounting of the conceptual system. The second figure depicts the progression over the five forms of disciplinarity as an unfolding of nested manifestations, in which each emergent form encompasses earlier forms. There is not a replacement of one form by another, but a development that is an advance, such that the inquirer has more choices to advance knowledge and can always regress to, or draw advantageous upon, an earlier form while continuing to progress to higher order and more challenging, complex forms. Meta-disciplinarity encompasses all forms and may be seen as a higher order and much more complex, integrated, and advanced form of mono-disciplinarity.

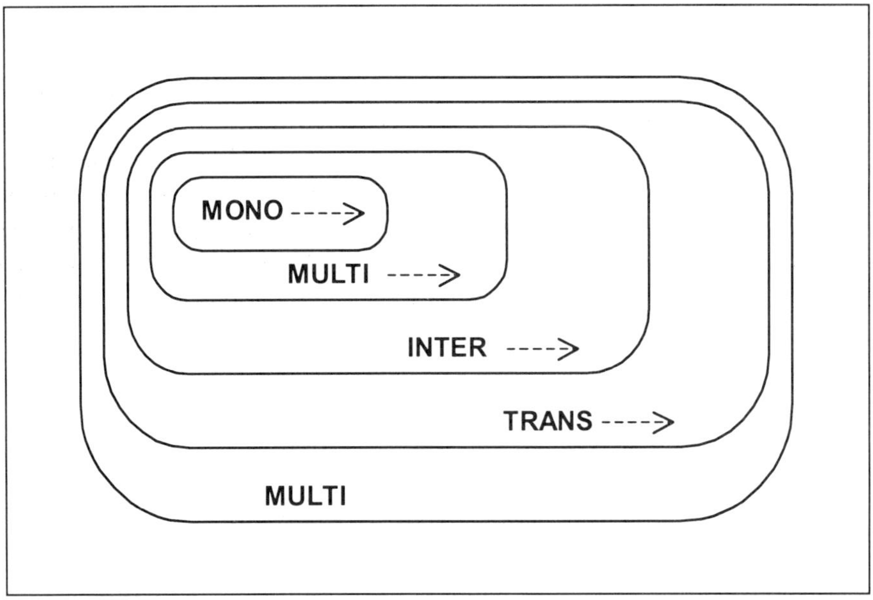

Figure 2. The embeddedness of disciplinarity.

Finally, it is equally important to view the scheme as a conceptual system of complementary forms, for in the more complex and systemic applications, the inquirer can move among them to reap the greatest benefits from the contributing disciplines and facilitate the advancement of knowledge.

5. DISCUSSION

When we speak of the meta- level, there is scantly much left but historical vestige to regress to anything mono-, multi-, inter-, and trans-. One might think of meta as a higher order emergent synthesis - a super mono-disciplinarity. In human history there is both a proliferation in specialization (mono- and multi-disciplinarity) as well as the turn toward generalization and unification (inter- and trans-disciplinarity). While there is certainly the necessity of more specialists as knowledge domains mushroom ever more beyond the comprehension of a single human being, there is concomitant and complementary need for more generalists to synthesize in parallel fashion.

Complexification (Casti, 1994) brings both differentiation within as well as unification beyond given disciplines. Although a cyclic process has been suggested earlier when examined transitionally from a side view, the growth of the disciplines is really a spiral-like process when viewed historically from above. Both volume and density of the body of information and knowledge expands, even though paradoxically, the horizons become clearer in the pursuit of how much we do not yet know and how much there is still to know. It is a never-ending process.

To design a system able to act on the processes of knowledge production is to design an inquiry system. The history of science may be viewed in this fashion. Churchman (1971) provides an informative and useful description of such a sequence of paradigmatic systems for inquiry, that is to say, broadly conceived research traditions in the history of Western science. Although granted the phrase "inquiring system" does convey an enlivened connotation—used to stress the dynamics and process of the search for knowledge—appearances of the phrases "disciplined inquiry," "inquiry system," and "system of inquiry" are not intended to diminish that vitality. In addition, Churchman contributes an important perspective to his history of inquiry systems by reminding us that our design and use of them must take into account future generations, else we fail to include fully the ethics of our pursuit.

We may speculate that with further globalization of information and knowledge, there will be a general enhancement of disciplinarity in all its forms. The Internet and World Wide Web, for example, are making it rapidly possible for an unprecedented proportion of humanity to develop more personalized forms of mono-disciplinarity that transcend physical limitations and traditional disciplines. The mutual presence formerly required between inquirers (co-researchers, mentor and apprentice) may be less necessary in coming generations, even though established traditions will be heavily used nevertheless, making full use of globalizing trends. Given a preliminary focus, the researcher can draw upon the global network of archives, experts, and related sources to cull, define, circumscribe, situate, contextualize, critique, evaluate, synthesize, theorize, and apply the focus. It is from these research-oriented activities that the disciplinarity of the inquirer emerges. Once established, the inquirer can offer his and her domain and expertise to others. With one's fellow inquirers, one can engage in multi-, inter-, and trans-disciplinary forms of inquiry. Finally, it is from the more collective pursuits that manifestations of meta-disciplinarity will become more prevalent by way of knowledge communities and cultures denuded of classical subject matter labels and boundaries among the disciplines.

REFERENCES

Bateson, G., 1979, *Mind and Nature: A Necessary Unity.* Dutton, New York.

Casti, J., 1994, *Complexification: Explaining a Paradoxical World Through the Science of Surprise.* HarperCollins Publishers, New York.

Churchman, C. W., 1971, *The Design of Inquiring Systems: Basic Concepts of Systems and Organizations.* Basic Books, New York.

Collen, A., 1994, Developing a systemic approach to human science research methodology. In *Information Systems Architecture and Technology '94* (M. Bazewicz ed.), Oficyna Wydawnica Politechniki Wroclawskiej, Wroclaw, Poland, pp. 30-38.

Collen, A., Minati, G., Penna, M., and Pessa, E., 1996, Describing transcultural activity in the framework of the systemic view. *Proceedings of the Third European Congress on Systems Science* (E. Pessa, M. P. Penna and A. Montesanto, eds.), Edizioni Kappa, Rome, pp. 881-885.

Folk, T., 1995, *Metapatterns Across Space, Time, and Mind.* Columbia University Press, New York.

Francois, C., (ed.), 1997, *International Encyclopedia of Systems and Cybernetics.* Saur, Munich.

Minati, G. and Collen, A., 1997, *Introduction to Systemics.* Eagleye Books International, Walnut Creek, California.

Linstone, H. et al., 1995, The multiple perspective concept. In *Design and Systems: General Applications of Methodology* (A. Collen and W. Gasparski eds.), Transaction Publishers, New Brunswick, New Jersey, pp. 177-242.

A new vision of the world transdisciplinarity. International Center for Transdisciplinary Research. Retrieved August 10, 2001, from http://perso.club-internet.fr/nicol/ciret/english/ visionen.htm.

Encyclopedia of World Problems and Human Potential. Retrieved August 12, 2001, from http://www.uia.org/uialists/kon/c0077.htm.

The Compact Oxford English Dictionary, 1991, Second edition. Oxford University Press, New York.

Management

Distributed Knowledge Management

A Systemic Approach

MATTEO BONIFACIO, and PAOLO BOUQUET
Department of Information and Communication Technologies, University of Trento
Via Somarive 10, I-38100 Trento, Italy

1. INTRODUCTION

Knowledge, in its different forms, is increasingly recognised as a crucial asset in modern organisations. Knowledge Management (KM) refers to the process of creating, codifying and disseminating knowledge within complex organisations, such as large companies, universities, and organisations for social and civil services. KM attracted a lot of attention and investments in the last ten years. However, the return from these investments is quite modest, and many managers and analysts are starting to manifest their disillusion.

Most projects of KM aim at creating large, homogeneous knowledge repositories, in which corporate knowledge is made explicit, collected, represented and organised, according to a single – supposedly shared - conceptual schema. These schemas, called for example knowledge maps, are meant to represent shared conceptualisations of corporate knowledge, and to enable communication and knowledge sharing across the entire organisation. The typical goal of these projects is to create a so-called Enterprise Knowledge Portal (EKP), namely a computer-based system that provides a unique access point to corporate knowledge.

In the paper, we argue that this approach reflects an objectivistic epistemology and a centralised paradigm of managerial control. Indeed, on the one hand, it presupposes that all contextual, subjective, and social aspects of knowledge can be eliminated in favour of an objective and general codification, and that this abstract and general knowledge can be shared and reused independently from the individual or community in which it was created. On the other hand, this assumption is coherent with traditional organisational models and paradigms of control, as knowledge is centrally managed (managers organise the process of knowledge extraction in analogy in analogy to what they do with operational work) with no loss of control (once it is made explicit, knowledge is no longer owned by individuals, and becomes property of the organisation).

We show that this approach is incompatible with the nature of knowledge, and that this way of managing knowledge explains, at least partially, why most KM systems are deserted by users. We propose a systemic approach, We propose a systemic approach, in which there is no attempt to reduce corporate knowledge to a single, objective representation. Subjectivity and sociality are taken as irreducible aspects of knowledge, but this is viewed as a potential source of value and richness, rather than as a problem to overcome. Instead of building a single, objective knowledge map, an organisation is thought of as a "constellation" of autonomous, local knowledge maps produced by individuals, communities, teams in their daily work. In this respect, KM becomes the problem of coordinating these multiple sources of knowledge in a distributed (that is, non centralised) way. In the last part of the paper, we also sketch a high level architecture which can support this distributed approach to KM from a technological point of view.

2. (IMPLICIT) ASSUMPTIONS IN CURRENT KM SYSTEMS

In the last ten years, many companies have been trying to design and develop systems to capitalise organisational knowledge. If we analyse the systems that have been implemented, we can see that, even if different technologies, tools and methodologies were adopted, they have a common high-level architecture and share some important features:

- the installation of corporate-wide intranets in order to ensure physical and syntactical accessibility to information (i.e., connectivity and shared formats);
- the design of a corporate language and of knowledge maps, which are used to represent corporate knowledge in a standard and common way and to create semantically homogeneous and ubiquitously accessible repositories;
- the creation/support of informal communities that represent the place where "raw" knowledge is produced through spontaneous and emerging social interaction of company peers. In practice, these communities are

materialised as "virtual communities" through the adoption of computer supported cooperative tools, such as groupware applications;
- the creation of a new role, the Knowledge Manager, whose goal is to support and facilitate the interaction across communities;
- the design of contribution processes which enable community members to explicit their tacit knowledge through the codification in the corporate language;
- the construction of EKP, which provides a simple interface through which people can contribute to the knowledge base, socialise, retrieve information.

These features are based on some general assumptions. In particular, we stress two of them:
- from a structural standpoint: though spread throughout the organisation, sometimes even at a implicit level (Nonaka, 1995), organisational knowledge is something that can (and should) be centralised within explicit, shared repositories;
- from a process standpoint: even though knowledge is produced at a peripheral level as a specific and concrete matter (the so-called tacit dimension of knowledge, see (Polany, 1966)), its use requires that it is transformed into an abstract (i.e. independent from the production context) and general (i.e. applicable in any context) object.

Even though many business operators claim that this architecture is the right answer to the needs of managing corporate knowledge, KM systems are often deserted by users, who instead continue to produce and share knowledge as they did before, namely through structures of relations and processes that are quite different from those embedded within KM systems (in (Bonifacio, Bouquet and Manzardo, 2000), a paradigmatic experience which embeds the assumptions of this approach was concretely illustrated using the case of a worldwide consulting company). In the next two sections we will propose an explanation of why this is the case, and argue that the features described above, far from being necessary elements of a KM system, represent a very clear choice in terms of what knowledge is and, moreover, on the nature of managerial control.

2.1 Epistemological assumptions

Despite the intention of supporting a subjective and social approach, the way most KM systems are designed reflects a marginal notion of sociality and embodies an objectivistic view of knowledge.

On the one hand, social aspects are still considered through a Hawthorn-like approach (Mayo, 1945), which is well rooted in western sociology. Essentially, this view considers sociality as a constraint to be dealt with in order to have operational and cognitive work done by people; people work better if some level of informality and sociality is preserved and guaranteed in working environments. Under this light, communities represent the place where socialization occurs, a way of satisfying the social factors needed by people in order to operate, learn and share knowledge.

On the other hand, the subjective nature of knowledge, although commonly recognized, is viewed more as a sort of primordial state rather then an intrinsic feature of knowledge. When knowledge is created, it is dependent on the context of production, such as the particular viewpoint of the individual or the daily practice of work, and thus not replicable. Therefore, this raw form of knowledge, called implicit by Nonaka and Takuechi (1995) and tacit by Polany (1966), must and can be "cleaned up" (objectified) from any contextual element, and be transformed into an abstract (avulse from the original context) and general (applicable in any similar situation) form.

2.2 Managerial assumptions

This epistemological view, which determines the organisational features of KM systems, is strongly related to a traditional paradigm of managerial control. This paradigm views management as a function grounded in the capacity of centralising the control on the company processes, such as resource and task allocation and monitoring. In particular, management allocates/distributes tasks and resources to employees and monitors the proper execution of tasks and use of resources. This traditional view on the managerial function can cope with KM only if an approach is taken which is compatible with the traditional paradigm; the processes of knowledge (resource) production and dissemination (tasks) need to be centrally driven (allocation) and controllable (monitoring). This condition is satisfied only if knowledge is thought of as an object, which can therefore be kept separate from the people that produced it. Otherwise, as far as knowledge remains embedded within subjective dimensions, it becomes a resource that falls outside the boundaries of managerial control.

It's hard to say whether it is the epistemological view that led to applying the traditional managerial paradigm to KM, or the other way around. Quite likely, they are in a circular relation in which the view on knowledge legitimates a managerial attitude, and the managerial attitude fits within the objectivistic view of knowledge. What is relevant for our purpose is that if the epistemological view becomes unrealistic, then the relationship between KM and the traditional managerial practices becomes problematic.

2.3 The role of technology

Technology plays a crucial role in KM. Many theories argue that technology cannot be considered a neutral matter with respect to organisational structures (see, for example, (Orlikowski and Gash, 1994). Technological architectures can shape organisational forms, and organisational forms can affect the concrete appropriation of technology. The experience with groupware applications in business processes re-engineering (generally considered a key technology for KM) illustrates very well the active role played by a flexible technology to stimulate organisational change. Therefore technologies, and in particular technologies for KM, far from being an independent variable, are a constraining and constrained force in play. From this perspective, the relevance of a KM technology is primarily to be judged on the basis of architectural considerations, rather then on technical or functional features. IT architectures represent and embed the assumptions on how an information environment is "factorised" (described as composed by basic elements) and "processualised" (described through the processes and interactions that are possible among the different elements). In other words, IT architectures can be seen as the informational lens through which an organization can be red, described and modelled. Therefore we can judge whether a technology is consistent or inconsistent, capable of shaping or being shaped by an organizational form only at an architectural level.

If now we consider KM systems from an architectural standpoint, we can observe that there is a discrepancy between social form and technological architecture. On the one hand, most organisational and sociological studies show that, as far as knowledge is concerned, the social form of most modern organisations is similar to a constellation of communities, each with its own languages, processes, tools, practices; in short, is a distributed system of local "knowledges" (see left hand side of Figure 1); on the other hand, most KM systems exhibit a completely different architecture, in which knowledge is "alienated" from its original locus (the community), and organised in a centralised knowledge base (KB), which is then used to feed the EKP (see right hand side of Figure 1). This process is sometimes supported by software tools, like text miners or content management tools, which allow knowledge managers to couple the knowledge map with existing and incoming documents. Of course, there are variations of this general architecture. For example, some systems allow forms of personalisation; a typical example is the use of personal assistants, whose goal is to support the single user (or group) to access in a personalised way to the corporate knowledge (a well-known example is the introduction of personal agents in the engine of Autonomy). However, from our perspective, this does not change the architecture in a significant way, and the process that leads to the creation of the corporate KB is basically the same.

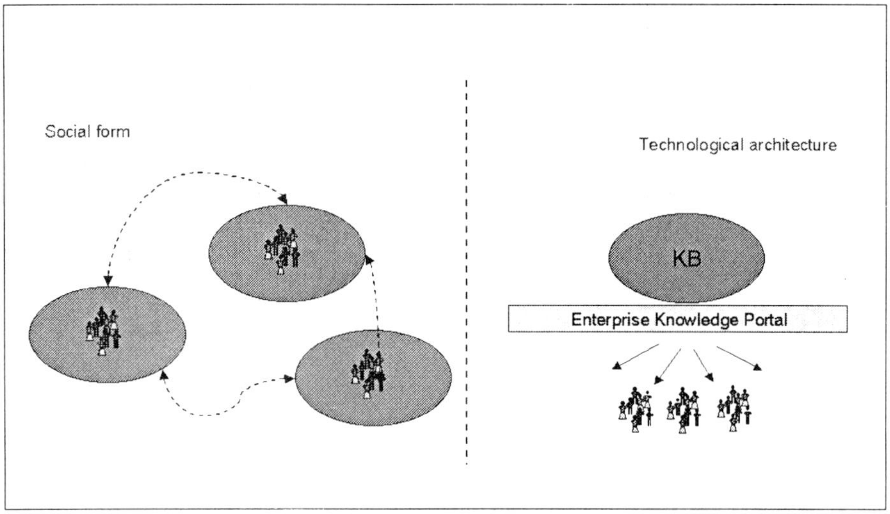

Figure 1. Social form vs. technological architecture in the traditional approaches to KM.

Therefore, if technology is not neutral (both from a cognitive and an organisational perspective), then adopting a different, non objectivistic epistemology requires not only to rethink the managerial function, but also to imagine new technological architectures that can support a different approach to KM.

3. A DIFFERENT EPISTEMOLOGY FOR KM

Most authors who stressed the subjective nature of knowledge argued also that meanings are not externally given; rather, individuals give meaning to situations through subjective interpretation. Interpretation is subjective, since it occurs according to some “internal” interpretation schema , not directly accessible to other individuals. These schemas have been called, for example, mental spaces (Fauconnier, 1985), contexts (Bouquet, 1998; Giunchiglia and Ghidini, 2001), or mental models (Johnson and Laird, 1992). Different interpretation schemas can produce very different interpretations to the “same” facts. Conversely, in order to produce similar interpretations, people need to share (to some extent) interpretation schemas, or at least to be able to make conjectures on what other people's schemas are. Internal schemas can be made partially accessible to other individuals only through language, since language is not just a means to communicate information, but also a way of manifesting an interpretation schema. Shared schemas are studied in different disciplines, and are called in many different ways, for example paradigms (Kuhn, 1970), frames (Goffman, 1974), thought worlds (Dougherty, 1992); the process of "viewing" and reasoning about other people's schemas was called perspective taking (Boland and Tenkasi, 1995) or double loop learning (Argyris and Schon, 1978). Since we are talking about organisations, and thus about a collective level, what is relevant for our purpose is that, without this inter-subjective agreement (or at least conjectured agreement), communication cannot take place, coordinated action is impossible, and meaning remains confined at an individual level (Weick, 1993). From an epistemological point of view, this approach leads to some significant consequences:

- knowledge is intrinsically subjective, as the meaning of any statement always depends on the context (Bouquet, 1998; Giunchiglia and Ghidini, 2001) or schema of the interpreter;
- at a collective level, groups of people can assume they share (or have a reciprocal view on) some part of their different schemas, and therefore they share (or understand) the other's meanings. Nonetheless meaning "lives" in the inter-subjective agreement of different individuals. Thus collective knowledge is social, since sociality is the precondition for something to be a meaning.

As a result, the concept of absolute knowledge, which refers to an ideal, objective picture of the world, leaves the place to the concept of local knowledge, which refers to different, partial, approximate, perspectival interpretations of the world, generated by individuals and within groups of individuals (e.g. communities) through a process of negotiating interpretation schemas. At an organisational level, each local knowledge appears as the synthesis of both a collection of statements, and the schemas that are used to give them a meaning. Local knowledge is then a matter that was (and is continuously) socially negotiated by people that have an interest in building a common perspective (perspective making for Boland and Tenkasi (1995) or single loop learning for Argyris and Shoen (1978)), but also in understanding how the world looks like from a different perspective (perspective taking for Boland and Tenkasi (1995) or double loop learning for Argyris and Shoen (1978)). Therefore, rather then being a monolithic picture of the world as it is, organisational knowledge appears as a heterogeneous and dynamic system of "local knowledges" that live in the interplay between the need of sharing a perspective within a community (to incrementally improve performance) and of meeting different perspectives (to sustain innovation).

4. A DISTRIBUTED ARCHITECTURE FOR KM SYSTEMS

The aim of this section is to draw some consequences on how an architecture for a KM system should be designed to be consistent with the distributed social form in which knowledge is created within organisations. As we said, the main limitation of a centralised architecture (Figure 1, right hand side) is not technological, but organisational. It creates a mismatch between social form and technological architecture, and this often produces a non acceptance by users. Our alternative proposal is based on three main ideas.

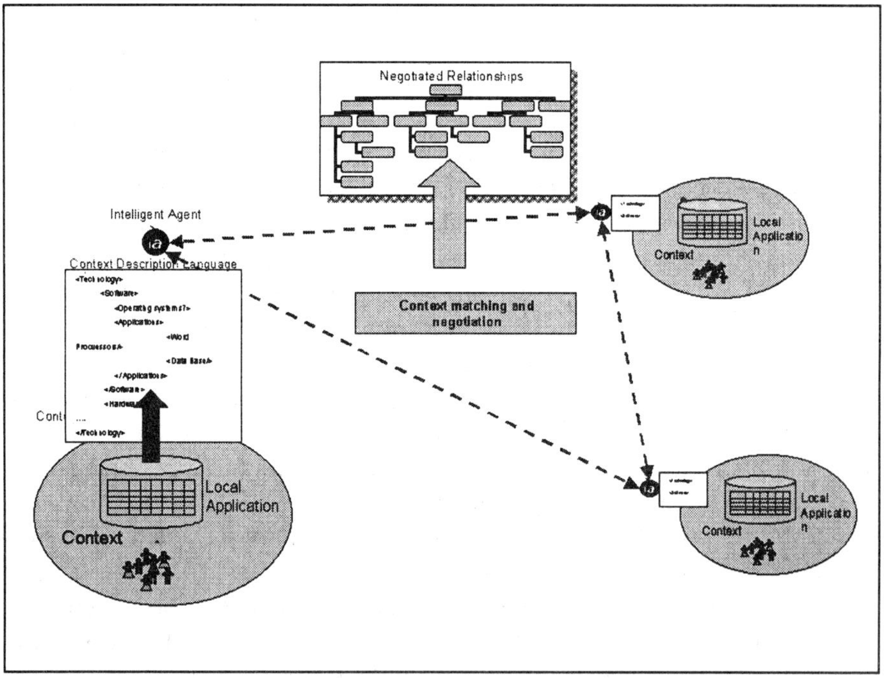

Figure 2. A distributed architecture for KM systems.

The first idea is that knowledge should be autonomously managed where it is created and used, namely within each community. Autonomy means, for example, that each community should be allowed to build its own local knowledge map, and choose whatever tools and processes are more appropriate to manage the life cycle of documents and other data. From an architectural point of view, this means that each community is allowed to build its own "local application" (see Figure 2) for managing local knowledge. This autonomy is a necessary condition for local intelligence to be exploited at its best, as it produces strong ownership on each community's knowledge.

The second idea has to do with the problem of coordinating many autonomous knowledge sources. Indeed, autonomy without coordination is almost useless for knowledge sharing in complex organisations. In traditional KM systems, coordination is achieved by creating common languages and schemas (see Section 2 above) which are then used to categorise all corporate knowledge; however, we already discussed the theoretical and practical limitations of such an approach. In a distributed system, coordination should be reached through interoperation rather than centralisation. Our way to semantic interoperation is based on two steps:

- on the one hand, the system must provide a way for each community to make (at least partially) explicit its own interpretation schemas. This explicit representation of a community's interpretation schema (or perspective) is what we call a context. A context can be "extracted" in different ways. Of course, a community can create its context from scratch (e.g., through a context editor): However, much more interesting is the possibility of extracting contexts in an automatic or semiautomatic way; for example, one can provide tools for translating category structures already in use in a community (e.g. local database schemas or taxonomies of content management tools) into the format of the context description language; or for extracting the context from the documents created or collected in a community's repository;
- on the other hand, each community must be enabled to create relations with explicit contexts of other communities. If we don't want to fall back to an objectivistic view of knowledge (e.g., by requiring that each local context is translated/mapped into a centrally built, shared knowledge map), we need to allow the creation of perspectival mappings from context to context. In our work, contexts are represented as directed acyclic graphs in which both nodes and arcs are labelled. Nodes represent concepts, and are labelled with terms from natural language; arcs represent relations between concepts and are labelled with a type (e.g., is-a or part-of). Intuitively, a mapping between two contexts is a "link" between a concept of the first context and a concepts of the second, for example, a link between the concept of "Venice" in a context and "the city of S. Marco" in another. Again, these mappings can be defined by hand, but a KM system should support the creation of automatic (or semiautomatic) mappings. This requires the ability to compare, in a

semantically relevant way, autonomously generated contexts. From a technological point of view, this can be done by implementing algorithms of context matching, for example, using graph matching combined with natural language processing techniques to suggest possible mappings).

However, context matching is not enough. For human beings, knowledge sharing is often the result of a social process, in which many different cooperative strategies are used. A typical example is a query-answer conversation, in which someone tries to understand what the interlocutor really means with a request (e.g., a question like "Do you have any information about Venice" can be interpreted in many different ways, depending, among other things, on the speaker's goals). This goes beyond the simple idea of context matching. The third idea underlying the proposed architecture is that the system should aim at to reproducing this social process. This is what we call meaning negotiation in Figure 2. In our architecture, meaning negotiation is implemented via communication protocols between autonomous agents. In the architecture a software agent is associated to each community. The role of agents is twofold. On the one hand, each agent "knows" the explicit context of its community; on the other hand, through a process of query-answer, agents can improve the simple idea of context matching. For example, during a search, an agent can ask other agents to provide more information about a concept in their contexts (to improve the result of context matching), can remember past interactions and reuse them for similar queries, can be redirected to other agents, can get suggestions by overhearers (i.e. other agents that overheard the request); and so on.

An important advantage of this architecture is that it reproduces the dynamic of social interaction at the system level. This provides a high degree of flexibility and scalability. An intuitive extension is the following. It is well-known that, in complex organisations, a crucial role is played by those who act as a bridge between organisational units, groups and communities. This role (sometimes called brokering) is well known in social network analysis and community management, as it enhances the ability of a social system to exchange knowledge across semantic boundaries (Wenger, 1998). In the proposed architecture, we can easily introduce brokers as a new type of agent whose contexts are "road maps" to the other contexts available in the system. Thus, brokers may facilitate the exploitation of synergies between communities that, even though working in different contexts, could fruitfully work together. Notice that brokers can become also repositories of "negotiatied relations" (see top of Figure 2), namely collect and remember relations that other agents have successfully negotiated in previous interactions and combine them in innovative ways.

5. CONCLUSIONS

The proposed architecture has many important implications, both from a managerial and technological perspective.

From a managerial standpoint, a distributed approach to KM poses fundamental questions on the nature of the managerial function. First of all, the supportive role played by knowledge managers and claimed by organisational experiences, needs to be brought from the aesthetics of a "nice to be" to a substantial acceptance of a loss of control. In the domain of knowledge, control is to be reinterpreted as a bottom-up process where emerging communities legitimate the "cognitive leadership" of knowledge managers rather then a top-down process of organising "cognitive activities" of people. Moreover, knowledge managers should be the internal expression of a community rather then being assigned from outside.

Second, managers should abandon the dream of having a unique and homogeneous materialisation of "knowledge" represented as an asset in some knowledge base, such as a balance sheet is for financial assets. Knowledge is intrinsically distributed, embedded and localised within the context of informal communities, where context is the fundament of meaning and sense making and thus, from an organisational standpoint, value.

Third, communities, when viewed as the ground where knowledge is created and lives, become centre of power and interest that negotiate organisational paradigms and resources. A community, that is a local knowledge, fights for its survival through the acquisition of resources in order to sustain the "evangelisation" of other communities according to its perspective. In this sense, a community represents not just an opportunity for its valuable knowledge, but also a threat for its natural tendency to create the conditions for its existence. Existence that, being based on knowledge, means forcing that knowledge to be "true".

Fourth, the emphasis on homogenisation as the main process to ensure control (the reduction of complexity through the elimination of differences) should leave the place to concepts such as "interoperation" and "coordination" among different and autonomous actors. In fact, autonomy and heterogeneity are not just intrinsic to the notion of knowledge, but are also the basis of innovation that has been described as the moment in which different perspectives meet, this way generating a discontinuity in traditional and incremental organisational learning paths (Boland and Tenkasi, 1995), (Argyris and Schon, 1978). The assumption of homogenisation as a KM fundamental process leads to the contradiction that knowledge is maximised when innovation (one fundamental organisational learning process) is minimised. On the one hand, managers should support the growth of different perspectives as a premise for innovation; on the other hand, they should seek for coordination providing processes and tools to support systematic interoperability.

From a technological standpoint, distributed architectures presuppose the explicit recognition of the distributed nature of knowledge. Distributed architectures sustain autonomy at different levels: technological (different groups may use different technologies), syntactical (different groups may use different information formats) and, most of all, semantic (different groups may generate different systems of meaning). From a group's or a community's perspective, a distributed system supports the exploitation and representation of a community's context that stands for the layer upon which a community's members produce and negotiate a common semantics. Contexts can be represented as local ontologies, taxonomies and, in general, as theories through which a community's members interpret their environment and make sense of organisational events. From this perspective, text mining tools, for example, instead of being used as a means to produce a corporate wide homogeneous classification of documents, can be seen as domain specific applications that are able to serve the specific needs of each community. Text miners become a community technology that, through use, circularly learns a community context from that community's documents, and learns how to classify documents on the basis of a community's context.

From an inter community standpoint, a distributed architecture facilitates knowledge sharing without assuming company wide shared meanings, and, as a consequence, naturally supports the genesis of innovation. Moreover, since it is more respectful of each community's interpretation schemas, it is more likely to be adopted and used by people. Indeed, it can cope with the need expressed by each community to protect and preserve its specific way to organise and use information.

If semantic interoperability ensures knowledge sharing without the need of a shared semantics, the reduced need of sharing tools and application is dramatically facilitates technological interoperability. Implementing a shared way to organise knowledge from a semantic perspective (semantic homogenisation) is a strong constraint on what technologies can be installed within an organisation. IT managers know how difficult is to balance group specific technological needs with the need to maintain a common information environment. As a consequence, technological choices tend towards general purpose content management platforms that offer a collection of differentiated functionalities in order to match some group's need, and at the same time offer a shared environment to ensure integration. The problems arise when group specificity grows enough to overcome the platform's ability to manage differences and, moreover, when companies are faced with situations that require the integration of different platforms (e.g., merge and acquisitions). Indeed, if one considers the huge investments required, centralised solutions manifest their irreversibility and a lack both in maintainability (for example a change in classification schemas issued by a local group requires a change in the overall structure) and scalability (platforms are scalable to the extent that the same platform is assumed to be used). We claim that distributed architectures, rather then being an expression of IT anarchy, are the best way to balance group specificity (each group can use the appropriate local technology) and shared environment (through interoperability), ensuring reversible investments (disinvesting a technology doesn't mean throwing away the entire system), maintainability (changes can be managed locally), and scalability (the addition of a component or technology is compatible by definition).

ACKNOWLEDGEMENTS

This work is part of the project EDAMOK (*Enabling Distributed and Autonomous Management of Knowledge*). The project is funded by the Provincia Autonoma di Trento.

REFERENCES

Argyris, C., and Schon, D. A., 1978, *Organizational Learning: a Theory of Action Perspective*. Addison-Wesley, Reading, Mass.

Boland, R. J., and Tenkasi, R. V., 1995, Perspective Making and Perspective Taking in Communities of Knowing. *Organization Science* **6** (4): 350-372.

Bonifacio, M., Bouquet, P., and Manzardo, A., 2000, A Distributed Intelligence Paradigm for Knowledge Management. In *AAAI'2000 workshop on "Bringing knowledge to business processes"*, Stanford University, March 2000.

Bouquet, P., 1998, *Contesto e Ragionamento Contestuale. Il Ruolo del Contesto in una Teoria della Rappresentazione della Conoscenza*. Pantograph. Genova (Italia).

Davenport, T., De Long, D. W., and Beers M. C., 1998, Succesfull Knowledge Management Projects. *Sloan Management Review*, Winter.

Dougherty, D., 1992, Interpretative Barriers to Successful Product Innovation in Large Firms. *Organization Science* **3**.

Fauconnier, G., 1985, *Mental Spaces: Aspects of Meaning Construction in Natural Language*. Bradford Books, MIT Press.

Fensel, D., Horrocks, I., Van Harmelen, F., Decker, S., Erdmann, M., and Klein, M., 2000, OIL in a nutshell. In *Proceedings of the European Knowledge Acquisition conference (EKAW-2000)* (Dieng *et al.*, eds.), Lecture Notes in Artificial Intelligence (LNAI), Springer-Verlag.

Giunchiglia, F., and Ghidini, C., 200, Local Models Semantics, or Contextual Reasoning = Locality + Compatibility. *Artificial Intelligence* **127** (2): 221-259.

Goffman, I., 1974, *Frame Analysis*. Harper & Row, New York.

Johnoson-Laird, P., 1992, *Mental Models*. Cambridge University Press, Cambridge.

Kuhn, T. S., 1970, *The Structure of Scientific Revolutions*. University of Chicago Press, Chicago.
Mayo, E., 1945, *The Social Problems of an Industrial Civilization*. Graduate School of Business Administration, Harvard University, Boston.
Nonaka, I., and Takeuchi, H., 1995, *The knowledge creating company*. Oxford University Press.
Orlikowski, W., and Gash, G., 1994, Technological Frames: Making Sense of Information Technology in Organizations. *ACM Transactions on Information Systems* **12** (2).
Polany, 1966, *The tacit dimension*. Routledge & Keel Paul.
Starr, S. L., and Bowker, G., 1999, *Sorting Things Out: Classification and its Consequences*. MIT Press.
Weick, E. K., 1993, Collective Mind in Organizations: Headful Interrelating on Flight Decks. *Administrative Science Quarterly* **38**.
Wenger, E., 1998, *Communities of Practice. Learning, Meaning and Identity*. Cambridge University Press, Cambridge.

The Role of Knowledge Management as a Key Tool for Companies to deal with the Turbulent Environment: A Survey

M. P. PENNA[°,#,*], E. CELANI[#], and N. S. BONFIGLIO[*]
[°]*Facoltà di Scienze della Formazione, Università di Cagliari - E-mail: npenna@hotmail.com*
[#]*LUMSA, Roma*
[*]*Facoltà di Lettere e Filosofia, Università di Pavia*

1. INTRODUCTION

The business environment in which companies are competing seems to have become more and more unpredictable, challenging them in many new ways; the high rate of change in several aspects of economic life, from technology to geopolitics, makes it virtually impossible for any firm to make medium or even long-term market forecasts (Nicolis and Prigogine, 1989). Corrective initiatives so far involved cost reductions, staff cuts and re-engineering processes, without questioning the traditional organizational model that ruled companies for generations (March and Simon, 1992).

We believe in a new approach to the problem of constituting an organization able to react and immediately adapt to environmental change. If companies want to survive and prosper, they have to find innovative strategies to deal with discontinuities in the economic context (Mintzberg, 1996); we retrace this need back to the process of communication between a company and its environment, in other words back to the dynamics of knowledge exchange. Most of the firms are in need of some kind of change, since they feel it's the only way to be competitive: only the acquisition of new skills an expertise will keep them alive. Learning can be incremental, if it tends to improve the existing strategy, or transformational, if its purpose is to put forward new strategies. The design of a learning organization (LO) includes both of the previous aspects of acquiring knowledge, but this double process has to be efficiently planned, implemented and controlled by a set of initiatives that goes under the definition of knowledge management (KM). An effective KM that takes place within the framework of a learning organization becomes the ultimate mean for companies to build and sustain a competitive advantage (Nonaka and Takeuchi, 1995).

The aim of this work is to photograph the standard of implementation of several KM initiatives in a company that defines itself a LO; we also wanted to investigate the nature of the connection between KM practices and innovation mechanism. The organization we chose is a large multinational operating in the service sector; it's based in New York, USA, therefore it tries to dominate its market in a considerable high-complexity context.

2. HYPOTHESES

Knowledge management initiatives can encourage innovation. Our research analyzes the dynamics of knowledge sharing/management in a learning organization and tries to put in evidence the relation between these dynamics and innovation. Therefore our hypotheses are:

1. Knowledge management and innovation management should be implemented regularly in a learning organization. Continuous change needs continuous learning: knowledge practices have to guide individual's job on a daily basis. In order to survive, a company has to manage its innovation potential through effective pre and post-implementation strategies.
2. There's a relation between KM and innovation management in a learning organization, therefore an innovation has to be seen, if not as a result, at least as a positive effect of the influence of KM on organizational processes.
3. A shared vision of goals and procedures is the condition of existence for a LO; all the individuals should deeply believe in the importance of knowledge for the future of their company, with no difference across departments and positions.

3. TESTING THE HYPOTHESES

In order to test the previous hypotheses we built a questionnaire containing 70 multiple-choice items that was administrated and collected by intranet and E-mail. The questionnaire was divided into 3 main sections:

knowledge management and innovation, knowledge management, innovation management.

3.1 Subjects

The questionnaire was given out to 40 employees of the company quoted in the Introduction during July and August 2001; by that time, they were all working in the company head office located in New York City, USA. The selection of the subjects was based on their position and on the department they were in; we chose people from every department and every position on the hierarchic scale (see Table 1). The average age was 26, females were 70% of the whole.

Table 1. Distribution of employees as a function of department/position

DEPARTMENT	POSITION		
	Director	Manager	Coordinator
Public Relations	1	2	8
Internship USA	1	2	5
Business	1	2	2
Sales & Services	1	1	3
Work & Travel USA	1	2	3
Work Abroad	1	2	2

4. RESULTS

We calculated the percentages for the answers (related to any single area of interest) pertinent to both dichotomous and ordinal items. In the following we will report the obtained results for each questionnaire section.

4.1 Knowledge management and innovation

The results relative to the first section of the questionnaire showed that innovation was considered as a key-value and a drive behind the growth of the organization. 67,5% of the sample believed that service innovation leads to success, while 62,5% feel that innovation can improve internal and external processes relevant to the organization. Everybody generally agreed on the fact that knowledge management considerably influences innovation; this result basically confirmed our second hypothesis.

As regards KM practices, we notice how most of them are seen as crucial for innovation: the Table 2 shows the percentage of the answers related to 8 KM practices (De Long, Davenport and Beers, 1996).

Table 2. Percentage of the answers relative to the KM practices

PRACTICES	ANSWER (%) Almost never	Some-times	Often	Regularly	N/A
Capture and reuse knowledge	0,0	37,5	52,5	47,5	0,0
Share "lessons learned"	0,0	15,0	37,5	40,0	7,5
Document expertise	5,0	25,0	37,5	32,5	0,0
Structure and map knowledge	0,0	15,0	45,0	40,0	0,0
Measure and manage the economic value of knowledge	0,0	15,0	60,0	25,0	0,0
Synthesize and distribute external knowledge	0,0	5,0	47,5	47,5	0,0
Use a technical infrastructure for knowledge exchange	0,0	12,5	40,0	47,5	0,0
Embed knowledge in services	0,0	5,0	35,0	60,0	0,0

New ideas come out from a creative recombination and reorganization of knowledge, we believe, and the employees we tested agree with our hypothesis. Innovation is a key-factor for growth in many firms, therefore they have to learn how to innovate better and faster than their competitors; innovation is a drive to efficiency as well: through process innovation in fact, companies can reduce costs and build closer relationships with customers and partners (Davenport, 1993).

4.2 Knowledge management

We tested our subjects about daily implementation of knowledge practices and we found out they are frequently executed: this fact gives a partial confirmation to hypothesis 1. The Table 3 contains the percentages of execution related to the same set of practices we used before (De Long, Davenport and Beers, 1996).

Some items of the questionnaire investigated the state of organizational knowledge in the company, with a stress on coding and circulation of information. 37,5% of the individuals reported that they documented «often» their know-how in a way that can be useful to the others, whereas 32,5% did it «regularly». This process is critical both for knowledge coding and for tacit knowledge to become explicit (Nonaka and Takeuchi, 1995).

Table 3. Percentages of execution of KM practices

PRACTICES	ANSWER (%) Almost never	Some-times	Often	Regularly	N/A
Capture and reuse knowledge	0,0	37,5	7,5	55,0	0,0
Share "lessons learned"	0,0	30,0	22,5	47,5	0,0
Document expertise	7,5	35,0	35,0	22,5	0,0
Structure and map knowledge	22,5	32,5	20,0	25,0	0,0
Measure and manage the economic value of knowledge	7,5	15,0	55,0	22,5	0,0
Synthesize and distribute external knowledge	0,0	22,5	55,0	22,5	0,0
Use a technical infrastructure for knowledge exchange	0,0	27,5	57,5	15,0	0,0
Embed knowledge in services	0,0	30,0	37,5	32,5	0,0

Even if technology made it possible for employees to virtually interact without being physically close by, our research shows how deeply they believe in face-to-face relations, when knowledge can find informal ways to flow. 60% of employees was thinking that their workplace promotes knowledge exchange.

The company we studied invests money and time to build people's know-how (regularly for the 35%, often for the 35%) following a steady schedule, and it rewards who acquires new skills or expertise (often for the 32,5%, regularly for the 25%). This strategy was traduced in a strong motivation to share knowledge: 52,5% agreed on the fact that people should pass on ideas or information that they believe may be of some use to someone else in the organization, and 32,5% strongly agreed.

4.3 Innovation management

A cooperative environment is the first condition for a learning organization to be able to make the most of any direct or indirect synergy between its members; an innovative behaviour emerges from a system that knows how to retrieve the information it's looking for and succeeds in reorganizing it creatively. 44% of the sample believed that cooperation leads to innovation, 18% believed it strongly.

Environmental complexity is hard to reduce, therefore a firm must continuously find new resources or use the ones it already has in new ways. We asked to the employees if the company had processes in place to identify new service ideas, and 55% answered «yes»; then we asked about processes in place to identify new process ideas, and 77,5% this time gave a positive confirmation.

Many organizations generate more ideas than they can handle, so we included some questions to investigate the process of discovering, classifying and financing new ideas. We found out that in our company the ideas were captured often for the 37,5%, regularly for the 12,5%, sometimes for the 42,5%; captured ideas were revisited and developed at some point in the future often for the 52,5%, sometimes for the 35%, regularly for the 12,5%. It's also useful to determine the value of an innovation fairly accurately both prior to implementation, (it happened often for the 37,5%, sometimes for the 35%), and after that (the 57,5% of individuals agreed on the fact that the organization views lack of success as a learning experience): this leads to an enrichment of personal and organizational knowledge (it happened often for the 45%, regularly for the 35%, sometimes for the 12,5%).

4.4 Relationships with "position" and "department" variables

The results of our testing didn't show any relevant difference between the answers of individuals characterized by different positions, thus giving a partial confirmation to the hypothesis 3. Yet, when we compared the same answers across departments, we found a surprising difference between the ones of Work & Travel USA (W&T) and the others. W&T seems not to be involved in any one of the corporate initiatives, revealing lack of motivation and integration within the system. This situation demonstrates how often opposite forces coexist in complex organizations, blocking a spontaneous synergy of components; without this synergy a set can't evolve in a system as "a new entity not reducible to the addition of its parts" (Minati, 1996).

5. CONCLUSIONS

Knowledge management improves both the innovation and adaptation factors. While the results of our questionnaire have turned out not to be linear at all, our subjects suggested many ways in which knowledge could or should be used to design, develop and implement new products and services.

The role of KM combines with care activities that allow and encourage people in generating ideas, support their diffusion, and store up value for the organization: KM is an effective set of approaches for the success of these initiatives.

Even though our testing tool has to be refined, it led us to interesting conclusions as regards the initial hypotheses. The company we observed was structured as a learning organization, and the everyday working experience seemed to reflect the directions of a knowledge management strategy: so our first hypothesis gets its confirmation.

The learning organization uses what it knows to evolve in new ways, in other words to innovate. Innovation is management's favourite evolution factor because it's the result of a creative reorganization of corporate knowledge. KM supports this process, and our subjects agreed with this statement: the relation between KM and innovation was perceived and verified by the majority of individuals (Hp. 2 gets validated).

The results of the comparison across departments, as we already said, showed us that not everybody is totally into the learning organization project: one department (W&T) is just a satellite of the company, and its knowledge gets dissipated. Some individuals act with no conviction, causing a weakening of the whole structure. All of these observations clash with our statements in hypothesis n. 3, confirming at the same time the complex and contradictory nature of the organizational model we studied. The risk for this LO is to mistake the mean with the real aim: a LO can be successful only if the support comes from the totality of its members.

We believe our questionnaire has been a precious tool, helping us to photograph the situation of a learning organization in its dynamic evolution. There are a lot of difficulties for a research project on KM to deal with, while there will probably never be an explicit mechanism to translate knowledge into innovation: we nonetheless hope knowledge management will become more effective and adaptable for companies to support the evolution process.

REFERENCES

Davenport, T. H., 1993, *Process Innovation: Reengineering Work through Information Technology*. Harvard Business School Press, Boston, MA.

De Long, Davenport, and Beers, 1996, *What is a Knowledge Management Project?*. Ernst & Young LLP Center for Business Innòvation, Research Note CBI311.

Forrester, J. W, 1969, *Urban Dynamics*. MIT Press, Cambridge, MA.

Gell-Mann, M., 1994, *The Quark and the Jaguar*. W.H. Freeman & Company, New York.

Heisenberg, W., 1958, *Fisica e filosofia*. EST, Milano.

Holland, J. H., 1995, *Hidden Order*. Helix Books, Reading, MA.

Kauffman, and Stuart, A., 1991, Antichaos and Adaptation. *Scientific American* **265** (2), New York. (It. trans.: Di Giuseppe and Longo, Anticaos ed evoluzione biologica. *Le Scienze* **47** (278), ottobre 1991, Milano).

March, J. G., and Simon, H. A., 1992, *Organizations*. 2nd ed. Blackwell, Oxford, UK.

Minati, G., 1996, *Introduzione alla sistemica*. Kappa, Roma.

Minati, G., 1998, *Sistemica – etica, virtualità, didattica, economia*. Apogeo, Milano.
Mintzberg, H., 1996, *La progettazione dell'organizzazione aziendale*. Il Mulino, Bologna.
Nicolis, G., and Prigogine, I., 1989, *Exploring Complexity: An Introduction*. Freeman, New York.
Nonaka, I., and Takeuchi, H., 1995, *The Knowledge Creating Company*. Oxford University Press, New York.
Penna, M. P., and Pessa, E., 1993, *Introduzione alla psicologia connessionistica*. Di Renzo, Roma.
Pessa, E., and Penna, M. P., 1994, *La rappresentazione della conoscenza*. Armando, Roma.
Taylor, J. R., and Van Every, E. J., 2000, *The Emergent Organization*. Lawrence Erlbaum Associates, Mahwah, NJ.
Waldrop, M. M., 1992, *Complexity*. Simon & Schuster, New York.

Systemic Approaches to Human Care

Balancing and Compensating Equilibrium Deficits in Beings Provided with Cognitive Processing Capabilities

GIANFRANCO MINATI[*], and ALBERTO RICCIUTI [*,#]
[*]*AIRS - Associazione Italiana per le Ricerca sui Sistemi,*
[#]*Attivecomeprima - Onlus (Breast cancer Association), Via livigno, 3 - 20158 - Milano, Italy*

1. BALANCING AND COMPENSATING

This paper refers to *Cognitive Science* (Benjafield, 1992) and *Systems Theory* (Minati, 1998b). The general topic is the ability for a system to use in different ways resources, totally or partially, non or no more available as expected by the system (for instance the ability to *move*, which is to *walk* for people having good availability of natural resources presumed to be used for that purpose *and* to use different resources in total or partial substitution of the unavailable ones. Related systemic topics in other domains as control and signal theory, cybernetic devices, computer science, dynamics, etc., where the subjects of *balancing and compensating* (introduced below) are of great interest, are not subjects of this contribution even if some common meaning and analogies may be identified. The focus in on the information processing capabilities of the system to manage processes of balancing and compensating. *Focus is on living systems supposed, at certain levels of complexity, provided with cognitive system, acting with cognitive models, cognitive resources and cognitive architectures.* Systems provided with simulated Cognitive Systems may also be of great interest such as in robotics when some physical damages occur and they may be dealt with by processes of balancing and compensating.

Emergence in Complex, Cognitive, Social, and Biological Systems, Edited by
Minati and Pessa, Kluwer Academic/Plenum Publishers, New York 2002

- ***Balancing*** is intended as a process referred to the *use* of two or more resources in such a way to distribute their contribution in order to get a state of equipoise, of equilibrium, by creating harmonic or proportional effects or consumption. An *unbalanced* use of resources may be an aware use of resources by the system to get a particular purpose. The inability to *design* an usage of available resources by the system may bring to inability to reach the system's purposes, being the problem related more to a non effective use of resources than to their quality, quantity, distribution, etc. An unbalanced use of resources may also bring to the destruction of the system, operating for some time *compensating*: in mechanical devices it happens for instance because of irregular consumption and friction. Balancing is an activity for systems provided with Cognitive Systems.
- ***Compensating*** is intended as a process referred to the use of one or more resources in substitution of others no more totally or partially available in order to reach the system's purposes. Compensating is often based on using unbalanced resources as referred to usual or expected configuration of resources in order to reach standard user's purposes. *It is a process of establishing a balanced usage of unbalanced resources expected to be used in a predefined, standard way, even substituting some, no more available, by others. It is matter of substituting standard, predefined functions with new uses* (*function* is intended a standardized way of using resources by the system. New ways of using resources may be established depending on the context, on specific needs, on unexpected situations, and may be extemporarily or becoming stable. In the later case the new uses may became new functions). It may take place spontaneously in a system having the configuration of resources no more as expected.

Balancing may be intended coincident with *compensating* for systems having not the ability to process information and to design uses. The embedded, pre-established way of using resources is constant because, for instance, no learning and designing abilities are available.

The two processes really are not rigidly distinguishable and they may take place interactively, simultaneously, becoming one part of the other. Resources are not balanced or unbalanced: it depends on the expected usage by the system. In fact what is unbalanced may become balanced and vice versa.

From an objectivistic point of view a balanced usage of two or more resources takes place when looking for the *optimum*, that's the trade off between resources consuming and achievement of the system's goal, that may be: to move in a specific environment, to manipulate, to detect and process signals, to control its temperature, etc.

It may happens that the system has standard, pre-established goals while the available resources are not as expected. In this case the system may (1) try to use the available resources in the standard way (that's usually a poor strategy, but the easy one because it doesn't need any change in the system's standard behavior; (2) look for different ways of using the resources available at the moment. In the later case the focus is on systems provided with cognitive models, able to allow Cognitive Processing (Anderson, 1985), to develop and apply strategies for using resources –balancing and, in case, compensating- even if dealing with resources differently than expected.

Balancing is a difficult task. A specific case takes place specially when balancing *deficits* in systems. Systems may have different reactions and behavior in front to a progressive or suddenly or permanent or non permanent reduction of their ability to perform tasks (and then to use resources related to the task).

Fundamental distinction must be done between systems assumed provided or not provided with *cognitive system* (Pessa, 2000), even if it may be simulated by some non-living systems (i.e. computer systems) (Allen, 1990). Balancing processes in human beings typically take place when resources, usually pair, become unbalanced for any reason: legs, feet, hands, arms, eyes, ears, fingers. Kidneys may became unbalanced because of illness, surgical remove, coupling with an artificial one. Left and right brain hemisphere may became unbalanced because of cerebral damages.

Before to focus on adaptive, intelligent systems and on their *strategies of using resources* thanks to their cognitive processing abilities, in order to balance deficits, let's consider the question: *is balancing a process always related to cognitive processing*?

Processes of balancing and of compensating may take place in non living systems also (i.e. mechanical) where cognitive processing is assumed to don't take place, being not even reproduced by simulation (i.e. computing systems). In this case where is a spontaneous, self-organized distribution of forces, weights, charges which may be assumed to be balancing and compensating processes. *In this case it is a physical phenomenon and not a cognitive one because cognitive processing in the system doesn't take place.* It is still possible to identify balancing and compensating processes because of the role of the observer as it happens for emergence processes where the role of the observer is crucial.

The reason to focus on processes manifested by systems provided with cognitive systems is that such processes may be decided, managed, regulated, designed, being also of different kinds, variable in time. For living beings at high level of evolution it means moving from the unconscious (cerebellum) to the conscious (brain) level.

It is interesting to note how unbalanced, compensated working in systems gives rise to unbalanced, non harmonic, non proportional, unusual and unexpected resources wear and consumption while in living systems, assumed provided with cognitive systems (then where cognitive processing takes place), may give rise to an *unusual energy and time consumption* used in compensation processes, in balancing unbalanced resources. In the latter case an abnormal use of non usually involved resources may take place, as mental, emotional, psychological, etc. in sum related to the cognitive system itself and not just driven by it. The cost of balancing-compensating processes must be considered in order to identify the trade-off and evaluate its convenience (for instance how much *stressing* is to try to walk for a patient having Multiple Sclerosis vs. to move by using a wheel chair).

The topic of considering balancing and compensating processes in systems provided with Cognitive Systems bring necessarily the attention to the classical discussion between empiricist and rationalist philosophers. This point is very well presented in the case of the discussion regarding blind people in (Ungar, 2000)[1].

The spatial understanding of blind people became a topic of study in the late 17th century, when the English philosopher Locke attempted to answer a question put to him by his friend Molyneux:

> "Suppose a man born blind, and now adult, and taught by his touch to distinguish between a cube and a sphere of the same metal, and nighly of the same bigness, so as to tell, when he felt one and the other, which is the cube, which the sphere. Suppose then the cube and sphere placed on a table, and the blind man be made to see: quaere, whether by his sight, before he touched them, he could now distinguish and tell which is the globe, which the cube?" (Locke, 1964)

Locke's answer to this question is negative. As an empiricist, he believed that our minds originate as a 'blank slate' (tabula rasa) and that all concepts we have are derived from our sensory experiences. Thus a congenitally blind person would only have tactile impressions of objects, and these could not automatically allow him to recognise the same objects by sight. It is only by integrating experiences from different senses that we build up abstract (amodal) concepts. In contrast to Locke, rationalist philosophers (e.g. Berkeley, 1975) argued that abstract ideas about the world (including spatial concepts) are present at birth: we recognise something as spherical, in any sensory modality, by matching it to a pre-existing concept of a sphere.

[1] http://www.lgu.ac.uk/psychology/ungar/research/papers/cmppf.html

The dichotomy of empiricism versus rationalism has continued to influence theories of the mind in the form of the famous "nature/nurture debate". However Millar argues that this way of thinking about the problem offers us little in either theoretical or practical terms (see (Millar, 1994), for a fuller discussion of this issue), and she suggests an alternative reading of the question, focusing on the empirical issue of how perception relates to knowledge. In other words, if one perceptual modality (i.e. vision) is missing, what (if any) effect does this have on our knowledge of the world? Here, the question is phrased in terms of information processing and directs us to a study of the nature of information handled by the various modalities, the way this information is processed and the way the resulting representations guide or influence spatial behaviour.

The focus of this paper is on information processing ability of the system having to balance and compensate. The problem of balancing and compensating is also real for the information processing activity itself, tacking place for instance in case of brain damages.

2. COMPENSATION AND STRATEGIES OF USE

As introduced above compensation is intended as a process of using unbalanced resources in order to cope with permanent or non permanent, partial or total, lack of availability of some resources normally expected available to the system. Let's focus on some details of the process of compensating.

If the available resources intended to be used in the process of compensating are considered just as *set*, the previous description introduced may be sufficient to understand the nature of the compensation processes. For instance they take place when a cooker substitutes non available ingredients with others, changing quantity and aspects of the cooking process (time, temperature, sequence of ingredients, etc.).

The available resources may instead be considered as a *system* (Minati, 1998a), being them in mutual, dynamic relationships, changing when *used*, in the cognitive space of the active observer (like a book by a reader, a violin by a violinist, an horse by a rider, the paints and paintbrushes by a painter, etc.). By considering resources not only as a set, but as a system, when used, when are part of the design produced by the user system, it is possible to assume compensation as *reorganizing, restructuring process* of them. Actually the interaction between resources and user may be divided in two types:

- ***Restructuring***, means to establish a new configuration of available resources introducing new links, new relationships among them in their set, i. e. introducing a *structure* on them. It is possible to say that the set of resources assumes a non standard, new structure imposed by user (new sequences, new hierarchy, new priorities, new combinations, etc.).

- ***Reorganizing***, means to establish new ways of using the structured set of available resources, by designing uses and not just applying standard, pre-established functions. *In reorganization processes the context itself may become a resource.*

Let's consider now the case related to human behavior, i.e. disable people, real interest of this contribution. We may focus on a particular aspect of the human behavior: the ability to walk. The ability to balance the use of the two legs is a very complex task, usually performed in unconsciously way, at the cerebellum level. Inability to properly walk may came from many reasons of different nature:

- physical damages to directly involved resources, such as broken bonds, feet's problems, legs having different length, pain in using muscles, knee's problems, blood circulation in legs, etc.
- neurological damages as it comes from pathologies as Multiple Sclerosis, brain damages, paralysis, damages to spinal cord, etc.
- mental, psychological problems inducing lack of equilibrium, insecurity remaining standing, great weakness, etc.

Human beings react to this kind of causes by using the cognitive system developing strategy of use of resources and designing scenarios, cognitive spaces, special worlds having their own (explicit and implicit, we may say symbolic and sub-symbolic) steps and rules.

The contribution presented in this paper regards author's experiences related to problems (lack of equilibrium, great weakness, reduced sensitivity) due to Multiple Sclerosis. The introduced approach may be assumed applicable to other cases.

3. USING THE CONTEXT

Focus must be on the prospective shift from *walking in the context*, as it usually happens for subjects having full availability of their resources, to *using of the context to walk.*

Usage of context to walk means ability to carry out in real time projects of movements including probabilistic considerations on steps, on reliability of supports available in the context.

Movement in human beings may take place:

- Having equilibrium, intended as when introducing the concept of balancing, in par. 1;

- Having *temporary* (for a significant period of time assumed to have an end), *unstable* (tacking place unexpectedly, for short periods of time) or *permanent* (from the birth or from a specific time during life) lack of equilibrium, f. i. because of an illness, a trauma, an accident etc., such as to require (?);
- use of proteases as extension of the body;
- ability to USE context as a source of equilibrium.

In the later case subjects *create* (they cannot avoid to do so because they have a Cognitive System) *projects of movement* by considering:

- reliability of supports;
- emergency actions in case of failure;
- designs of paths of supports.

That's in case of availability of light and in non blind subjects.

Considering subjects that are able to see, in case of lack of light they must:

- Remember and process the scenario, modeling it, hypothesizing how to use it;
- Use tactile redundancy.

In case of blind subjects additional balancing information must be considered, such as the acoustic ones.

A system, human or artificial, may have two different goals:

- to move to know the context;
- to use the context to be able to move, *creating hypotheses and learning about it*, if the context is not known in advance.

The two approaches are usually simultaneously assumed by the system.

Cognitive processing about the environment in which to walk and used to walk as source of supports and reference points, takes place in terms of:

- memorizing about configuration, scenario, possible dangers, noises and techniques to be used to carry weights, to take objects from the ground, from high levels, etc.;
- context representing;
- designing, figuring projects of steps by using available supports;
- ability to react to unexpected situations;
- hypothesizing about unknown area of the scenario;
- how to process non empiric, direct experiences about the context (f. i. verbally communicated, or coming from seeing somebody else walking having equilibrium problems);
- ability to use prostheses as canes, crunches;
- how to cope with temporary loss of ability to see, important source of reference point (f. i. while washing the face, sneezing, etc.);
- ... and many others depending on the pathology, i.e. on levels of ability to use body's and external resources.

4. COGNITIVE PROCESSING

In the framework just introduced it is interesting to consider the *amount of information processing* required to compensate the lack of equilibrium, modeling the context and adopting a *dynamic use of models* (Minati, 2001a, 2001b, 2001c). As introduced in literature for *dynamic use of, dynamic and non dynamic, models*, reference is to the ability to select among the available ones, which ones to consider in order to establish the more effective use. It particularly applies in this case because of the different kinds of information to be, often in parallel, processed: visual, acoustic, tactile, internal. And they may be processed by using different simultaneous models continuously interacting and bringing the system to take decisions. The cognitive processing is in this case expected to suggest behavioral models to be adopted in order to move, to make specific actions like to walk, to dress, to wash the face closing the eyes, to blow the nose, to use the restroom, etc. Results of the cognitive processing based on a dynamic use of models (Allen, 1990; Minati, 2001a) are sequence of actions to perform tasks as the ones just listed. As *systemic use* it may be intended the ability to select among results coming from information processing by using *simultaneously* (Minati, 2001b, 2001c) different models, making parallel processing, to be combined and considered, for instance to carry something, to speak and walk, to take something difficult to be reached, etc.

- Results must be used following the strategy of combining them, of using *errors* and not just to avoid them (learning that a sensation processed and interpreted as before the illness gives arise to wrong information to be adjusted tacking count of others sensations, see the case of handwritten digits recognition by using DYSAM (Minati, 2001b, 2001c) of considering previous preprocessed situations (where learning processes took place), of considering probability, of generating hypotheses and abductive reasoning(typical cases are to button up, to take a shower, etc.).
- Decisions must be not only locally and context-dependent, but depending also on the expected or learned scenarios.
- Processes already made by the Cognitive System regarding emotions, affection, attention, perception, inferential system and language, must be consciously or unconsciously considered.

Many different illness may create such a situation to be compensated by cognitive processing. People around subjects having such problems must help them to do by themselves (i. e. activating cognitive processing and physical actions) more than to avoid them to do by assuming their inability to perform a task. *Subjects having more processing power and many cognitive models available have more ability to walk, to do.*

Rehabilitation cares must help ill people to combine physical and Cognitive Rehabilitation, by suggesting and inducing usage of behavioral model verbally and through mental images regarding the everyday life.

Applications in medical care are related to teach patients how to compensate lack of equilibrium *and* to take advantage from rehabilitation. To compensate by using Cognitive Models must *not* substitute physiotherapy focusing *only* on mental processes, but a kind of Cognitive Rehabilitation must be established during Physical Rehabilitation.

5. ADDING "INTELLIGENCE" TO PROSTHESES

To support both usability of the context and rehabilitation processes (Physical and Cognitive) intelligent interactive tools (proteases, shoes, crutches, wheelchairs, etc.) able to adapt, to learn, to correct and to track (registering detected information, available on demand both on line and off line) patients behavior, may be designed. Thanks to recorded information cognitive models and the cognitive space of disables may be assumed *and corrected by the therapist.*

Applications of computerizes systems able to detect movement, to keep track of movement and of the use of devices as well, are available in fields such:

- security, for controlling, detecting crossing borders;
- animal control (both animals free in the environment and animals in farm), for controlling, movement, temperature, etc. References to *pedometer technology* may be found at[2];
- dancers, for controlling artistic movements. References may be found at[3]. References to "3D-accelerometer-based Drancing-A suit, driven by PC control software *DranceWare*" may be found at[4].

It must be reminded that not only discrete movements may be detected but also *dynamics* (Harrington, 1994; Harrington, 1995; Verplaetse, 1996), for instance by using *accelerometers*. References to accelerometers technology may be found at:

a) Putnam W., Knapp R. B., Input/Data Acquisition System Design for Human Computer Interfacing[5].
b) Analog devices, Products and Datasheets – Magazines, Accelerometer news[6].

[2] www.afimilk.co.il, www.tdm.it,www.alfalaval.com
[3] http://www.genmars.com/adrian/photos/ted11/
[4] http://www.art.net/Resources/dtz/archive/DanceTech97/0677.html
[5] http://ccrma-www.stanford.edu/CCRMA/Courses/252/sensors/sensors.html
[6] http://www.analog.com/publications/magazines/accel_news/accel.html

The possibility to take trace of the way by disables use prostheses and tools must be completed with knowledge of the Cognitive Models and behaviour they have in mind. This may happened through a typical psychological inquiry based on questioners to be interpreted by the therapist.

6. CONCLUSIONS

Focus of the paper has been on the importance of cognitive processing in (disable) human beings having reduced or damaged resources, f. i. to walk, as due to reduced equilibrium as it happens in some cases of Multiple Sclerosis. Compensation processes must take place not only physically, but cognitively also. Therapists must focus on physiotherapy having

- not only the goal to re-establish previous situation as for problems expected to evolve recovering in time, non permanent, non degenerating, having temporarily decreasing of functionality,
- not only the goal to retard decreasing in degenerating pathologies, trying to reduce damages coming from lack of normal activity too,
- but the goal to teach patients how to use at the best the remaining resources, avoiding accidents due to misuse of them and tracking everyday activity by getting information from prostheses and tools of support.

Physiotherapy intended not only as rehabilitating activity, but supportive in using damaged and reduced resources must take in count Cognitive Processing (Anderson, 1985). This is the joint activity supposed to be accomplished by the physiotherapist and the psychologist working in team on each patient.

Cognitive Processing allows patients to realize their own model of the scenario, of the environment, of the physical and human context. Depending on models carried out by patients they'll be able to establish degrees of freedom and ties: the configuration, the system of them will be their life.

A scenario poor of freedom and full of limits and ties will generate only costs for societies and low quality level for patients. Well identified degrees of freedom are able to allow patients to be productive, both intellectually and socially, reducing cost for society and improving life quality. That's has been and is experienced by the author. It must be pointed out as for Cognitive Processing considered in the just mentioned cases is not intended *Cognitive Rehabilitation* devoted to re-establish cognitive functions (i.e. memory, logical functions) perturbed by accidents as ictus or cranial trauma, but the process of inducing new behavioral models more appropriate to a new physical condition. Intelligent tools able to take trace of usage and psychological inquiry procedures based on questioners may help the therapist to realize the cognitive word in which the disable lives. It must also pointed out the relation between the approach just introduced and the *FELDENKRAIS METHOD®*[7] based on strong mind-body interaction (intended as a whole), awareness and ability to full of self physical area of the body and movements.

REFERENCES

Allen, R. B., 1990, User model: theory, methods, and practice. *International Journal of man-machine Studies* **32**: 511-543.

Anderson, J., 1985, *Cognitive Psychology and its Implication*. 2nd Edition, W. H. Freeman and Co., New York, N.Y.

Benjafield, J. G., 1992, *Cognition*. Prentice-Hall, Englewood Cliffs, NJ, USA

Berkeley, G., 1975, *Philosophical Works on Vision*. Dent, London.

Harrington, M. E., Daniel, R. W., and Kyberd, P. J., 1995, Gesture Recognition of Arm Movements using Accelerometers. *The European Context for Assistive Technology* (I. Placencia, Porrero, and R. Puig de la Bellacasa, eds.), IOS Press, pp. 432-435.

Harrington, M. E., Daniel, R. W., and Kyberd, P. J., 1994, A measurement system for the recognition of arm gestures using accelerometers. *Journal of Engineering Medicine* **209** (H2): 129-34.

Locke, J., 1964, *An Essay Concerning Human Understanding*. Fontana, Glasgow.

Millar, S., 1994, *Understanding and representing space: theory and evidence from studies with blind and sighted children*. Oxford University Press, Oxford.

Minati, G., 1998b, *Sistemica*, Apogeo, Milano, Italy.

Minati, G., 2001a, *Esseri Collettivi*, Apogeo, Milan, Italy (Collective Beings, in progress).

Minati, G., 2001b, Experimenting with the DYnamic uSAge of Models (DYSAM) approach: the cases of corporate communication and education. In: *Proceedings of the Forty Fifth Meeting of the International Society for the Systems Sciences (ISSS)*, July 8-13, Asilomar, California, USA.

Minati, G., 2001c, The Dynamic Usage of Models (DYSAM). In: *Proceedings of The Second Italian Conference on Systems Science*. Nov. 9-10, 2001, Castel Ivano, Italy, published by Kluwer Academic, 2002, USA.

Minati, G., Pessa, E., and Penna M. P., 1998a, Thermodynamical and logical Openness. *Systems Research and Behavioral Science*, **15** (3).

Pessa, E., Penna, M. P., and Minati, G., 2000, Collective Phenomena in Living Systems and in Social organization. *Chaos & Complexity Letters*, Italy.

Ungar, S., 2000, Cognitive Mapping without Visual Experience. In *Cognitive Mapping* (R. Kitchin and S. Freundschuh, eds.) Routledge, an imprint of Tasylor & Francis Books.

Verplaetse, C., 1996, *Inertial proprioceptive devices: Self-motion-sensing toys and tools*. IBM, http://www.research.ibm.com/journal/sj/mit/sectione/verplaetse.txt.

[7] http://www.feldenkrais.com/index.html

Emergence, Individuality, Therapy

A Systemic Vision of Homeopathy

MAURIZIO TRIONFI
Via Quarto dei Mille, 10 - 25128 Brescia, Italia
E-mail: maurizio.trionfi@tin.it

1. INTRODUCTION

The appearance of a systemic vision is a recent cultural event whose origins can be traced back to the years between 1940 and 1950.

The purpose of such a vision springs from concrete working needs: "when we proceed from the enunciation of the elementary properties of physical particles to the way in which they are organized, giving birth to the unthinkable complexity of reality, we realize that we lack conceptual, interpretative and operative tools". These words were used by L. von Bertalanffy to synthetically express the limitations of a reductionistic and physicalistic vision.

This paper is not about The Theory of Systems in a narrow sense, as it has been characterised, over the years, by an approach relying mainly on mathematics and engineering; it will be dealing with Systemic Vision, which is particularly useful when describing phenomena concerning Open Systems in non-necesseraly mathematical environments.

This kind of vision is convincingly suitable for Homeopathic Medicine, which, it is possible to say, is intrinsically systemic; since its birth, its approach has been connected with globality producing individualization.

In homeopathic writings we can find such concepts as Order-Disorder, Globality-Identity, Peripheralization, Suppression = Abolition - Respect of Variety, Dynamism, Teleology etc.

Even if these concepts were, those times, expressed in a way which was not wholly self-conscious, but explicit in their deep meaning, the use of a

Emergence in Complex, Cognitive, Social, and Biological Systems, Edited by
Minati and Pessa, Kluwer Academic/Plenum Publishers, New York 2002

modern concept based vocabulary allows us a reinterpretation of homeopathic texts and events.

Systemic Vision is a new paradigma whose aim is to offer a deeper awareness of the mechanisms ruling the behaviour of Complex Systems; as such, it can be placed in a

cross-cultural position where exchange of information and meaning between seemingly distant environments can take place.

We wish to remind you that we will be referring to the concept of System as: "A number of elements dynamically interacting and which are organized towards an objective". This very general definition, which can be applied to all Open Systems, be they biological or not, will be used with special reference to psychobiological organization.

The systemic vision of homeopathy allows us to state, and if it is the case re-organize, such concepts as "disease" and "therapy" and gives us inspiration for a wider and deeper medical debate, even in territories which are not exclusively homeopathic.

2. HOMEOPATHIC PROVING AS A SIMULATION MODEL

2.1 (Opening a " Black Box" which can think)

The stages leading to the birth of homeopathy are generally well-known. In a very synthetic way: Samuel Hahnemann has the intuition that farmaco-dynamic substances (poisons) can produce symptoms in a healthy man if they are repeatedly taken in sufficient quantities and repetitions. This stage, consisting in the *production of symptoms in a healthy man* is called "Proving", that is Experimentation, and it is used to obtain a picture of unbalance induced in a system through the administration of an identified pathogenic substance.

This phase of the process towards knowledge can be placed in a totally scientific prospective: it is the production of an experiment.

In figure 1. you can see a graphic model of the stages of experimentation; It consists in:

a) induction-production of symptoms in the single prover
b) collection of data from a number of provers
c) compilation of Pure Materia Medica (subdivision and rewriting of findings)

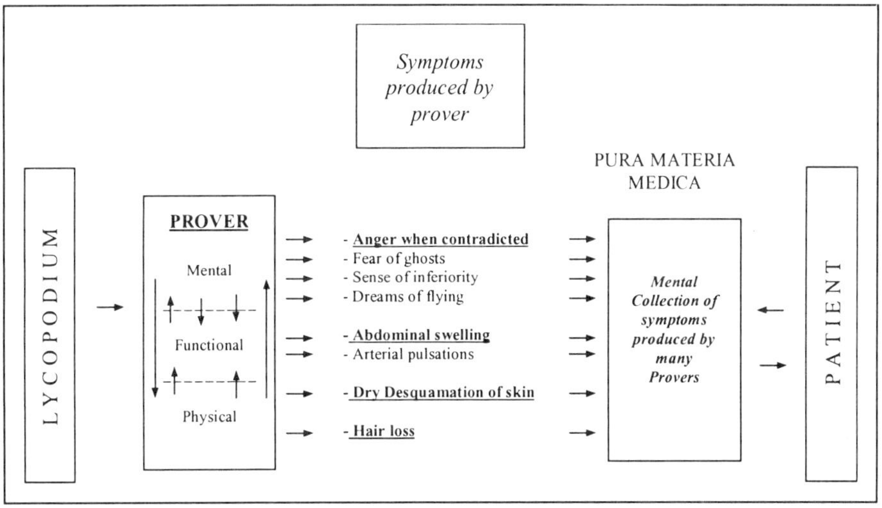

Figure 1. Model of homeopathic experimentation (Proving).

NOTA BENE. *Homeopathy is the only science-methodology which has Proving as its initial and structural reference unit. As a consequence of this, no drug can be defined as "homeopathic", unless it has gone through this experimental phase. Every other formulation can be defined, if it is the case, with the term " homeopathically diluted".*

There are two points that I would like to emphasize in my paper:

1. Unlimited extension of the panorama of usable substances

Lycopodium, which we are using as an example, is a lichen-derived homeopathic remedy whose chemical composition is totally incapable of producing symptoms in a healthy man (and so it is used as an excipient), unless it is homeopathically treated, that is, it is diluted and dynamized.

In the Pure Materia Medica of Lycopodium we can find more than 8,000 symptoms

found in various provers or in subsequent Proving sessions. Some symptoms are very frequent, others less well represented.

The experimentation with this remedy corresponds to the second cycle of Hahnemanian Provings, that is when he turned from the use of raw, pathogenic substances to the use of diluted, dynamized substances to avoid the crudest, sometimes dangerous toxic effects. The unexpected, surprising result was that even inert substances, if they are treated this way, acquire the ability to evoke symptoms (which means that theoretically every substance can).

2. *From the knowledge of substances to the knowledge of provers*
(From the concept of Energy to the concept of Information).

The second decisive point is the fact that, while when using pathogenic substances (poisons like Belladonna, Stramonium, Mercury, Arsenicum etc) all Provers produce symptoms as a necessary consequence of chemical intoxication, on the contrary, with diluted-dynamized substances only subjects who are sensitive to that pharmacological information become able to produce symptoms*.

This is how we can introduce the concept of Information, not chemically compulsory and above all separated from quantity: this is crucial both to understand the symptom triggering mechanism and to explain the possible answer of Prover (and patient!). The most coherent metaphor takes us towards the concept of language.

An example may be of help.

If a person is moved by sad news, we can say this person is sensitive or has a kind soul; in other words, noxa has shown who this person is, what makes him/her as he/she is. The same piece of news doesn't produce any reaction in a different person, and we can then say that this person is

* Avoiding the problems connected to the biophysical mechanisms underlying this activity, we can only assume, in the present state of our knowledge, that variations induced in the homeopathic remedy (electromagnetic, vibrational, etc) may "talk to" and so "direct" the biological solvent of which every mono or pluricellular biological individual is structurally made of. The modulations of the solvent could induce inhibitions or stimulations in the various metabolic processes. Water is the pervading medium of every living being and it is, in a certain sense, "the place of relationships", the one that can be found in very different structures, both inside and outside cells. The possibility to inhibit or facilitate certain relationships (both metabolisms and distant communications between components), by H_2O modulation, is an explanation of the global (i.e. systemic) effects that we happen to verify during the execution of a Proving or, to the contrary, in the "restitutio ad integrum" during therapy. A further supporting element for this hypothesis is the fact that the use of remedies with known chemical structure (e.g. sodium chloride, calcium carbonate, potassium carbonate) will affect expicitly the use of sodium and chlorine, rather than calcium, as well as the potassium in the organism under examination. The relevance of these events is well known if we compare the patient's symptoms with the physiology or physiopathology of these elements, known in modern times. It is, therefore, not a single symptom, but a whole group of processes , linked to the primary biophysical functions of the system. It is possible to see by intuition how remedies which are seemingly simple in their chemical structure can induce deep transformations in the biological modus operandi. Our lack of knowledge of the intimate mechanisms governing these events, but which happen constantly, doesn't prevent us from using such a powerful instrument to re-balance the system under examination; on the other hand we have been using acetylsalicylic acid for over a century even if we didn't know prostaglandins and, in a different field, we are extremely efficient in our management of electricity (we can even direct a single electron), even if its essence is unknown to us.

insensitive or cynical. A third person, who can't speak the language in which the information is being conveyed won't have any reaction, but this depends on the fact that the message was only noise for him/her. In this model, the quantity of energy used to produce sound has no importance; what is important is which kind of sound has been produced, that is if this sound is (or isn't) understandable . I can yell and shout: that person will never be able to understand my orders, suggestions or advice. He/she will be still, as if listening to unmeaningful noise. The Prover (or patient) receiving a totally inadequate "drug-information" won't have any change on his axis of balance (or unbalance). Things are very different when my language is understood by the listener: a single speech can move hundreds of millions of people.This is not quantitative energy, but " command energy", that is the minimum amount necessary to convey information fit for that system. Therefore the remedy-information makes explicit 1) the language the person "speaks" and 2) later on, the person him/herself. Before the arrival of that piece of news the two people who have been able to receive the message looked the same; later on they showed characteristics which differentiate them: we can now distinguish two different individuals.

If we had compared them on an anatomic table we could not have differentiated them and it is very likely that this would have been true even if we had investigated their microscopic or ultramicroscopic components; in fact, the deeper we go the more a cell is identical to itself no matter if it belongs to a hamster or to an elephant; finally, the carbon in my brain is the same as carbon found in graphite, diamonds or fulleren.

What individualizes two different beings and unequivocally distinguishes between them is the different way in which they function and react even when there is a structural identity of their components.

In a sense, we can legitimately say that the hardware section of biological organisms is essentially the same while possible software is infinite. *The problem, then, is not what constitutes elements but the way in which they relate.*

Which relations are promoted, which ones are inhibited?

It's exactly thanks to the diversity of relationships and connections, even if with identical components, that we can build the three hundred thousand species of Coleopterons and finally pick a single one among them all. No complex being is totally identical to another even if the constituting atoms are. A similar concept can be applied to all possible single-individuals of the Homo sapiens genus which can be known individually in this way (that is by making their individual characteristics emerge).

3. HOMEOPATHIC REMEDY AS A PASSWORD TO FUNCTIONAL STRUCTURE

The Proving model we have previously described is a *simulation* informing us a) about Prover receptivity and, secondarily, b) about its functional structure. We can then say that the remedy, but only if it is diluted and dynamized, acts as a key opening the expressive possibilities of that individual from the psychological, somatic and functional point of view.

This is a procedure which is perfectly comparable with the technique which has been used in recent times to investigate the so-called "Black boxes".

We know that when we open it we risk destroying it. Our chance to know the way it works, what it does, what it is for, to imagine an internal structure, is to give an input, to later verify what sort of reaction we will have as "output". The correct stimuli have to be used to elicit an understandable answer. It is evident that, if we apply a noise input to a light bulb, we will never be able to understand what its function is: only an adequate electric tension can enlighten us. On the other hand even the microscopic fragmentation of the object would not bring us to any kind of knowledge.

Taking an example from a different area we will never be able to know what a person thinks or feels if we observe him/her from the point of view of an anatomic table: he/she must be alive. As a consequence of this, a deeper knowledge can only be attained if all the functions of the system are whole. The history of medicine can confirm that an absolute majority of therapeutic knowledge derives from physiology and physiopathology, while very little comes from anatomy

The group of symptoms resulting from Proving makes up a "*model of experimental unbalance*", induced by the remedy under examination. The deeper our knowledge of the global image of the proving, the more precise therapeutic reproducibility we will have: if we see a great quantity of symptoms when we produce unbalance, taking into consideration only one of them does not correspond to what happened during the experimentation.

Therefore, on a therapeutic level, I will need a greater number of symptoms so that adhesion is formed, in the most perfect way it is possible, between what has happened and what I am going to treat in my patient. If I don't do this, therapeutic failure will ensue.

4. THE " IMAGO" OF REMEDIES, OR INDIVIDUALIZING RELATIONSHIPS

Every remedy produces an image which is its own and corresponds to a departure from a condition of balance which is specific for the subject under examination.

We will call this configuration "imago", and it can be described as a group of line linked knots : the knots are the symptoms and the lines are the relationships between them. The final picture will be a coherent one. In any particular configuration symptoms will have a precise position in relation to each other, and they can't be arbitrarily separated; if they are, the information content of their mutual positions will be lost. It is necessary to underline that this "relational configuration" cannot be represented fully on a bidimensional sheet; Time, too, has a fundamental role and what we have in practice is a switch from a static, quantitative vision of "symptoms taking place" to a " cinematic" vision of symptomatological evolution .

This corresponds to a much more adhesive description of what happens in a living being during the process of departure from and return to a state of balance (see present disease, recent, remote and constitutional pathology …).

The final graphic, even if a very simplified one, can be represented thus

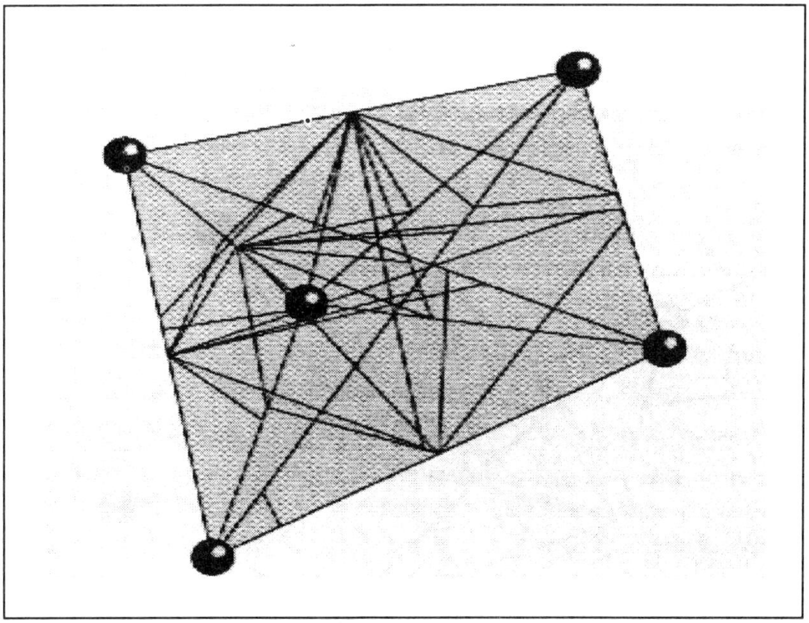

Figure 2.

The same symptom, but in a different picture, has a different position, that is, it has specific relationships with other symptoms and thanks to its position, on one side it will acquire a sense of belonging (to that system), on the other it individualizes, that is it gives its contribution to the picture it belongs to. In other words, it acquires and gives individuality.

Therefore the homeopathic remedy, during the phase of Proving, brings to light an "individuality of relationships" between symptoms rather than single events; that is why the same remedy, during therapy, will be able to cure an ordered and coherent group of symptoms and not a single one.

To make a concrete example, we can say that a symptom like "fear of examinations", is connected, in the Silicea subject, to the feeling of not being able to express its own abilities, in the Calcarea Phosphorica subject to a feeling of impermanence and difficult communication of acquired notions, in the Gelsemium (Imago ...) subject to excessive watchfulness and the fear of a real Truce of all acquired data; in this subject, examination anxiety appears also with a strong gastro-enteric movement and heavy sweating. In the other two subjects somatic effects are much less evident.

As we can see, it is never a single symptom, but a syndrome, that is a group of symptoms coming together: sensations, functional bodily expressions and, at the end of the processes, eventual real somatic damage. The fear of examination, a symptom which is seemingly an isolated one, generically defined as "anxious", will show, when analyzed more in depth, different motivations and expressions in different individuals.

In this model, which has experimental objective reality and can be reproduced every time we perform a Proving with the same substance, symptoms are connected by internal coherence which cannot be eliminated. In fact, this coherence is a structural datum of functionality and identity, its importance is fundamental and we cannot do without it, just as we can't, on a different level, operate without structural elements such as potassium, carbon, nitrogen and hydrogen in a biological individual.

5. CYBERNETIC REGULATION OF A BIOLOGICAL ORGANISM

When a patient shows the disorder whose global imago (psichosomatic) I, as a therapist, have, I know that the prescription of a suitable remedy will allow a homeostatic reaction (homeopathic aggravation), which will bring the system back to its normal functions. *We can then say that my therapeutic act consists in prescribing the system a global picture of its own state of disorder in as precise a way as possible (Law of Similars = The most suitable remedy = Homeopathic Simillimum*). This modus operandi is perfectly coherent with the autoregulation schema of all systems, which is here reproduced.

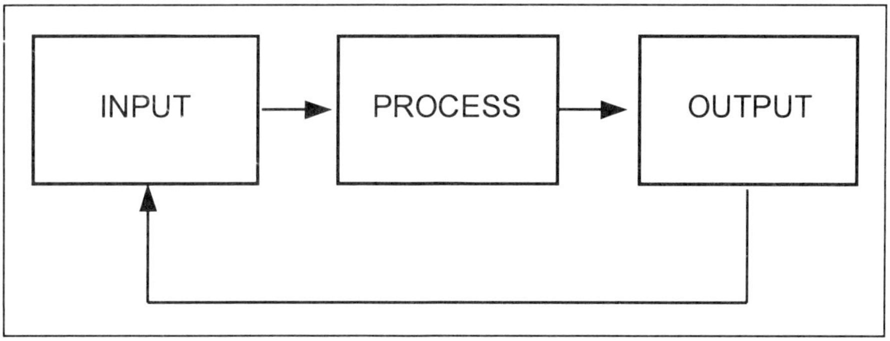

Figure 3.

The above-mentioned schema is the founding reference of every cybernetic regulation.

The work of a homeopathic therapist consists then in using the entire axis of adaptive counterreactions since they are the essence of the functioning of all biological organisms.

It is in this sense that homeopathy is a "natural" therapy: it doesn't use techniques or drugs that act in a partial way, that is on single subsystems, thus avoiding an increase of entropy in the system. "Naturalness" has nothing to do with the fashion of nature: it identifies the global individual as a therapeutic aim. The objective is not the use of softer, less aggressive drugs (naturopathy, fitotherapy etc.) but the establishment of a model of disesase and healing which explicitly admits the concept of entropy reduction in the system under examination.

This modus operandi belongs almost exclusively to unicist homeopathy whose desideratum terapeuticum consists in modifying the global functioning of the system: not the disappearance of one or more local symptoms. To reach this objective the re-prescription of the dysfunctioning global imago has as its aim the "awareness" by the system of its own dysfunctional state in order that the necessary correcting processes may take place (see output re-prescribed on input).

We are then very distant from any symptomatological cosmetics.

6. CONCLUSION: FROM DISEASE AS A WHOLE TO HEALING AS A WHOLE

In systemic theory and vision the peripheralization mechanism is well known: it is the ability to transfer internal unbalance conditions, especially when these affect the SP (Sensitive Point) or the CP (Critical Point), to more superficial, hierarchically less important subsystems.

By Critical Point, the decision centre of a system is meant, that is the special point towards which positive and negative retroaction rings converge.

Therefore, when "partial" therapies are used, the ones which cannot reach the Critical Point, in a majority of cases, two kinds of situation will be found:

- the first: relapse into symptoms (and symptoms are only the expression of dysfunctioning processes)
- the second: the apparent disappearance of superficial symptoms and their re-appearance, even if seemigly modified, in more important organs or functions.

In the first case the relapse into symptoms shows that despite the "therapeutic" attempt to cancel disorder, the system constantly resets itself on the functioning it had been able to adopt as a compromise between its internal resources and the stressful external conditions.

In the second case, on the other hand, the disappearance of symptoms shows that the total depletion of system energy has become chronic and the disorder which was more superficial earlier has now become deeper. The subject is objectively more ill, even if this will be clearly shown only later on.

From this perspective we have taken into account also the internal time of the system as well as the modalities, and the entropy levels, underlying the system. The disappearance of a single symptom does not correspond to a desideratum terapeuticum, but sometimes to its contrary, if the general wellness indicators are not respected. In the systemic vision, effectiveness is a concept which needs precise specifications, linked to the system under examination, not to any individual system, and also clearly determined objectives.

On this premise we can state:

Healing consists in a better functioning of the system, a prelude to higher degrees of expressive freedom; i.e. the system under examination, starting from its own characteristics and not following abstract conceptual models it doesn't belong to, expresses its reality in our worlds, evolving (Health) and thus becoming able to elaborate what we can call its own "poetics" (from poiesis = creation).

In the absence of a theory of healing, universally expressed and accepted, to which these final lines are referring and trying to give an answer, we maintain that a single symptom, inside a constellation, becomes a solitary indicator, absolutely unsuitable for the description of the whole.

Homeopathy, and specifically unicist homeopathy, represents an organized, hyerarchized theory of biological events and beings.

In the light of these considerations, it appears as a subject rich in experimental and therapeutic experience, useful for the definition of the concept of disease, and consequently, of therapy towards healing as a whole.

REFERENCES

A.A.V.V., Comunicazioni personali negli incontri dell'A.I.R.S. (Ass. Italiana Ricerca sui Sistemi), 1986 - 1999.

Allen, T. F., 1982, *Encyclopedia of pure Materia Medica*. Jain Publishers, New Delhi.

Ashby, W. R., 1971, *Introduzione alla cibernetica*. Einaudi, Torino.

Von Bertalanffy, L., 1983, *Teoria generale dei sistemi*. Mondatori, Milano.

Ball, P., 1999, *H_2O una biografia dell'acqua*. Ediz. Mondolibri, Milano.

Hahnemann, S., 1977, *Organon*. Edium, Milano.

Haken, H., 1983, *Sinergetica*. Boringhieri, Torino.

Kent, J. T., 1978, *Lezioni di omeopatia*. Edium, Milano.

Kent, J. T., 1994, *Omeopatia. Come collaborare con il proprio Medico per una prescrizione ottimale* (a cura di M. Trionfi), Scalvi Editore, Brescia.

Minati, G., 1998, *Sistemica*. Apogeo, Milano.

Systemic Support Therapy for Cancer Patients During and After Chemotherapy

ALBERTO RICCIUTI
Attivecomeprima – Onlus (Breast cancer Association) – Milano
AIRS - Associazione Italiana per la Ricerca sui Sistemi – Milano

1. INTRODUCTION

The antitumoral chemotherapy, that is the use of medicines aimed at the destruction of neoplastic cells, is regarded as a cornerstone in tumor therapy. This therapy is nevertheless burdened with such frequent and intense side effects, that serious pain is induced in the patients and that doctors are often forced to space further out in time the treatments, to reduce the dosage, or even to interrupt the therapy. Besides, this is lived by the patients as a reduction of their life expectancy.

2. THE PROBLEM OF SIDE EFFECTS IN ONCOLOGICAL THERAPIES

In cases in which chemotherapy is administrated, oncologists usually face especially those side effects – nausea and vomit, myelodepression (leukopenia, thrombocytopenia, anaemia), mucositis (stomatitis and general suffering of mucosae, candidosis), alopecia – that handicap the administration itself. The current therapeutic approach to face such toxic effects is basically aimed at opposing the symptoms: anti-emetics against nausea and vomit, dietary devices and sodium bicarbonate rinsing of the oral cavity against stomatitis, anti-mycotics against candidosis, anti-diarrhoeic in

Emergence in Complex, Cognitive, Social, and Biological Systems, Edited by
Minati and Pessa, Kluwer Academic/Plenum Publishers, New York 2002

case of dysentery, hemopoietic growth factors and transfusions in case of myelodepression and so on.

Moreover, a syndrome – the complex causes of which are still being discussed about and which is often underestimated by the doctors – known as "fatigue" and characterized by malaise and such a loss of energy that it is considered to be the main obstacle to normal daily activities and to an acceptable quality of life, is anyway associated to cancer and/or to its therapies by the patients (Curt Gregory, *et al.*, 2000). Such syndrome, well documented in scientific literature with an estimated prevalence between 60% and 96%, seems to be a common characteristic of all patients affected with cancer, but especially of those who undergo chemotherapy and/or radiotherapy and of patients treated with different modifiers of the biological response such as interferone-a and interleukines (Vogelzang, 1997). The symptoms of "fatigue", besides, tend to gradually raise during oncological therapies reaching their maximum after some weeks, and last, sometimes for a long time, after the therapies have ceased. The treatment of this complex syndrome is still very problematical, and it basically consists of psychostimulating drugs, anti-depressants, cortisone-like medicines and epoetine alpha, which, raising the levels of hemoglobin, proved effective in reducing the symptoms of "fatigue", apart from the effects of the oncological therapies (Portenoy Russel, 1999).

On the whole, therefore, the support therapies currently administrated to oncological patients are certainly effective from a strictly farmacological point of view, but it is not always so from a clinical point of view, due to the complexity of different oncological situations and of all that escapes from our knowledge. Such therapies seem mainly aimed at opposing the direct and immediate effects of an organ-specific poli-toxicity, using valuable drugs which in turn can be accountable for further side effects. In other words, support therapy seems generally directed towards chasing the symptoms rather than preventing them, substantially following a chemotherapically caused-damage "repairing logic" on "single components" (i.e.: tissues, organs, functions, etc.) of the organisms. This logic is useful in some cases, but certainly insufficient in others.

In spite of the current support therapies, chemotherapy remains therefore up to now a very oppressive therapy for the patient, reducing the quality of life to levels often judged barely tolerable, inducing many cancer patients to refuse it.

3. THE MEANING OF THE CURRENT RESORT TO NONCONVENTIONAL THERAPIES

At least mainly for these reasons, those who resort to nonconventional medicine practices (alimentary integrators, vitamins, phytoterapeutics, dietary prescriptions, mind-body reharmonization activities) has been raising for many years in every country of the world, involving on average 30-50% of citizens. This is further confirmed by some recent researches on the use of such methods by oncologic and hematologic patients that report figures of 52-65% for Germany, 7-54% for North America, 22% for Australia. Moreover, 41% of patients report having been given advice to use such therapeutic support by the doctors themselves. Anyway, only 50% of the patients has informed their doctors in charge (Kappauf *et al.*, 2000; Schraub, 2000) about this choice - even though it generally was not caused by unsatisfaction towards conventional treatments -, this being followed by all the possible consequences (possible interactions among medicines, conflicts in doctor-patient relationship, etc.). Such therapies are vaguely sensed by the patients as being effective in facing the global need for support of the organism, including a relationship with the therapeutist successful in mobilizing hope (Paltiel *et al.*, 2001; Tagliaferri *et al.*, 2001) and, in particular on the physical level, in counteracting the toxicity of the therapies, the reduction of immune defenses, and some symptoms related to the disease itself (Eguchi *et al.*, 2000).

All this expresses a need for global support on behalf of the patients, whom cannot and must not be ignored, and who above all have the right of not being left in loneliness and uncertainty, in such a dramatic time of their life in which the uneasiness due to the therapy itself adds up to that of cancer.

The problem is that, among the so-called support therapies, the offer is very wide and it is hard to understand first and then choose. Many of these therapeutics proposals are related to models of health and illness that are defined by adjectives like 'global', 'holistic', integrated', 'natural', and so on, which are words that, unlike the abstruse tecnical terms used by our medicine, are generally more comprehensible, both because they belong to the common language and because they are referred to much-discussed themes from an emerging psycho-bio-ecologic culture proposing to center the attention (and not only in medicine!) on the human person, her relationship with nature, the diseased rather than the disease, the theme of freedom of cure, of doctor-patient relationship, and so on. All that, in a nutshell, represents a value. Moreover, in a patient affected with cancer, all this expresses just that need for attention to the person and to her condition of global suffering that is indipendently searched for through the different complementary support cures, during and after the oncological therapies.

The problem, though, lies in the fact that those who propose such complementary cures often make reference to theoretical models of health and illness which use concepts and languages other than those of our traditional medicine, introducing strong conflict elements in patient-to-oncologist communication with serious damage for the diseased, who does not feel anyway suitably cured in the global dimension he is looking for.

In fact, those who refer to so-called 'holistic' or 'global' models, do it not only using a language far from our traditional medical culture, but also believing that they can exceed the limits – of which, by the way, our medicine itself has long been aware (see the Di Bella case) – of an approach oriented more towards the disease than the diseased, opposing to the solidity of knowledge and methods belonging to traditional medicine the frailty of a knowledge referring to 'globality' in too vague terms from a theoretical point of view and only empirical from a practical one.

Without entering upon their efficacy, some method problems must be pointed out with regard to such complementary therapies:

a) they too often follow an exclusively 'repairing' logic, which is not negative in itself, but sometimes insufficient;
b) they flank oncological therapies with prescriptions sometimes more meant to cure cancer in an alternative way rather than to support the organism during oncological therapies;
c) they often use theoretical concepts and languages, that, due to their distance from those currently shared by the scientific community, make the desirable dialogue and integration, which has long been searched for by different parties, harder and harder;
d) the lack of a common objective and language prevents from evaluating whatever effectiveness of different therapy instruments and methods.

Therefore, for such reasons, it is now not only useful but necessary *to systematize the general medical support therapy and to promote its development and its application as a major part of oncological therapies*, given some indispensable content and method premises:

a) it is not to appear as an 'alternative' cancer therapy method, at least because this goes beyond the general medicine support during oncological therapies objective;
b) it is to be able to join together the 'objectifying clinical view' of the current medical tradition with the 'anthropological clinical view' emerging today as a need for attention to the person as a whole – both by the patients and by many doctors;

c) it is to use, for obvious opportunistic reasons, languages and concepts acquired and shared by the scientific community, making communication between patients and doctors easier, and not harder;
d) it is to allow verification of its results through instruments that are accepted or acceptable by the scientific community.

4. TOWARDS A SYSTEMIC APPROACH TO SUPPORT THERAPY IN ONCOLOGY

What must start to be considered, is the fact oncological therapies and chemotherapeutics in particular cause toxic effects at a level that is well deeper and wider of the organ-specific one mentioned above. These effects (including "fatigue"), or at least their intensity, can indeed be thought of as due to disturbing systemic actions developing at a level of important cell- and tissue-biochemical homeostatic regulation processes that involve the entire organism as an "autopoietic system", that is self-producing.

Such is, in the definition given by Maturana and Varela (1980) the trait that specifies every living system as an organized unit in the form of a network of processes producing (transforming and destroying) components which, realizing the network itself that produces them, form a unit out of it. Such "organization scheme" – defining the configuration of the relationships among system components production processes and thus necessary and sufficient to define any system as living – is materialized, in physical space, in the different structures characterizing the different living organisms as we know them, including human beings. "Structure" therefore means the physical materialization of the organization scheme, that is the physical components of the system, their shape, chemical composition, etc. Such structure has the characteristics of a "dissipative structure" (Prigogine and Stengers, 1979) thus every autopoietic system is closed at the organization level and open at the thermodynamical level.

Therefore, what we call life can only be grasped - in a living system - in the unceasing activity of the processes producing each single component that maintains the invariance of the organization scheme, that is what grants the system autonomy, unity, and individuality. But the ability to contact the environment, continuously carrying out those structural internal changes pertaining to self-maintenance, specifies necessarily every living system as a "cognitive system" and living, as a process, a cognition process (Maturana, 1980). Thus, what we call 'mind' is not an entity but a process. This fundamentally important aspect - initially caght by Bateson (1979) and described in a more subtle way by Maturana and Varela - leads the way to exceeding the cartesian dualism that, albeit its importance in our civilization's scientifical and tecnological development, has long started to show those limits that, for what regards medicine, the various holistic, naturalistic and so on approaches try to exceed, often incoherently.

These Authors and Capra's effective synthesis (Capra, 1996) are beginning to be referred to, searching for an anthropological foundation of medicine and of the ethical issues in which it is struggling today (Galimberti, 2000).

For our purpose it is important, therefore, to be able to pay particular attention to the *connections*, or the component production processes, and not only to the component themselves as subsystems, with their shape, chemical composition, etc.

Considering that, as Maturana states, "all that is said is said by an observer" in his interaction with the observed object – and that, therefore, every *explanation* of any phenomenon consists of a reformulation of the observed phenomenon which shows how its components generate it through their interactions and connections –, it is always necessary to be clearly aware of the difference between what belongs to the observer's descriptive domain and what belongs to the constitutive organization of the phenomenon that is being explained. "The explanations, that is, take place operatively in a meta-domain with respect to what they intend to explain" (Maturana, 1990). This observation is important since, for example, in medicine we have to do with what we call "organs" and with what we call "functions". While the former belong to the structural organization of the organism as an autopoietic system, or to its *phenomenological domain*, the latter belong to the *domain of the descriptions* of that system, or part of it, that the observer – he too as an autopoietic system – believes he can make, by interacting with the observed system on the basis of his perceptive abilities.

This specification is not merely an academical one, but it has its practical importance since our decisions about medical interventions on the diverse functions of the organism follow the model - or models - that we have of them, that is the type of description we made of such functions.

5. THE LOGICAL BASIS OF GENERAL MEDICAL SYSTEMIC SUPPORT THERAPY

On the basis of acquired and shared knowledges it is possible to point out some important connections among the processes that involve the entire organism in maintaining its structural organization (Fig. 1).

In fact, for what regards antitumoral chemotherapeutics it is expedient - just like for other medicines (antibiotics, hormones, anti-inflammatory) - to take into account - together with the known toxic effects on single components - their *disturbing effect on those self-organizing, self-repair, and self-defense processes that specify the ability of the organism to autonomously maintain, in physical space, its structural unity organization*, of keeping, that is - in a more common language - in good not only short-term, but middle- and long-term health.

Figure 1 describes the disturbances – occurring during oncological therapies - of some fundamental processes that allow maintaining the structural organization of the organism as an autopoietic system through time.

The biological processes, obviously, involve single components of the organism (cells, tissues, organs, etc.), of which it is always possible to describe the involvement in purely physical terms. In doing so, though, we obtain – at the best – a merely physical description that is unable to grasp the phenomenon in its biological dimension, which concerns the connections among processes in the autopoietic framework. The "objectifying clinical view" (Foucault, 1963) of western modern medicine and the diagnostic instruments it consequently adopted, allow to grasp the status but not the processes; the consequent description of the organism, that is, rises from a data 'set' grasping the status of its single components, mainly observed in different moments. Let us think, for example, of what a radiography, a glycemia measurement, an hormonal dosage are. These observations are very precious to a doctor but they cannot grasp, like photograms in a film, the 'process' that puts a sense to the whole set of observations.

But *an autopoietic network is a system of connections among processes producing components, that is to be considered within the complex development of the self-regulating and self-regenerating dynamics which grant the invariance of its structural organization on the global level.* It is when the connections among these component producing processes (transformation and distruction) are altered, that the single components - regenerated in turn by those same processes – undergo a damage that can either be reversible or irreversible.

As it can be seen, firm conceptual instruments already exist, on which a therapeutic approach can be founded, taking into account some of the most important processes specifying the biological system of the organism in its unity, and trying to oppose the disturbing effects of oncological therapies, first of all chemotherapy.

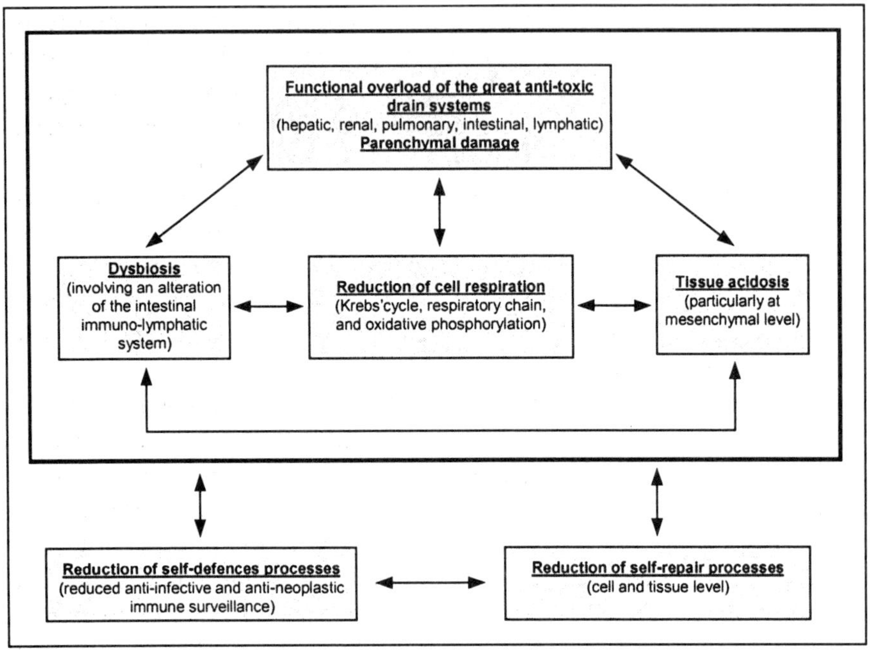

Figure 1. The web of autopoietic processes affected by chemotherapy.

The processes that must particularly be carried out – that is the alterations caused by oncological therapies -, without considering therapy, concern what in Fig. 1 is indicated as:

- "***Functional overload of the great anti-toxic drain system***"; it is the overload of the fundamental processes of removing toxic substances (drugs and substances involved in their metabolism, and toxins developed consequently to the action of the drugs themselves) by the organs and the apparati that are supposed to carry out this function, i.e.: liver, lungs, intestine, lymphatic system.
- "***Dysbiosis***": it represents a disorganization, caused by many factors, of the physiological bacterial flora in the intestine gradually leading to a progressive and serious dysfunction of the entire organism and of the digestive processes on the whole. The local immune tolerance has long been known to depend mainly on the functional integrity and on the synergy of complex structures. The latter play a major role in the immuno-lymphatic system that links all the mucosae of the organism (Mucosa-Associated Lymphoid Tissue = MALT). The chemotherapeutics, besides altering the delicate equilibrium of the intestinal flora because of their antiblastic action, directly damage epithelial cells in all mucosae, hinder their physiological regenerative turnover, and alter, especially at a gastroenteric level, the enzimatic

production and thus the correct absorption of macro and micronutrients. This allows the development of a pathogenic flora, first aerobic (Enterobacteria, Proteus, Kleibsiella, Staphylococci, pathogenic Escherichia Coli, etc.) and then anaerobic (Clostridia, Fusobacteria, Peptococci, Actinomycetes, etc.). Consequently, the reduction of the energetic metabolism in colon epithelial cells involves a reduction of cell respiration, along with the progressive establishment of an actual organ damage. Besides, the consequent increase in decomposition processes causes the production of different endotoxic substances (ammonia, skatole, amines, indole, phenol, fuselic alcohol) which further overload the various emunctory organs, particularly the liver.

- "***Reduction of cell respiration***"; the complex energy production process, through degradation of food to its most simple components (aminoacids, mono- and disaccharides, fatty acids, and glycerol), from these to acetyl-Coenzyme A, and eventually to water and carbon dioxide, yields large amounts of ATP (adenosine triphosphate). Its synthesis is known to take place in mitochondria through a complex series of biochemical reactions mainly identified in Krebs' cycle (tricarboxilic acid cycle or citric acid cycle) and in the respiratory chain – a complex electrochemical process occurring on the mitochondrial membrane and requiring its perfect integrity. In patients under chemotherapy and/or radiotherapy, production and accumulation of exogenous and endogenous toxins (medicines, intermediate metabolism products, free radicals) occur on one hand, and a lack of those substances which are necessary to their neutralization and elimination on the other. An easy example is the chelating effect of many mentioned drugs, inhibiting the bioavailability of enzymes which depend functionally on the presence in the substrate of various oligominerals. Especially in case of tissue acidosis the latter lack or are biologically unavailable, and this seriously compromises the redox reactions in the respiratory chain and in the oxidative phosphorilation.
- "***Tissue acidosis***"; our metabolism normally produces an amount of acids slightly higher than that of alkali, and these acids must be continuously neutralized or removed. In order for hematic pH to be kept constant, the exceeding acids that cannot be eliminated accumulate in intercellular spaces within that particular tissue called mesenchyme, rich in fibroblasts, neurovegetative terminations, proteins, and mucopolysaccharides. Such tissue has a colloidal structure that during the day – because of the progressive accumulation of acid catabolites – tends to a gel state (semisolid, viscous). During the night, instead, mesenchyme tends – because of the inflammatory hydrolisis caused by hyaluronidase - to a sol state (semiliquid and much less viscous), allowing transport and removal of acid catabolites and the consequent shift of pH towards

alkalinity. An excessive gelation of mesenchyme, occurring under effect of drugs (chemotherapeutics, antibiotics, hormones, non-steroid anti-inflammatory substances, analgesics, etc.) or/and of an improper diet (excess of acidifying foods, like refined carbohydrates and animal-derived foods, and chemical polluting substances), thus involves a reduction or a blockage of those processes that allow the removal of acid toxins from tissues.

These processes reciprocally interact following a network scheme that, in turn, involves a "*reduction of self-defense processes*" (reduced anti-infective and anti-neoplastic immune surveillance) and a "*reduction of self-repair processes*" (cell and tissue level), of those processes, that is, which grant the maintenance of the structural organization of the entire organism as an autopoietic system.

Therefore, the therapeutic strategy aims at *maintaining the organization of the autopoietic network through administration of drugs and substrates stimulating and supporting the autonomy of the self-repair processes of structural components determining it* (Ricciuti, 2001). This is the rational suggesting the prevailing – but not exclusive – use of natural medicines, that is of biomolecules carrying adequate biological informations that are consistent to the nature of the system. Very briefly, these are specifically: drugs containing 'probiotics' that can correct the dysbiosis and support the immune surveillance of the entire organism; biocatalysts and intermediate products of Krebs' cycle and of the respiratory chain that can regulate cell metabolism and the production of energy at a mitochondrial level; adequate nutritional substrates allowing the correction of the excessive acidification of the mesenchyme and thus the correct anti-toxic drain of the entire organism. The support of these processes is a premise that cannot be given up in order to a correct expression of the self-defense and self-repair-functions that grant the entire organism autonomy, unity, and individuality, which are implicit consequences of autopoiesis. Such a typology of therapeutic intervention – because of its characteristic purposes of regulating the self-organization processes of the organism as an autopoietic system - involves the use of many of the mentioned instruments in 'periphysiological' dosages, that is containing biomolecules in concentrations similar to those physiologically present within the organism itself, and therefore suitable to carry on functions of 'modulation' of the informative network forming the delicate web of the autopoietic processes.

6. THE INNOVATIVE VALUE OF A SYSTEMIC APPROACH IN ONCOLOGY

The last fundamental point that must be underlined is that such a support care logic in oncology gives the concrete chance to use, within the doctor-patient relationship, a cognitive model centered on the person and not on the disease. And this has a profoundly innovative value both on a cultural and a practical level.

In fact, such a model gives the doctor the chance to enrich his bedside manner with new and fertile communicative and operational contents, which on one hand stimulate the patient to participate in a more trusting way to the oncological therapies themselves, and on the other hand offer a satisfying answer to the issues that induce so many patients to either refuse chemotherapy or look for treatments other than oncological therapies.

Moreover, a cognitive model centered on the person, helps the establishment of the 'therapeutic alliance' between doctor and patient that is the fundamental premise to allow the latter to find an answer and a concrete content to the need for hope that every cancer patient desperately looks for. Besides this, it also allows recognizing as a *primary objective of the medical praxis the restoration and the strengthening of the natural self-repair and self-organization processes of the organism, by supporting and stimulating personal resources*. Recovery, from this point of view, belongs more to the set of possible 'outcomes', rather than to that of the 'objectives' (Ricciuti, 1998), and – because of its preventive value implied by such a therapeutic strategy – is likely to be more easily reachable and lasting.

REFERENCES

Bateson, G., 1979, *Mind and Nature: a Necessary Unity*. Dutton, New York. (it. tr.: *Mente e Natura*. Adelphi, Milano, 1984).

Capra, F., 1996, *The web of life*. Doubleday-Anchor Book, New-York. (it. tr.: *La rete della vita*. Rizzoli, Milano, 1997).

Curt Gregory, A., *et al.*, 2000, Impact of Cancer-Related Fatigue on the Lives of Patients: New Findings From the Fatigue Coalition. *The Oncologist* **5**: 1-8.

Eguchi, K., *et al.*, 2000, Current status of cancer patients' perception of alternative medicine in Japan. *Supportive Care in Cancer* 8 (1): 28-32.

Foucault, M., 1963, *Naissance de la clinique. Une archéologie du regard médical*. Presses Universitaires de France, Paris. (it.tr.: *Nascita della clinica. Il ruolo della medicina nella costituzione delle scienze umane*. Einaudi, Torino, 1982, (1a ed. 1969)).

Galimberti, U., 2000, Etica e medicina. In *Orme del sacro*, Feltrinelli, Milano.

Kappauf, H., *et al.*, 2000, Use of and attitudes held toward unconventional medicine by patients in a department of internal medicine / oncology and haematology. *Supportive Care in Cancer* **8** (4): 314-322.

Maturana, H. R., 1980, Biology of Cognition. In *Autopoiesis and Cognition. The Realization of the Living*, D. Reidel Publishing Company, Dordrecht, Holland. (it.tr.: Biologia della cognizione. In *Autopoiesi e Cognizione*, Marsilio, Venezia 1985).

Maturana, H. R., 1990, *The Biological Foundations of Self Consciousness and the Physical Domain of Existence*, Humberto R. Maturana. (it.tr.: *Autocoscienza e realtà*. Raffaello Cortina, Milano 1993).

Maturana, H. R., and Varela, F. J., 1980, Autopoiesis. The Organization of Living. In *Autopoiesis and Cognition. The Realization of the Living*, D. Reidel Publishing Company, Dordrecht, Holland. (it.tr.: Autopoiesi. L'organizzazione del vivente. In *Autopoiesi e Cognizione*, Marsilio, Venezia 1985).

Paltiel, O., *et al.*, 2001, Determinants of the use of complementary therapies by patients with cancer. *J. Clin. Oncol.* **19** (9): 2439-2448.

Portenoy Russel, K., 1999, Cancer-Related Fatigue: Guidelines for Evaluation and Management. *The Oncologist* **4**: 1-10.

Prigogine, I., and Stengers, I., 1979, *La Nouvelle Alliance*, Gallimard, Paris. (it. tr.: *La Nuova Alleanza*, Einaudi, Torino 1981).

Ricciuti, A., 2001, *Il sostegno medico generale alle donne durante e dopo la chemioterapia*. Attivecomeprima-Onlus, Milano (internal document used in refresher seminars on the activities of "Attivecomeprima" – Breast cancer Association).

Ricciuti, A., 1998, Le speranze, le preoccupazioni e la relazione terapeutica del medico personale. In *...e poi cambia la vita*, FrancoAngeli, Milano.

Schraub, S., 2000, Unproven methods in cancer: a worldwide problem. *Supportive Care in Cancer* **8** (1): 10-15.

Tagliaferri, M., *et al.*, 2001, Complementary and alternative medicine in early-stage breast cancer. *Semin. Oncol.* **28** (1): 121-134.

Vogelzang, N. J., 1997, Patient, Caregivers, and Oncologist Perceptions of Cancer-Related Fatigue: Results of a Tripart Assessment Survey. *Seminars in Hematology* 34 (3-2): 4-12.

A Systemic Approach to the Modelling of Epileptic Phenomena

P. L. BANDINELLI [°,#,^], C. PALMA [#,^], M. P. PENNA [*,^,+], E. PESSA [#,*,^], and R. PETRONI [#]

[°] *Dipartimento di Salute Mentale ASL Roma «E», c/o Servizio Psichiatrico di Diagnosi e Cura Azienda Complesso Ospedaliero S. Filippo Neri. Via G. Martinotti 20, 00135 Rome, Italy*

[#] *Cattedra di Intelligenza Artificiale. I° Facoltà di Psicologia, Università degli Studi di Roma «La Sapienza», Via dei Marsi 78, 00185 Rome, Italy*

[*] *Dipartimento di Filosofia- Sezione Psicologia, Università degli Studi di Pavia Piazza Botta 6, 27100 Pavia, Italy - E-mail : eliano.pessa@unipv.it*

[^] *AIRS - Associazione Italiana per la Ricerca sui Sistemi Viale Jenner, 10 - 20159 Milano, Italy*

[+] *Facoltà di Scienze della Formazione, Università degli Studi di Cagliari*

1. INTRODUCTION

This contribution deals with a model of brain electric activity, designed to explain the occurrence of epileptic phenomena. The macroscopic features of these latter seem to be a consequence of a sort of global synchronization of neural activity, due to the presence of long-range correlations between neural outputs, in some way analogous to the ones associated to critical points of second-order phase transitions. As is well known, it is very difficult to build models of biological systems exhibiting such correlations, owing to the fact that biological systems are far more complex than simple physical systems which are the subject of the theory of phase transitions.

Such a circumstance suggested us to adopt a general strategy in order to model in a realistic way emergent phenomena (and epilepsy is just an emergent phenomenon) in biological systems. Such a strategy made resort to two different modelling levels:

Emergence in Complex, Cognitive, Social, and Biological Systems, Edited by Minati and Pessa, Kluwer Academic/Plenum Publishers, New York 2002

a) a microscopic level, in which the individual system components are described in a detailed way, taking into account as much as possible the constraints imposed by biological data; at such a level the behavior of the whole system cannot be forecast in advance, but must be studied only through computer simulations; the eventual occurrence of long-range correlations can be detected only through a monitoring of suitable global quantities (such as, e.g., crosscorrelation functions) describing model behavior;
b) a macroscopic level, in which the system behavior is described through (hopefully simple) phenomenological evolution equations ruling suitable global variables; these equations are amenable to a detailed mathematical analysis, letting us forecast and classify the qualitative features of system behavior; the eventual occurrence of long-range correlations is a logical consequence of suitable theorems or of analytical computations done on evolution equations.

In principle, the macroscopic phenomenological equations should be derived from the microscopic level descriptions through rigorous procedures, such as the ones of statistical mechanics. However, the practical implementation of this idea, which should let us build an exact theory of emergent phenomena, appears as impossible. Namely, in their present form the methods of statistical mechanics can be applied with some success only to very simple systems, made of identical components and satisfying suitable conservation or symmetry principles, such as the assemblies of interacting particles. On the contrary, biological systems are characterized by the fact that each component is endowed with its individuality, and that we lack conservation or symmetry principles.

In such a situation the only possible way for studying emergent phenomena is to make resort to phenomenological macroscopic equations derived from the microscopic level through approximate procedures. Then, the macroscopic equations cannot be longer be considered as the exact counterpart of microscopic ones. However, they are nonetheless very useful, as their analysis suggests:

1. what global quantities should be considered as the order parameters (in the sense of Haken, 1983) associated to system's phase transitions, and, whence, what global quantities should be monitored during the computer simulations of microscopic behaviors;
2. how emergent phenomena could derive from the behavior of individual components and what microscopic features are relevant to this regard;
3. how to refine the statistical analysis of microscopic level in order to obtain more reliable macroscopic phenomenological equations.

Thus, the building of approximate macroscopic equations triggers a process of actions and back-reactions regarding the analyses of microscopic and macroscopic levels, leading to a progressive improvement of our description of emergent phenomena. It is to be remarked that this latter will never become exact and objective, as the mutual influences between microscopic and macroscopic descriptions depend, in a crucial way, on subjective choices made by the researcher (see, to this regard, Minati et al., 1998). In this paper we will describe shortly the first step of this process of mutual influence in the case of a model of emergence of epileptic phenomena. As regards, instead, the results of computer simulations of the model itself, they will be presented elsewhere.

2. NEUROPHYSIOLOGICAL ASPECTS OF SEIZURE PROPAGATION

Epilepsy is a greek term that means «get hold of somebody». It denotes the clinical aspect of an electrophysiological disfunction of cortical neurons, consisting of an ictal, abnormal, ipersynchronous (of high voltage and low frequency), or paroxystic discharge (of high frequency). At the beginning the discharge is generally localized only within a restricted area of cortical neurons.

Several types of clinical expressions have been observed, depending both on the cortical area originarily interested, and on the possibility that abnormal discharge can be spread to deep encephalic structures. Partial seizures initially affect only a small part (focus) of the cerebrum (most commonly the temporal or frontal lobes). Complex partial seizures are subdivided into two categories: 1) simple partial onset followed by impairment of consciousness that may or may not be associated with automatism; 2) impairment of consciousness at onset, with or without automatism. Tonic-clonic seizures may or may not be preceded by a prodromal period during which an attack is anticipated, often by an ill-defined vague feeling.

The propagation of epileptic activity in brain tissue, and the phase transition from simple to generalized seizures, are interpreted as an evidence for a neural nonlinear chaotic dynamics (sensitively dependent upon initial conditions of the process), deterministic but unpredictable. The involved factors are related to both local structural properties (architecture and excitatory-inhibitory properties of synaptic network) and variations in neurotransmitters or ions concentration.

From the electrophysiological point of view, intracellular recording of cortical pyramidal cells within the epileptic focus show large paroxysmal depolarization shifts (PDS). These are longer than an action potential, and indeed give rise to a burst af action potentials which propagate down the axon. PDS is triggered by very large excitatory postsynaptic potentials (EPSPs) generated in the dendritic tree of pyramidal cells because of excessive excitatory input and/or insufficient inhibitory input, but is itself composed of an envelope of calcium action potentials.

The PDS is then followed by one and then several depolarizing afterdischarges which look like brief replications of the PDS. Finally the cell enters a phase of continuous depolarization and high-frequency action potential generation (tonic phase). In clinical major motor epilepsy the tonic phase is generally followed by a clonic phase in which the neuron undergoes intermittent afterdischarges accompanied by bursts of action potentials. Gradually the membrane repolarizes and the seizure stops.

Immediately after the PDS the cell is hyperpolarized for several hundred ms, during which the membrane conductance is greatly increased and the cell is profoundly inhibited. At least two mechanisms are likely to contribute. The first is a recurrent IPSP. However, the probable transmitter for the recurrent inhibition in the cortex is GABA. Another mechanism is probably the Ca-activated K^+ conductance. During PDS neurons release large amounts of K^+ into the extracellular space. This can depolarize neighboring neurons and increase their excitability. During epileptiform activity the $Ca2^+$ movement is so great that extracellular $Ca2^+$ concentration approaches zero mmol/L.

As a conclusion, the factors involved in priming and spreading of epileptic activity are related to both local and global structural properties of synaptic network:

1. architecture and excitatory-inhibitory properties
2. variation in neurotransmitters or ions concentration.
3. homeostatic properties of glial cells for $[K^+]$ and $[Ca^{2+}]$ in ECL

To know how epileptic activity spreads in brain tissue is very important both for a diagnosis and a classification of different kinds of ictal activity, and for the understanding of the mechanism of transmission of electrical discharges within central nervous system. As regards the modelling of microscopic features of such a phenomenon, the first obvious possibility is the one of resorting to artificial neural network models. Such a possibility was effectively explored in the papers by Kudela *et al.* (1997, 1998, 1999), and Trabka *et al.* (1988a,b). However, all these authors introduced artificial neural networks including only synapses between neurons, without taking into account the role eventually played by other features of biological neural networks, such as the fluidodynamics of Extracellular Liquid (ECL), and homeostatic properties of glial cells.

3. THE MICROSCOPIC MODEL

This model was inspired by a systemic approach. Neural system was viewed as a complex system constituted by three interacting subsystems (neurons, ECL, and glial cells) in such a way as to describe emergent phenomena as a result of these interactions. A list of these latter includes: neuron-neuron interactions, interactions between single sites of ECL, interactions between glial cells, neurons-ECL interactions, ECL-glial cells interactions, glial cells-neurons interactions. Each of them is ruled by the laws specifying the dynamics of the single subsystems, which will be described in the following.

3.1 ECL dynamics

ECL was modelled as a 2-dimensional grid where every site is located at the cross of a row and of a column and whence is individuated by two integer coordinates. Besides, every site is allowed to contain different ion species. The definition of the neighbourhood of a given site depends on the species of ion taken into consideration. More precisely, given a particular ion species, whose concentration in the site of coordinates i, j is denoted c_{ij}, we considered as belonging to neighbourhood D_{ij} of the site of coordinates i, j all sites with coordinates h, k, lying in the second-order square neighbourhood of i, j, such that $c_{hk} > c_{ij}$.

We assumed that the concentration of a given ion species in a chosen ECL site should be modified in such a way as to favour the tendency towards an equidistribution of concentrations within the whole ECL. The updating algorithm for each ion species concentration is parallel, works on a discretized time scale and starts from c_{ij}, i, j as input variables. The first step requires the computation of the minimum value of the ion concentration of the chosen ion species within the sites lying in the neighbourhood of the considered site. Let us denote such a value as $min\ A$. Then we define the upper limit on the amount of ions that can be transferred to the site i, j as:

$$Q_{trsf.le} = (\min A - c_{ij}) / 2 .$$

At this point for each site of coordinates l, k belonging to the neighbourhood we compute an ownership coefficient, denoted as x_{lk} and defined by:

$$x_{lk} = c_{lk} \Big/ \sum_{rs} c_{rs} \qquad r, s \in D_{ij} .$$

Finally, the amount of ion concentration received from the considered site by its neighbouring side of coordinates l, k is given by:

$$x_{lk}\, Q_{trsf.le}\,.$$

Owing to the fact that we are interested in the simulation of a seizure and whence in the structural description of a injured glial area, as it appears after the acute phase, our model allows for the possibility that the ECL (with some difficulty) continue to flow through the channels obstructed by the damaged glial cells (*ballooning*). The relevant geometry of 2-dimensional cellular space containing ECL, glia and neurons is represented in the Figure 1. This means that, instead of simulating the effective growth of the glial cells during insult and the consequent narrowing of the channels open to ECL flow, we will suppose that the geometry will be unchanged and the narrowing effect will be represented through the variation of a coefficient in the diffusion equation.

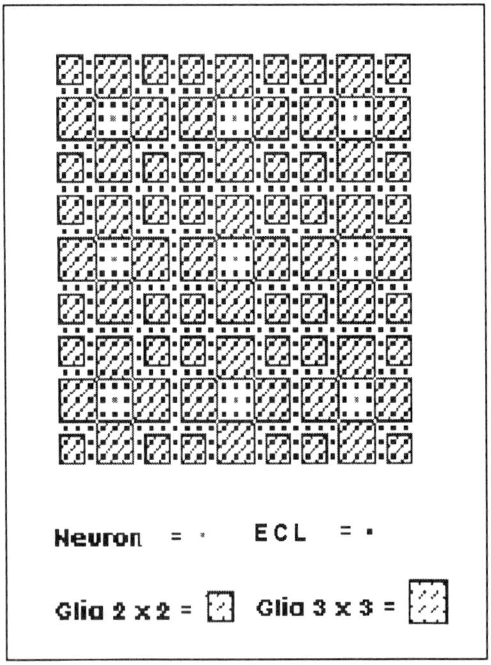

Figure 1.

3.2 The glial subsystem

Withn our model, the glial subsystem consists of two types of cells, both square-shaped. Each cell of the first type is in contact with 3×3 ECL sites and four of these cells surround each neuron (see figure 1). On the other hand, the cells of the second type (astrocytes) are just in contact with 2x2 ECL sites and border only on other glial cells. In this way, astrocytes cover a large part of the ECL but, in normal conditions, don't block the free movement of ions in the ECL; furthermore, the different dimension of the two types of cells fits with the biological variety in the brain. So, the cell spatial distribution results very simple but plausible from a biological point of view. Finally, each glial cell can interact only with seven others neighbours.

Generally, in spite of the equilibrium mechanism, potassium concentration $[K^+]$ is not homogeneous within ECL, and, consequently, different glial cells are characterized by a different amount of stored potassium too. However K^+ can flow from a cell to another, across the gap-junctions; in this way the glial network provides another spatial distribution mechanism for K^+. Within our model the glial cell G_{ij} can get potassium only from its seven neighbours G_{hk}, provided that $[K^+]_{hk} > [K^+]_{ij}$. In addition, at the next step $[K^+]_{ij}$ must be not greater than the minimum $[K^+]_{hk}$ among the sites that made a positive contribution to G_{ij}. Like for ECL algorithm, let be

$$\min A = \min_{h,k \in D} \left\{ [K^+]_{h.k} - [K^+]_{i,j} \right\}$$

As $[K^+]_{ij}$, at the next step, must be not bigger than $[K^+]$ in the site corresponding with *min A* , the total amount of potassium transferred to G_{ij} is

$$\Delta[K^+]_{i,j} = \frac{\min A}{2} - \eta_{i,j} \qquad \left(0 < \eta_{i,j} \leq \frac{\min A}{2} \right)$$

where $\eta_{i,j}$ denotes the gap-junction efficacy of the cell; when $\eta_{i,j} = (\min A)/2$ the contact is interrupted.

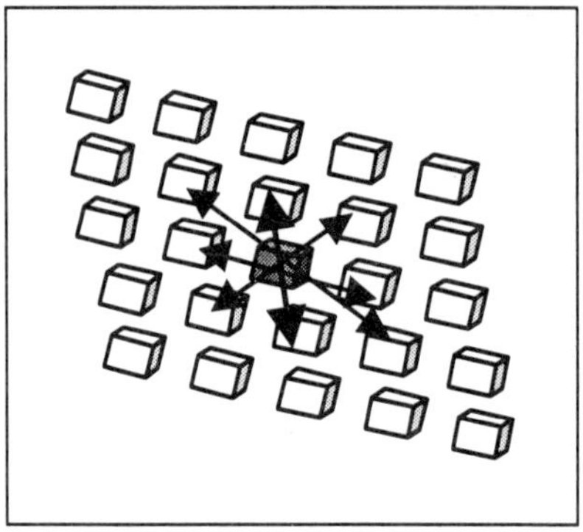

Figure 2.

3.3 The neuronal subsystem

The neuronal subsystem consists of a Cellular Neural Network (CNN). Each cell can interact only with its neighbours (first order interactions; see Figure 2).

Moreover, each neuron can also interact directly with its ECL neighbouring sites and indirectly with its glia neighbouring sites. So, the neural activity depends on neighbouring neurons and on some chemical substances present within ECL. sites near the neuron. To this regard, in the present model we will take in account only K^+ and Glutammate (Glu) concentrations. The activation function of a single neuron is:

$$x_{i,j}(t+\tau) = \tanh\left[a(K^+) \cdot \sum_{h,k \in D} \varphi_{h,k}(Glu) \cdot w_{i,j;h,k} \cdot x_{h,k}(t) \right]$$

where:

$x_{i,j}$ is the neuron state (output);

$w_{i,j,h,k}$ is the synaptic efficacy;

$\varphi_{i,j}$ is a function that depends on the glutammate concentration within ECL;

α is a parameter modelling the K^+ concentration influence on neuron activity.

The synaptic efficacies are obtained from a previous learning phase and are kept fixed; during the subsequent network activity.

3.4 Interactions between neurons and ECL

In our model $[K^+]$ and [Glu] in ECL sites near the neuron influence the neuron activity in the following way:

- $[K^+]$ modifies the slope of the activation function; an increase in $[K^+]$ increases the firing frequency through the following law of dependence of α on $[K^+]$:

$$a(K^+) = \frac{2}{1 - e^{-\alpha \cdot [K^+]}}$$

- [Glu] acts on the synapse in an excitatory way, but doesn't affect directly the synaptic efficacy; it amplifies this latter through a function $\varphi_{i,j}([Glu])$; this function takes into account the variation ΔGlu from a typical value $[Glu]_0$; more precisely, we supposed that:

$$\varphi(Glu) = A \cdot \left(1 - e^{-\Delta Glu / B}\right)$$

$$\varphi'(Glu) = \frac{A}{B} e^{-\Delta Glu / B}$$

In the model, each neuron is endowed with a fixed amount of *Glu*; after *N* fires the neuron exhausted its reserve.

The *Glu* near a synapse is recovered, in a time τ, during an inactivity period, in an exponential way:

$$[Glu(t)] = G * (1 - e^{-t/B})$$

On the other hand, neuron firing frequency modifies $[K^+]$ and $[Glu]$ in ECL. sites near the neuron in the following ways:

$$[K^+(t)] = [K^+]_0 + \kappa * (1 - e^{-t/B})$$

$$[Glu(t)] = [Glu(t)]_0 \, \gamma * (1 - e^{-t/B})$$

where $[K^+]_0$ and $[Glu(t)]_0$ are the equilibrium values in ECL and the symbols κ and γ denote terms directly proportional to neuron firing frequency.

3.5 Interactions between ECL and glial cells

An unimpaired glial cell captures a portion of the K^+ ions exceeding the equilibrium value $[K^+]_{eq}$. However, as stored potassium increases, glia ability to capture potassium declines. When a saturation value $[K^+]_{sat}$ is reached the glia is unable to capture further potassium. At each time step each glial cell captures a portion of K^+ according to the formula:

$$\Delta K^+_{i,j} = \sum_{\lambda=1}^{N} \mu^{\lambda}_{i,j} (Kecl^{\lambda}_{i,l} - Keq)\left(Ksat - \frac{1}{N} Kglia_{i,j} \right)$$

where the symbols used denote, respectively:

- $\Delta K^+_{i,j}$ the total potassium captured, in one step, from the glial cell of coordinates i , j ;
- λ the specific ECL site in contact with the glial cell of coordinates i , j ;
- $N = 4$ (fist type of glial cell) or 9 (second type of glial cell) the total number of sites in contact with the glial cell of coordinates i , j ;
- $\mu_{i,j}$ the absorption efficacy;
- $Kecl_{i,j}^{\lambda}$ the potassium concentration in the λ-th site in contact with the glial cell i , j ;
- Keq an equilibrium value in ECL.;
- $Ksat$ the saturation value in glial cells;
- $Kglia_{i,j}$ the total potassium concentration in the glial cell i , j .

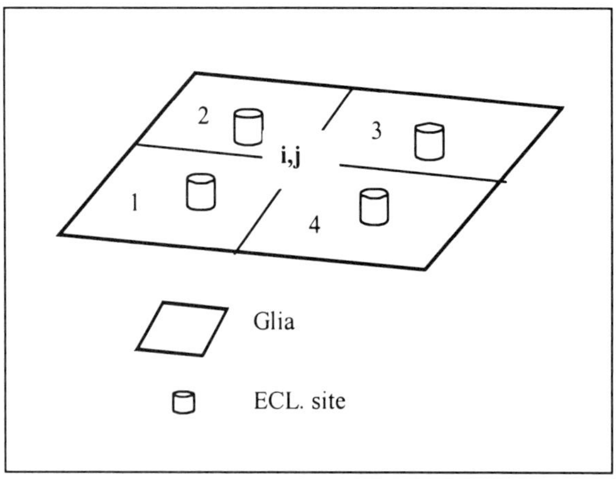

Figure 3.

3.6 Interactions between glial cells and neurons

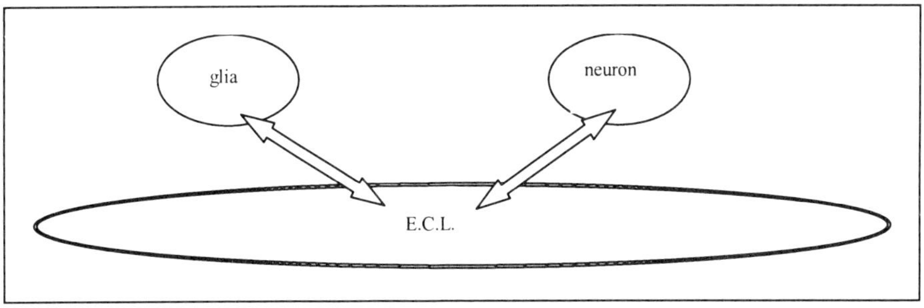

Figure 4.

Glial cells and neurons interact indirectly by means of the ECL, as shown in the Figure 4. In fact, the neuron firing modifies the ions concentration in the ECL and, consequently, glia activity. On the other hand, processes driven by glia (especially K^+ absorption, diffusion and reintroduction in the ECL.) modify the ionic concentrations in the ECL, thus influencing the neuronal activity.

4. INJURY

The model, during the physiological behaviour, can prevent a local excessive increase in $[K^+]$ by means of two mechanisms:

- Diffusion in the ECL
- Glial cells captures, stores and, if necessary, passes to its neighbours the exceeding K^+ ions

An injury that damages the glial cells in a bounded district can reduce the efficacy of these mechanisms or quite destroy them. We modelled the effect of an injury principally through a variation of some model parameters:

- the injury produces a scar and, consequently, in a delimited region, damaged glial cells reduce or quite obstruct the ionic flow in the ECL. In the model, this effect is simulated through an increase of $\varepsilon_{i,j}$, this symbol denoting the difference between (*min* A)/2 and the actual amount of potassium, received in a ECL site of coordinates i, j from neighbouring sites; if $\varepsilon_{i,j} = (min\ A_\varepsilon)/2$ the flow is interrupted.
- damaged glial cells reduce their ability to capture K^+ ions from the ECL. If $\mu_{i,j} = (min\ A_\mu)/2$ the glial cells quite lose this ability.

- damaged astrocytes reduce their ability to redistribute K^+ ions within neighbouring glial cells. The parameter that controls this mechanism is $\eta_{i,j}$. If $\eta_{i,j} = (min\ A_\eta)/2$ the gap-junction is destroyed.

Each of these effects can act by oneself or go together with others.

5. A MACROSCOPIC MODEL OF EPILEPSY

From the foregoing description of our microscopic model it is possible to extract a macroscopic model, useful to forecast the appearing of epileptic emergent phenomena and to individuate what macroscopic variables should be monitored during the computer simulations in order to detect the presence of these phenomena. The procedure used to obtain the macroscopic model will consist in taking a *continuum limit* of a suitably simplified version of the previous microscopic model. The adopted simplifications were the following ones:

- the model was working within a 1-dimensional space, instead of a 2-dimensional space;
- the factor $[1 - exp(-t/B)]$ was neglected and put, in a long-time perspective, equal to unity.

As regards the continuum limit, it consists in allowing all entities (neurons, glial cells, ECL sites) to be distributed in a continuous way in all points belonging to a geometric segment and in letting all time steps Δt and spatial coordinate steps Δi tend to zero in such a way as to keep fixed the relationship $\Delta i = v\ \Delta t$, where v denotes a given, constant, signal propagation velocity.

Such a procedure, when applied to our microscopic model together with the previous simplifying assumptions, gave rise to the following system of macroscopic equations:

$$\partial \xi(x,t)/\partial t = -\ \xi(x,t) + \tanh\{[2/(1+\exp(-\alpha K(x,t)))] \int A\,[1-\exp(-(G(x,t)-G_0)/B)]\, w(x,x')\, \xi(x',t)\, dx'\}$$

$$\partial K/\partial t = D_{ECL}\,(\partial^2 K/\partial x^2) + \kappa_0\,(\partial \xi/\partial t) - \mu\,(K - Keq)\,(Ksat - Q)$$

$$\partial Q/\partial t = \mu\,(K - Keq)\,(Ksat - Q)$$

$$\partial H/\partial t = D_H\,(\partial^2 H/\partial x^2) + (K - Keq)$$

$$\partial G/\partial t = -\ G + \gamma\,(\partial \xi/\partial t)$$

where $\xi(x, t)$ denotes the output acitvity at time t of the neuron located in the position of coordinate x, K the space-time dependent potassium concentration in ECL, Q the space-time dependent potassium concentration in glial cells of the first type, H the same concentration in glial cells of the second type (astrocytes) , G the glutammate concentration in ECL, and $w(x, x')$ a two-point function describing a suitable synaptic efficacy field. The other symbols denote model parameters.

By looking at these equations, it is easy to acknowledge that they have a homogeneous stationary equilibrium solution given by:

$$\xi = G = H = 0 \ , \quad K = Keq \ , \quad Q = Ksat \, .$$

A straightforward linear stability analysis around this equilibrium solution (done by neglecting the useless equation for H) showed that two of the eigenvalues of the Ljapunov matrix had the values $+i$ and $-i$, the symbol i denoting the imaginary unit, whereas the other two were the solutions of the following quadratic equation:

$$\lambda^2 - \lambda\,(\mu\, Keq - D_{ECL}\,\mathrm{k}^2 - \mu\, Ksat) + D_{ECL}\, k^2\, \mu\, Ksat = 0 \ ,$$

the symbol k denoting the wavenumber of the perturbations around the equilibrium state, connected to their spatial wavelength l by the well known relationship:

$$k = 2\,\pi / l \ .$$

A tedious algebraic analysis of the previous equation shows that it has two real roots, both of the same sign. In turn this sign coincides with the one of the expression:

$$\mu\, Keq - D_{ECL}\, k^2 - \mu\, Ksat \ .$$

It is immediate to acknowledge that, if the following condition is satisfied:

$$Keq < Ksat \ ,$$

then the expression is negative for every value of k . This implies that the equilibrium state previously quoted is *stable*, so that the neural activity is very small and cannot grow. What happens, however, in presence of an injury? As we have seen in the previous section, it implies the reduction of the ability of glial cells to capture potassium, so that the concentration of this substance within ECL can increase, leading, in turn, to an increase of *Keq* beyond the value of *Ksat*, so as to invalidate the previous stability condition. It is possible to see that, in this case, the equilibrium state becomes unstable with respect to spatial perturbations, whose wavenumber satisfies the condition:

$$k^2 < \mu (Keq - Ksat) / D_{ECL} \ ,$$

that is to perturbations with *high wavelength*. The instability, then, will occur necessarily under the form of *long-range correlations* between neural activities, such as the ones observed in epileptic phenomena. Our macroscopic model, therefore, was able to forecast the occurring of an emergent phenomenon in presence of an injury. As a further consequence we obtain that the macroscopic quantities to be monitored at every time step during the computer simulations of microscopic model behavior should be the following ones:

a) the average neural activity;
b) the average potassium concentration in ECL
c) the spatial crosscorrelation function of neural activities.

6. CONCLUSIONS

The previous analyses showed how our program can effectively work. From a detailed microscopic model of a biological system we extracted a macroscopic model able to forecast when and why emergent phenomena can take place, in particular the occurring of an epileptic phenomenology. Besides, we obtained precise indications about the macroscopic quantities to be monitored when simulating on a computer the behavior of the microscopic model. In turn, the results of these simulations should be the starting point for a revision, and whence of an improvement, of the macroscopic model. And, again, the improved model should be the basis for undertaking new computer simulations, and so on. Actually, this procedure seems the only one available for dealing with emergence in complex biological systems.

REFERENCES

Heinemann U, Lux H. D., and Guttnich, M. J., 1977, Extracellular free calcium and potassium during paroxysmal activity in the cerebral cortex of the cat. *Exp brain res* **27**: 237.

Trabka, W., Trabka, J., and Mikrut Ż., 1988a, Changes in synaptic potentials and axonal delays as factors causing epileptic instability of neuronal networks. *Neurol Neurochir. Pol.* **22** (5): 429-38.

Trabka W., Trabka J., and Mikrut Z, 1988b, Study of the mechanisms of the development of epileptic activity in neuronal networks using a simulation model. *Neurol. Neurochir. Pol.* **22** (5): 421-28.

Kudela, P., Franaszezuk, P. J., and Bergey, G. K., 1997, A simple computer model of excitable synaptically connected neurons. *Biological Cybernetics* **77**: 71-77.

Kudela, P., Franaszezuk, P. J., and Bergey, G. K., 1998a, A neural network model of propagation of burst activity. *Neural Modeling of Cognitive and Brain Disorders Workshop*. University of Maryland, College Park, MD, June 4-6, 1998.

Kudela, P., Franaszezuk, P. J., and Bergey, G. K., 1998b, A neural network model of traveling seizure activity wave. *Epilepsia*, **39** (6): 3.

Kudela, P., Franaszezuk, P. J., and Bergey, G. K., 1999, Model of the propagation of synchronous firing in a reduced neuron network. *Neurocomputing* (26-27): 411-418.

Synergetics and Systemics

Some Comments about Determinism and Classical Physics

LUIGI GALGANI
Dipartimento di Matematica, Università di Milano
Via Saldini, 50 - 20133 MILANO, Italy

There is no doubt that classical mechanics has always been associated with a strict deterministic vision, according to the celebrated sentence of Laplace. A reading of the popular writings of the beloved great Boltzmann gives a vivid impression of how deep an influence such a vision can have on the psychological attitude of people concerning life.

In order to find a different attitude in science one can go back to the work of Goethe, where the role of the individual is somehow vindicated. Due to the fact that Goethe was strongly contrasting Newton, one might imagine that ``ordinary scientists" did not take him into serious consideration; it is thus quite interesting to notice that his work was actually appreciated by Helmholtz. A considerable appreciation was also manifested by Heisenberg, the creator of quantum mechanics; this in fact seems to be quite understandable, due to the role that the observer plays in quantum mechanics too.

More generally, quantum mechanics is usually interpreted as providing a vision of the world which is not so strictly materialistic as that of classical mechanics. In the present short note I intend to point out that even classical physics, when extended to include electrodynamics and in particular the dynamics of charged point particles, seems to require a formalization which corresponds to a vision of the world much subtler than the traditional one.

Essentially this goes as follows. When trying to get a mathematical description of a charged point particle which be consistent with the presence of the electromagnetic field "created" by the particle itself and governed by Maxwell's equations, one meets with severe difficulties, and one is somehow obliged to provide some independent prescriptions defining the limit

Emergence in Complex, Cognitive, Social, and Biological Systems, Edited by Minati and Pessa, Kluwer Academic/Plenum Publishers, New York 2002

dynamics, so that one is actually providing some new theory. This was done in a consistent way by Dirac in the year 1938, but such a new theory presented strange somehow paradoxical features; this was manifested in the fact that the Dirac equation for a point particle involves the third derivative of the position, in such a form that the generic solution presents an absurd "runaway" character; one has thus to impose form outside a suitable prescription on the "final" motion which selects the good, "natural" motions. In Dirac's words: "We must restrict ourselves to those solutions for which the velocity is constant during the final period when the particle is left alone", and: "We must merely impose the condition that these solutions are the ones that occur in Nature".

Moreover: "We now have a striking departure from the usual idea of mechanics. We must obtain solutions of our equations of motion for which the initial position and velocity of the electron are prescribed, together with the final acceleration, instead of solutions with all initial data prescribed" ... And finally: "This will lead to the most beautiful feature of the theory" ... although ... "It would appear here that we have a contradiction with elementary ideas of causality".

In some joint work with Andrea Carati, it was shown that, in the case of a scattering by a barrier, the "final" prescription of Dirac does not select in a unique way the initial acceleration leading to nonrunaway motions, and that the different allowed initial accelerations determine whether the particle will be reflected by the barrier or go beyond it; so we have here a classical analogue of the tunnel effect. In such a way the initial acceleration (or rather the initial electromagnetic field, because the latter can be shown to determine the initial acceleration) turns out to play the role of a hidden parameter. Moreover we are here in presence of a kind of nonlocality, because the domain of definition of such a hidden parameter turns out to depend on the height of the barrier, even if the barrier is located in Tokyo and the particle in Milan. It is thus easy to show that in such a model one can have a violation of Bell's inequalities.

In conclusion, I hope I was able to communicate the information that classical physics appears to require a formalization which is akin to a form of determinism and causality somehow subtler than the familiar one of Laplace.

REFERENCES

Carati, A., and Galgani, L., 1999, *Nuovo Cimento* **B114**: 489-500.

Carati, A., and Galgani, L., 2001, *Foundations of Physics* **31**: 69-87.

Dirac, P. A. M., 1938, *Proc. Royal Soc.* (London) **A167**: 148-168.

What is Emergence?

ELIANO PESSA
Dipartimento di Psicologia, Università di Pavia, Pavia, Italy; E-mail: eliano.pessa@unipv.it

A general theory of emergence should first start taking into account that this concept cannot be defined in an *objective* way. The word "emergence" refers to a relationship between an observer, the models which he is equipped with, and certain results of observations, or of measure operations, in turn dependent on his mental schemata and on his technological apparatuses. Therefore the emergence can be defined only *relatively to a given observer*, whose features should be specified in a suitable way. For the purposes of the present discussion we will assume that the observer is a theoretical physicist, endowed with all actual tools of this discipline and with the best computing facilities so far existing. This means that the emergence which we will speak of will be associated to some discrepancy between the forecastings, whose attainment was the goal of the models introduced by the observer, and actual behaviors evidenced within models themselves.

Once given such an intuitive definition of emergence, it allows for a number of further subcases and specifications. The best categorization of these latter appears the one given by Crutchfield (see Crutchfield, 1994; cfr. also Pessa, 1998). Here we will take into consideration, as the most interesting one, the subcase denoted by Crutchfield as *intrinsic emergence*. The emergent behaviors belonging to the latter category are associated to the following features:

1. they are *compatible* with the model used by the observer; for instance, they are solutions of model's equations or are produced by computer simulations of it;
2. their occurrence, however, *cannot be foreseen in advance* on the only basis of the adopted model;
3. they are *macroscopic*, that is their occurrence persists despite changes in the observational scale.

Such features appear to be typical of most phenomena commonly classified as "emergent" and associated to the behaviors of many complex systems in the domains of physics, biology, economics, psychology, and so on. For this reason the study of this kind of emergence appears as one of the most important actual goals of Systemics.

To this regard, the first question to be answered is: do exist mathematical models allowing for intrinsic emergence? Happily, the answer is positive. Besides, these models can be classified into two main categories: the *ideal* models of intrinsic emergence and the ones which are *not ideal*. Between the main features of the former, we will quote the following:

a) they derive from general principles;
b) they work typically in a infinite volume limit;
c) they are focussed on ground states and on asymptotic limits.

On the contrary, not ideal models of intrinsic emergence are:

d) based on phenomenological considerations;
e) relying on the existence of suitable boundary conditions;
f) focussed on metastable states and on transient dynamics.

Between the ideal models of intrinsic emergence the most celebrated are the ones based on the mechanism of *Spontaneous Symmetry Breaking* (SSB) in Quantum Field Theory (cfr. Itzykson and Zuber, 1986); to the category of not ideal models belong, instead, neural networks, cellular automata, and all models of Artificial Life.

A second fundamental question is related to the delimitation of the class of ideal models. The importance of this question stems from the fact that, so far, the only models of emergence amenable to a rigorous mathematical analysis are the ones belonging to this class. Thus, ideal models appear as the only existing candidates for the building of a rigorous mathematical theory of emergence. To this regard we remind that a number of physicists (cfr. Anderson and Stein, 1985; Umezawa, 1993) made a very strong claim, which sounds like: *the only possible ideal models allowing for intrinsic emergence are the ones based on SSB in Quantum Field Theory.*

Here we will shortly present a number of points which constitute the rationale for such a claim:

- Quantum Field Theory is the only theoretical framework allowing for the existence of different representations of the same physical system, which are not unitarily equivalent one to another; this means that there exist qualitatively different descriptions of a same system which cannot be reduced one to the other, as they correspond to different "physics"; such a circumstance evidence how Quantum Field Theory is the only existing framework allowing for the existence of different *phases* of the same physical system; only within it, therefore, we can formulate a theory of *phase transitions*;

- SSB within Quantum Field Theory is associated to a property sometimes denoted as *generalized rigidity*, consisting in the fact that, once a system chose one between the many equivalent ground states resulting from SSB, the perturbations tending to destroy such a choice are counteracted by the occurrence of long-range correlations (associated to the so-called *Goldstone bosons*) which contribute to keep stable the ground state previously chosen; in other word, the system is endowed with a sort of *macroscopic coherence*, which is associated to measurable physical effects; such a circumstance cannot occur, in principle, within classical systems of particles obeying Maxwell-Boltzmann statistics, where the relaxation towards a ground state is associated to a weakening of long-range correlations (cfr. Rumer and Ryvkin, 1980); nonetheless, there exist notable exceptions to this latter circumstance, as evidenced by the pioneering work of Fermi, Pasta and Ulam in the Fifties (see Carati and Galgani, 2001), who showed the existence of long-range correlations in classical systems of particles; actually, however, it is still unclear to what extent such exceptional phenomena threaten the validity of the claim quoted before.

What is actually known about ideal and not ideal models of intrinsic emergence gives rise to a number of further interesting, but unfortunately still unanswered, questions, such as:

- what is the relationship between ideal and not ideal models of emergence?
- could we generalize the mechanism of SSB?
- how to deal with metastable states within ideal models?
- how to describe within an ideal model the formation of finite volume subdomains, associated to suitable boundaries?

As regards the first question, it can be split into two subquestions. The first one deals with the problem of *simplification* : if we simplify in some way an ideal model, should we obtain a not ideal model still allowing for intrinsic emergence? Such a question is related to the one of the usefulness of neural networks as simplified, even if approximate, models of field theories. Would the answer be positive, we could use neural networks as approximate, and easier to manage, implementations of Quantum Field Theory. In this important field of research the work is still in progress.

The second subquestion deals with the problem of *understanding*: given an apparently emergent behavior produced by a not ideal model, could we explain it as due to the fact that this model is really nothing but a simplification of some other ideal model? In other words, could it happen that, when we observe on our computer screen an emergent behavior produced by a deterministic program, we were actually doing quantum physics without being aware of it? On this topic we still lack an investigation, which, however, is urgently needed if we want to build an effective theory of emergence.

All these arguments assume, of course, that we have a precise definition of what is Quantum Theory and, in particular, Quantum Field Theory. Unfortunately, this is a sort of *punctum dolens*. The opinions on this subject still differ from a physicist to another and many questions are unanswered. All we can say is that we know what is *actual* Quantum Theory, developed for particular, and limited, purposes, in turn determined by the history of physics in the Twentieth Century. We have no idea about the *possible* Quantum Theory, that is a general theory of the interaction between an observer and his environment. However, without such a knowledge, it will be impossible to formulate a better theory of emergence.

REFERENCES

Anderson, P. W., and Stein, D. L., 1985, Broken symmetry, emergent properties, dissipative structures, life. Are they related? In *Self organizing Systems: The emergence of Order* (F. E. Yates, ed.), Plenum Press, New York, pp. 445-457.

Carati, A., and Galgani, L., 2001, Theory of Dynamical Systems and the Relations Between Classical and Quantum Mechanics, *Foundations of Physics*, **31**:69-87.

Cruchtfield, J. P., 1994, The calculi of emergence: Computation, dynamics, and induction, *Physica D*, **75**:11-54.

Itzykson, C., and Zuber, J. B., 1986, *Quantum Field Theory*. McGraw-Hill, Singapore.

Pessa, E., 1998, Emergence, Self-Organization, and Quantum Theory. In *First Italian Conference on Systemics* (G. Minati, ed.), Apogeo, Milano, pp.59-79.

Rumer Yu. B., and Ryvkin, M. Sh., 1980, *Thermodynamics, statistical physics, and kinetics*. Mir, Moscow.

Umezawa, H., 1993, *Advanced Field Theory. Micro, Macro, and Thermal Physics*. American Institute of Physics, New York.

Quantum Field Theory and System Theory

GIUSEPPE VITIELLO
Dipartimento di Fisica, Università di Salerno, 84100 Salerno, Italy
INFN Gruppo Collegato and INFM Salerno, Italy - E-mail: vitiello@sa.infn.it

1. INTRODUCTION

First of all I would like to thank the Organizers of this very interesting meeting for giving me the opportunity to participate to it and to learn so many things from the System Theory which I find extremely stimulating for my work in Quantum Field Theory (QFT). I would like, indeed, to extract from known facts in QFT some peculiar features which could have a possible connection with a systemic approach.

What I have been thinking in listening to some of the talks is in fact that there must be a connection between a systemic approach and the generation of *collective modes* in QFT. I will try to illustrate this point by referring to an example which I like since it is intuitively quite understandable. It is the example of the crystal.

I will closely follow in my short exposition a couple of sections of the first chapter in a recently published book of mine (Vitiello, 2001). Detailed mathematical formalism can be found in the existing literature which is there quoted. However, there is no need for such a formalism for what I would like to communicate to you here.

So let me start by observing that in Physics it is not enough to investigate what things are "made of". Listing elementary "constituents" is a crucial, necessary step, but it is only one step (the naturalistic level). We want to know not only what things are made of, but *also* "how all of it works": we are interested also in the dynamics. In short: we are interested *in structure and in function*. To a physicist, having a detailed list of constituents does not mean to know *enough* about the system under study. The study of the *structure of matter* is not simply making up a catalog of the elementary constituents of matter; it is knowledge of the evolution in space and in time of the elementary constituents and of their interactions (the forces). Thus when we physicists say "matter", we do not even think of the mere collection of elementary constituents; we always refer to the full dynamical description of matter. This is what we mean by the *dynamical* understanding of Nature.

In most physical systems the interplay among the elementary constituents and their interactions is so intricate that it is not even possible to make a *complete* list of the constituents *without* knowing how they work all together in the system. As we will see, the same concept of constituent is meaningless outside the *dynamical* knowledge of the system. The possibility of a clear definition of structure the constituents and of function (how they behave) is thus reduced down to the normalsize point that we can no longer make a sharp distinction between them.

In the crystal example I am going to present the notions of structure and function indeed merge each into the other. And for that the *quantum* aspect of the formalism is crucial. I discuss the crystal since it provides a relatively simple example of some dynamical mechanism described by QFT. I am not interested here in properties of some specific crystal system. My only interest is in discussing the dynamical generation of order and the merging of structure and function.

As is well known, when atoms (or molecules) of some kind sit in some lattice sites we have a crystal. The lattice is a specific geometric arrangement with a characteristic lattice length; I am thinking of a very simple situation which is enough for what I want to say. The crystal lattice may be deformed in several ways under the action of specific, external boundary conditions. Different *phases* are associated to different lattice deformation states. An extreme lattice deformation is one where the lattice is actually destroyed, e.g. by melting the crystal at high temperature. Once the crystal is broken (melted), one is left with the constituent atoms. So the atoms may be in the crystal phase or, e.g. after melting at high temperature, in the gaseous phase. We can think of these phases as the functions of our structure (the atoms): the physical properties and behaviors of the system are manifestly different in each of these phases and thus we can talk of the crystal function and of the gaseous function.

In the crystal phase one may experimentally study the scattering of, say, neutrons on phonons. The atoms in the crystal sites are continuously vibrating, and these vibrations manifest in the form of elastic waves which propagate all over the crystal. The vibrating atoms interact among themselves by means of the elastic waves and are thus correlated by them over large distances. It is such a long range correlation which keeps the atoms in the crystal ordering. The elastic waves can be mathematically described as *fields* and the phonons are the quanta of these elastic fields (or waves). The phonons are true particles living in the crystal. We observe them indeed in the scattering with neutrons. As a matter of fact, they are the same thing as the elastic waves. In quantum theory, to any wave can be associated a corresponding "quantum" which behaves as particle; in this way "wave description" and "particle description" become *complementary* descriptions. Thus phonons propagate over the whole system as the elastic waves do and therefore act as long range correlation among the atoms (for this reason they are also called *collective modes*). The atoms in the crystal do not behave anymore as free atoms. They are "trapped", like in a net, by the long range correlation mediated by the phonons. The phonons (or the elastic waves) are in fact the messengers exchanged by the atoms and are responsible for holding the atoms in their lattice sites in a *stable* configuration.

Thus we are led to conclude that the list of the crystal constituents includes not only the atoms but also the phonons. Including only the atoms, the list of the constituents is not complete. However, when one destroys the crystal one is left only with the atoms; one does not find the phonons! They disappear! Thus the phonons are "confined" to exist only "inside" the bulk crystal.

On the other hand, if one wants to reconstruct the crystal in its stable configuration after having broken it, the atoms one was left with e are not enough: one must supplement with the long range correlation fields (or quanta of the elastic waves, phonons) which tells the atoms to sit in the special lattice one wants (cubic or whatever). One needs, in short, to supplement the ordering information which was lost when the crystal was destroyed.

Exactly such an ordering information (the crystal function) is "dynamically" realized in the phonon particles (structure). Thus, the phonon particle only exists (but really exists!) as long as the crystal exists, and vice-versa. The function of being a crystal is identified with the particle structure!

Now you can object: "But how it can happen that the stable ordering information (the crystal function) is dynamically realized in the phonon particles (structure)?". I will try to clarify this point. Let me, however, first observe that this is the point where the *quantum* aspect of the theory is essential. Such a thing cannot happen in a *classical* theory. It cannot happen in a nonlinear dynamical theory as far as it is a *classical* theory. Nonlinearity plays its (?)role, but the theory must be a *quantum* theory. No way out. Let me also observe that the description of the crystal in terms of phonons is a mathematically well formulated, experimentally well tested physical theory. Moreover, there is no hope of building up a stable crystal without the *long range* correlation mediated by the phonons: if you try to fix up atom by atom in their lattice sites, holding them by hooks, you will never get the coherent orchestra of vibrating atoms playing the crystal function. Of course, one can also get results of practical interest by treating the crystal as a n-body problem. If, however, one wants to go beyond the phenomenological approach, then he must realize that the crystal is not the output of the assembling atoms by short range forces (*hooks*). On the contrary, it is the macroscopic (systemic!) manifestation of the long range correlation among the atoms born out of the microscopic quantum dynamics.

But the crystal, is it a classical system or a quantum system? It is both! It is a macroscopic quantum system. In a trivial sense, any physical system is a quantum system since it is made by atoms which are quantum objects. But this is a trivial statement, indeed, and it is not in such a trivial sense, that the crystal is a quantum system.

As I said, you cannot think of the crystal without thinking of the phonons. These are quantum particles and the system property of being a crystal is identified with them. It is therefore the *macroscopic* property of being a crystal that finds its identification in the *quantum* phonon mode: *although the crystal property is a classically observable property, it can only be described in terms of quantum dynamics.*

What one means by saying that the crystal is a macroscopic quantum system is that large scale properties of such a system cannot be explained without recourse to quantum dynamical mechanisms. Therefore, those physical properties of the crystal, specific to the nature of being a crystal, appear to be "classical" not because the Planck constant h is taken to be zero, but exactly because h is not zero, i.e. because the quantum dynamics is at work. The crystal is thus *at the same time* a classical system and a quantum system.

If you want, you can put this in different words: quantum theory is not restricted to the explanation of microscopic phenomena. The crystal is only one of the many examples to which such a conclusion applies.

To better understand this point, I recall that the phonons are boson particles. This means that as many of them as one wants can be put in the same state with the same quantum numbers, e.g. same energy, same momentum, etc.. Since they are massless, their lowest energy state is a zero energy state. Therefore we can collect a large number of phonons in the system lowest energy state (the ground state) without changing its energy: it will remain the system ground state, even with a very large number of phonons *condensed* in it. Since they have the same quantum numbers we say that they are *coherently* condensed in the ground state. Needless to say, this is also the root of the crystal ordering and of its stability. Let me observe that at a classical level it would be impossible that the state of the ordered crystal has the same energy of (is degenerate to) the state without ordering (where there are no phonons condensed in it). This can happen because we are considering the quantum mechanism of condensation of massless quanta. It is such a mechanism which allows the existence of crystals, ferromagnets, superconductors, superfluids, and other ordered quantum systems, which are in fact *ordered* and at the same time *highly stable* systems. Their high stability would not be allowed by classical laws of thermodynamics: they are macroscopic quantum systems, indeed.

Let me mention in this connection that the ordering I am talking about is, in the Schrödinger words in his book of 1944 *What is life*, the "dynamical" ordering, not the "statistical" ordering which emerges in Statistical Mechanics when one considers the "average" behavior of many particle systems.

In order to identify the phonon properties of the system ground state (thus the system crystal property) one does not need to monitor the quantum numbers of all the condensed phonons since, as just said, these quantum numbers are coherently the same for all the phonons. This means that it is not the small scale observations which are needed in order to identify the crystal state: it is here that the transition to large macroscopic scale is made possible. This happens just because of the *coherence* of the crystal state. Stated in different words, the phonons are *collective modes* of the system.

On the other hand, since a large number of them is condensed in the ground state, small fluctuations in their number due to quantum excitation processes are irrelevant to the system's macroscopic behavior: thus we reach the well known statement that a system behaves classically when quantum fluctuations are irrelevant. However, notice that the sense of such a statement is not that the Planck constant is zero for the classical system, but that the ratio between the quantum fluctuations and the condensate phonon number is negligible. If the Planck constant would be taken to be zero, there would be no condensate (which is a quantum effect) and the whole picture would collapse.

Of course, a careful analysis would require us to consider many more details in the above reasoning. However, I will not discuss them to make the story short. Nevertheless, there are a couple of further points which deserve to be mentioned.

The first is that the coherent condensation often proceeds through the formation of coherent domains. The minimum size of these domains is called the coherence length. Only at a further stage, the domain boundaries may open so that domains merge into larger coherent regions and the transition to the ordered state in this way occurs. This means that order can only appear on a scale larger than the coherence length: order is an "intrinsically diffused" property of quantum origin.

Another point is the interaction of a classical system with a quantum system. This is a crucial question in Quantum Mechanics (QM), which still waits a full understanding. All the measurement problematic (the measuring apparatus is typically a classical system) in QM has to do with such a question. In the case of macroscopic quantum systems, say of the crystal, we see that a classical probe may interact with the quantum modes which macroscopically manifest themselves in the coherent crystal structure. The crystal atoms act as a whole (as a crystal indeed) since they are trapped by the long range correlation like in a net, and cannot behave anymore as free atoms. In this way the classical probe interacts with such a macroscopic "net" (the crystal collective modes).

Of course, there are other examples of macroscopic quantum systems. The laser is one of these. We know all of it from Professor Haken work, so I will not insist talking about the laser system. Further examples of macroscopic quantum systems are superconductors, superfluids, ferromagnets, and in general every system showing an ordered pattern "quantum dynamically generated". In all these cases the ordering appears as a "system" property, or better it is "the" characterizing "systemic" property. This is my question: In all these cases, is QFT describing the dynamics "generating" the system?

We have seen that in the study of the crystal system one must allow for the possibility of the description of several stable "phases" (e.g. the gaseous phase, the crystal phase). These clearly present different physical properties and behaviors. They are therefore said to be physically inequivalent, or also, more technically, unitarily inequivalent. Without such a possibility one would not be able to handle the "ordering information". Let me observe that the formalism of QM does not allow for the existence of unitarily inequivalent phases. This is called the von Neumann theorem in QM. Thus, in QM it would never be possible for the system to go from one phase to another (breaking or melting the crystal, or reconstructing it are processes called "phase transitions"). What one needs is QFT, where, instead, there exist many, physically inequivalent stable phases associated with a given physical system.

This is why one needs quantum *fields*. Sometimes, the difference between QM and QFT is not enough appreciated. Or sometimes QFT is simply ignored. Therefore, I would like to stress that the dynamical generation of ordered states of minimum energy (i.e. ground states presenting long range correlation) is a characteristic feature of QFT, not present in QM. If any sensible connection exists between System Theory and quantum theory, such a connection involves QFT, not just QM.

I want to mention also that, in ordered systems sometimes vortices, domain walls, dislocations and other "defects" appear. These are extended objects of quantum origin whose behavior, however, manifests classical features: they are described indeed as solutions of nonlinear classical equations. QFT thus provides the basic dynamics out of which macroscopic, i.e. classically behaving, systems are generated.

Since the concepts of dynamical generation of order and of constituent particles acting as messengers of the ordering information (like the phonons in the crystal) are very general concepts, the question now can be asked: Do these concepts and the related mathematical formalism of QFT apply also to living matter and to the brain?

Living matter is no more than matter in the "living phase". The crystal is matter in the "inert phase". Certainly the living phase of matter is much more complex than the inert phase; it is in no way comparable to a crystal. However, a world where the basic dynamical laws ruling the physics of matter in the inert phase would not apply also to matter in the living phase would be a really crazy world.

Just as the crystal is the output of the quantum dynamical ordering of the atoms, functional features of the brain, systemic features of living matter, such as ordered patterns, sequentially interlocked chemical reactions, etc., may result as the output of dynamical laws underlying the rich phenomenology of molecular biology.

Moreover, living systems are also characterized by *plasticity*, namely they can adapt to many different boundary conditions. The multiplicity of responses of the system to many external inputs or perturbations is eliminated by *coherent* responses (functional stability). Such an *adaptive* character of living matter is one of its most typical features, and one of the major sources of differentiation with respect to inert matter. Living matter appears to be characterized by several organization "levels" and presents an evolution ("story"), which appear impossible to explain simply in terms of structural facts. And here, again, structure and function appear indistinguishable.

The question of the possible connection of QFT and System Theory appears thus to be relevant also in connection with the much broader area which includes living matter systems.

REFERENCES

Vitiello, G., 2001, *My Double unveiled*. John Benjamins, Amsterdam.

Index